AF361862

Advances in Aquaculture Ecology Research

Advances in Aquaculture Ecology Research

Editors

Xiangli Tian
Li Li

MDPI • Basel • Beijing • Wuhan • Barcelona • Belgrade • Manchester • Tokyo • Cluj • Tianjin

Editors
Xiangli Tian
Ocean University of China
China

Li Li
Ocean University of China
China

Editorial Office
MDPI
St. Alban-Anlage 66
4052 Basel, Switzerland

This is a reprint of articles from the Special Issue published online in the open access journal *Water* (ISSN 2073-4441) (available at: https://www.mdpi.com/journal/water/special_issues/Aquaculture_Ecology).

For citation purposes, cite each article independently as indicated on the article page online and as indicated below:

LastName, A.A.; LastName, B.B.; LastName, C.C. Article Title. *Journal Name* **Year**, *Volume Number*, Page Range.

ISBN 978-3-0365-7974-0 (Hbk)
ISBN 978-3-0365-7975-7 (PDF)

Contents

About the Editors . . . . **vii**

Xiangli Tian and Li Li
Editorial: Advances in Aquaculture Ecology Research
Reprinted from: *Water* **2023**, *15*, 1629, doi:10.3390/w15081629 . . . . **1**

Wan Zhang, Wen Zhao, Jingjing Zou, Jie Wei, Shan Wang and Dongpeng Yin
Comparative Biology of *Daphniopsis tibetana* from Different Habitats under Seawater
Acclimation
Reprinted from: *Water* **2023**, *15*, 34, doi:10.3390/w15010034 . . . . **5**

Mingyang Wang, Yang Liu, Kai Luo, Tengfei Li, Qingbing Liu and Xiangli Tian
Effects of *Bacillus pumilus* BP-171 and Carbon Sources on the Growth Performance of Shrimp,
Water Quality and Bacterial Community in *Penaeus vannamei* Culture System
Reprinted from: *Water* **2022**, *14*, 4037, doi:10.3390/w14244037 . . . . **17**

Xiuhong Zhang, Jiajia Wang, Chengwei Wang, Wenyang Li, Qianqian Ge, Zhen Qin, et al.
Effects of Long-Term High Carbonate Alkalinity Stress on the Ovarian Development in
Exopalaemon carinicauda
Reprinted from: *Water* **2022**, *14*, 3690, doi:10.3390/w14223690 . . . . **39**

Kun Wang, Nan Li, Zhaohui Wang, Guangjun Song, Jing Du, Lun Song, et al.
The Impact of Floating Raft Aquaculture on the Hydrodynamic Environment of an Open Sea
Area in Liaoning Province, China
Reprinted from: *Water* **2022**, *14*, 3125, doi:10.3390/w14193125 . . . . **55**

Jiahao Liu, Ce Shi, Yangfang Ye, Zhen Ma, Changkao Mu, Zhiming Ren, et al.
Effects of Temperature on Growth, Molting, Feed Intake, and Energy Metabolism
of Individually Cultured Juvenile Mud Crab *Scylla paramamosain* in the Recirculating
Aquaculture System
Reprinted from: *Water* **2022**, *14*, 2988, doi:10.3390/w14192988 . . . . **75**

Hanmo Song, Yan Liu, Jingyu Li, Qingli Gong and Xu Gao
Interactions between Cultivated *Gracilariopsis lemaneiformis* and Floating *Sargassum horneri*
under Controlled Laboratory Conditions
Reprinted from: *Water* **2022**, *14*, 2664, doi:10.3390/w14172664 . . . . **89**

Yilin Yu, Jiwu Wan, Xiaochen Liang, Yuquan Wang, Xueshen Liu, Jie Mei, et al.
Effects of Protein Level on the Production and Growth Performance of Juvenile Chinese Mitten
Crab (*Eriocheir sinensis*) and Environmental Parameters in Paddy Fields
Reprinted from: *Water* **2022**, *14*, 1941, doi:10.3390/w14121941 . . . . **101**

Yiming Xue, Li Li, Shuanglin Dong, Qinfeng Gao and Xiangli Tian
The Effects of Different Carbon Sources on the Production Environment and Breeding
Parameters of *Litopenaeus vannamei*
Reprinted from: *Water* **2021**, *13*, 3584, doi:10.3390/w13243584 . . . . **129**

Jingyu Li, Guohua Cui, Yan Liu, Qiaohan Wang, Qingli Gong and Xu Gao
Effects of Desiccation, Water Velocity, and Nitrogen Limitation on the Growth and Nutrient
Removal of *Neoporphyra haitanensis* and *Neoporphyra dentata* (Bangiales, Rhodophyta)
Reprinted from: *Water* **2021**, *13*, 2745, doi:10.3390/w13192745 . . . . **145**

Jing Yuan, Chuansong Liao, Tanglin Zhang, Chuanbo Guo and Jiashou Liu
Advances in Ecology Research on Integrated Rice Field Aquaculture in China
Reprinted from: *Water* **2022**, *14*, 2333, doi:10.3390/w14152333 . . . . **159**

About the Editors

Xiangli Tian

Xiangli Tian is a professor at the Laboratory of Aquaculture Ecology at the Fisheries College in Ocean University of China. He was honored with the Distinguished Young Scientists of Shandong Province and New Century Excellent Talents of the Ministry of Education awards. He obtained his Ph. D in the field of aquaculture from Ocean University of China in 2001. He carried out his postdoctoral study in Flinders University of South Australia from 2001 to 2002. He was a visiting scholar at Auburn University, USA from 2008 to 2009. His primary expertise focuses on Aquaculture, Aquaculture Ecology and Aquatic Microbial Ecology. For Aquaculture and Aquaculture Ecology, his research has an emphasis on the optimization of structure, function and management of aquaculture ecosystems. Moreover, his recent research interests are mainly focused on the structure and function of microbial communities in aquaculture ecosystems and the application of probiotics. He has served as the PI for more than 50 grants, including for the National Natural Science Foundation of China, National Key R&D Program of China, Natural Science Funds for Distinguished Young Scientists of Shandong Province, etc. He has published more than 200 peer-reviewed papers in English and Chinese and six books (in Chinese).

Li Li

Li Li is an associate professor in the Laboratory of Aquaculture Ecology at the Fisheries College in Ocean University of China. She obtained her Ph.D. degree from Auburn University, USA, in 2012 and was honored as a "Graduation Marshal" of the university. She has been studying and working in the field of aquaculture ecology for over 20 years. Two of her main research interests include: (1) traceability of aquatic products using multiple technologies, including stable isotope analysis, trace element analysis, phospholipid fatty acid analysis, etc.; and (2) water quality management principles and technologies in aquaculture systems. She has served as the PI for more than 10 grants and has published more than 70 peer-reviewed papers, of which 60 were published in English. The papers were published in journals such as *Reviews in Aquaculture, Aquaculture, Aquaculture Research, Food Chemistry, Food Control, Marine Pollution Bulletin*, etc. She has also published two book chapters and obtained two patents.

Editorial

Editorial: Advances in Aquaculture Ecology Research

Xiangli Tian * and Li Li *

Key Laboratory of Mariculture, Ministry of Education, Ocean University of China, Qingdao 266003, China
* Correspondence: xianglitian@ouc.edu.cn (X.T.); l_li@ouc.edu.cn (L.L.)

Citation: Tian, X.; Li, L. Editorial: Advances in Aquaculture Ecology Research. *Water* **2023**, *15*, 1629. https://doi.org/10.3390/w15081629

Received: 12 April 2023
Accepted: 17 April 2023
Published: 21 April 2023

This Special Issue describes the advances in the last decades in the research fields of individual ecology of commercial aquatic organisms, the ecology of aquaculture systems, interactions between aquaculture activities and the environment, the structure and function of the microbial community, principles of environment management in aquaculture ecosystems, etc. We collected ten valuable contributions focused on advances in aquaculture ecology research. All of the authors are from China, which is the largest aquaculture country and contributes more than 60% of global aquaculture production.

Aquaculture is one of the fastest-growing human activities, which not only provides high-quality food for human beings but can also pose a potential risk to the surrounding environment. The multiple outbreaks of golden tide caused by *Sargassum* have attracted lots of attention. Song et al. studied interactions between cultivated *Gracilariopsis lemaneiformis* and floating *Sargassum horneri* [1]. Results of the study could provide important references for mariculture management to reduce golden tide outbreaks. Wang et al. evaluated the impact of floating raft aquaculture on the hydrodynamic environment of an open sea area in Liaoning Province, China by establishing depth-averaged two-dimensional shallow water equations and three-dimensional incompressible Reynolds-averaged Navier–Stokes equations [2]. The work provides a good reference for other studies on aquaculture in open sea areas.

Much research has been performed to develop bioremediation technologies to reduce environmental influences from aquaculture and ensure the sustainability of aquaculture activities. Li et al. selected the appropriate seaweed species for bioremediation of aquaculture wastewater [3]. Results of the study demonstrated that the seaweeds *Neoporphyra haitanensis* and *N. dentata* are likely to be used as efficient and environmentally friendly remediation tools.

The bio-floc technology has been developed in the recent decade and is considered an environmentally friendly technology in aquaculture. Carbon sources are added in biofloc systems to increase the carbon-to-nitrogen ratio (C/N) and thus promote the growth of heterotrophic bacteria [4,5]. Water-soluble carbon sources such as molasses need to be applied frequently, which increases the management effort. Two collected papers investigated the production environment in biofloc systems [4,5]. The papers evaluated the effects of *Bacillus pumilus* BP-171 and different carbon sources, i.e., poly-3-hydroxybutyrate-co-3-hydroxyvalerate (PHBV) and molasses, on water quality, bacterial community and the production of *Litopenaeus vannamei* in culture systems. Both papers demonstrated that adding carbon sources or probiotics could affect the water quality and microbial community. The PHBV, which is insoluble biodegradable polymers and simple to manage, is a good alternative for a water-soluble carbon source.

Effects of various environmental factors, including temperature, carbonate alkalinity and protein levels in compound feeds on commercial aquatic species including ridgetail white prawn (*Exopalaemon carinicauda*), juvenile mud crab (*Scylla paramamosain*) and Chinese mitten crabs (*Eriocheir sinensis*) were comprehensively investigated in three collected papers [6–8]. Yu et al. comprehensively analyzed the protein requirements of juvenile Chinese mitten crabs in a rice–crab co-culture system and provided important references for the optimization of the feeding strategy in the rice–crab co-culture system [6]. Liu et al.

evaluated the optimal temperature range for juvenile mud crabs in terms of growth, molting, energy metabolism, antioxidant capacity and stress response [7]. Results of the study could provide guidance for crab management in aquaculture and support the design of recirculating aquaculture systems for the species. The saline–alkaline water areas in China is about 46 million hectares and the government encouraged the land to be reclaimed into fishponds [8]. The saline-alkaline water usually has high carbonate alkalinity and pH, which limits the growth of most aquatic species. Zhang et al. explored the effects of long-term high carbonate alkalinity stress on ovarian development and revealed the genes and pathways involved in the ovarian development of *E. carinicauda* under long-term high carbonate alkalinity stress [8]. The study demonstrated that *E. carinicauda* is an excellent candidate species for aquaculture in saline-alkaline water as this species could tolerate the saline–alkaline stress.

The *Daphniopsis tibetana* is an important food source for marine fish and shrimp during the nursery period. Zhao et al. evaluated the biology of *D. tibetana* from three lakes in Tibet or on the genetic difference between wild-type and seawater domesticated *D. tibetana*, which provides important information for the large-scale cultivation of *D. tibetana* [9].

In China, we have explored a wide diversity of polyculture applications, both marine and freshwater. It is very important to understand the underlying biological processes of various polyculture models. We collected one review paper from Yuan et al., who systematically reviewed the advances in ecology research on three major integrated rice field aquaculture models in China including rice–fish, rice–crab and rice–crayfish co-culture systems [10]. Integrated rice field aquaculture is one of the main freshwater aquaculture systems. The progress in ecology research on theories, biological studies, models and eco-engineering techniques were systematically reviewed in the paper, which could help aquaculture scientists to further study ecology in integrated aquaculture systems.

The fastest-growing aquaculture achieved high and predictable yields in the past decades; however, the industry is also facing numerous challenges in the long term, such as environmental pollution, excessive resource consumption, etc. The mission of aquaculture ecology is to lay an ecological foundation for the sustainable development of aquaculture. The Special Issue on "Advances in Aquaculture Ecology Research" is closed, but the research on aquaculture ecology is still being rapidly developed.

Author Contributions: Conceptualization, X.T. and L.L.; writing-original draft preparation, X.T. and L.L.; writing—review and editing, X.T. and L.L. All authors have read and agreed to the published version of the manuscript.

Data Availability Statement: Not applicable.

Conflicts of Interest: The authors declare no conflict of interest.

References

1. Song, H.; Liu, Y.; Li, J.; Gong, Q.; Gao, X. Interactions between Cultivated *Gracilariopsis lemaneiformis* and Floating *Sargassum horneri* under Controlled Laboratory Conditions. *Water* **2022**, *14*, 2664. [CrossRef]
2. Wang, K.; Li, N.; Wang, Z.; Song, G.; Du, J.; Song, L.; Jiang, H.; Wu, J. The Impact of Floating Raft Aquaculture on the Hydrodynamic Environment of an Open Sea Area in Liaoning Province, China. *Water* **2022**, *14*, 3125. [CrossRef]
3. Li, J.; Cui, G.; Liu, Y.; Wang, Q.; Gong, Q.; Gao, X. Effects of Desiccation, Water Velocity, and Nitrogen Limitation on the Growth and Nutrient Removal of *Neoporphyra haitanensis* and *Neoporphyra dentata* (Bangiales, Rhodophyta). *Water* **2021**, *13*, 2745. [CrossRef]
4. Wang, M.; Liu, Y.; Luo, K.; Li, T.; Liu, Q.; Tian, X. Effects of *Bacillus pumilus* BP-171 and Carbon Sources on the Growth Performance of Shrimp, Water Quality and Bacterial Community in *Penaeus vannamei* Culture System. *Water* **2022**, *14*, 4037. [CrossRef]
5. Xue, Y.; Li, L.; Dong, S.; Gao, Q.; Tian, X. The Effects of Different Carbon Sources on the Production Environment and Breeding Parameters of *Litopenaeus vannamei*. *Water* **2021**, *13*, 3584. [CrossRef]
6. Yu, Y.; Wan, J.; Liang, X.; Wang, Y.; Liu, X.; Mei, J.; Sun, N.; Li, X. Effects of Protein Level on the Production and Growth Performance of Juvenile Chinese Mitten Crab (*Eriocheir sinensis*) and Environmental Parameters in Paddy Fields. *Water* **2022**, *14*, 1941. [CrossRef]

7. Liu, J.; Shi, C.; Ye, Y.; Ma, Z.; Mu, C.; Ren, Z.; Wu, Q.; Wang, C. Effects of Temperature on Growth, Molting, Feed Intake, and Energy Metabolism of Individually Cultured Juvenile Mud Crab *Scylla paramamosain* in the Recirculating Aquaculture System. *Water* **2022**, *14*, 2988. [CrossRef]
8. Zhang, X.; Wang, J.; Wang, C.; Li, W.; Ge, Q.; Qin, Z.; Li, J.; Li, J. Effects of Long-Term High Carbonate Alkalinity Stress on the Ovarian Development in *Exopalaemon carinicauda*. *Water* **2022**, *14*, 3690. [CrossRef]
9. Zhang, W.; Zhao, W.; Zou, J.; Wei, J.; Wang, S.; Yin, D. Comparative Biology of *Daphniopsis tibetana* from Different Habitats under Seawater Acclimation. *Water* **2023**, *15*, 34. [CrossRef]
10. Yuan, J.; Liao, C.; Zhang, T.; Guo, C.; Liu, J. Advances in Ecology Research on Integrated Rice Field Aquaculture in China. *Water* **2022**, *14*, 2333. [CrossRef]

 *water*

Article

Comparative Biology of *Daphniopsis tibetana* from Different Habitats under Seawater Acclimation

Wan Zhang, Wen Zhao *, Jingjing Zou, Jie Wei, Shan Wang and Dongpeng Yin

Key Laboratory of Hydrobiology in Liaoning Province, College of Fisheries and Life Science,
Dalian Ocean University, Dalian 116023, China
* Correspondence: zhaowen_1963@163.com; Tel.: +86-159-4266-0511

Abstract: In this paper, we used experimental ecology methods and third-generation transcriptome sequencing to see the differences in growth, development, and reproduction of three strains of Daphniopsis tibetana Sars, 1903 from different locations in Tibet (Lake Namukacuo, NMKC; Lake Pengcuo, PC; and Lake Zigetangcuo, ZGTC). We also wanted to determine if the genes had changed after seawater-domesticated D. tibetana was reared in a laboratory. The results showed that at 15–16 ppt salinity and 15 ± 0.5 °C, the NMKC strain exhibited the highest survival rate of 26 d, and the ZGTC strain had the lowest survival rate at 53 days of culture. The body length was observed in NMKC (153.6 ± 12.1%), followed by PC (136.4 ± 16.1%), and then ZGTC (86.2 ± 7.6%). Combined, wild-type and seawater-acclimated D. tibetana produced 7252 DEGs, of which 4146 were up-regulated and 3106 were down-regulated. DEGs were subjected to gene ontology enrichment analysis. The DEGs were mainly enriched in single-organism localization, transporter activity, macromolecule localization, and organic substance transport. The Kyoto Encyclopedia of Genes and Genomes enrichment analysis was also performed and revealed that the RNA transport, protein digestion and absorption, and protein processing in the endoplasmic reticulum pathways were highly enriched. The data mined can provide a reference for follow-up research.

Keywords: *Daphniopsis tibetana*; biology; different habitats; population growth parameters; third-generation transcriptome sequencing

Citation: Zhang, W.; Zhao, W.; Zou, J.; Wei, J.; Wang, S.; Yin, D. Comparative Biology of *Daphniopsis tibetana* from Different Habitats under Seawater Acclimation. *Water* **2023**, *15*, 34. https://doi.org/10.3390/w15010034

Academic Editors: Xiangli Tian and Li Li

Received: 15 October 2022
Revised: 12 December 2022
Accepted: 17 December 2022
Published: 22 December 2022

1. Intraducation

Daphniopsis tibetana Sars, 1903 belongs to Daphniidae and is a rare saltwater cladoceran [1] that is mainly distributed in high-altitude areas such as Tibet, Qinghai, and Xinjiang in China. *D. tibetana* preferentially live in low temperatures. Moreover, they are well-suited for living at high altitudes in cold and nutrient-poor saline water bodies. *D. tibetana* also plays an important role in the study of cladoceran biology [2].

With the rapid development of mariculture, the demand for live farmed animals is increasing. China and other countries have conducted several series of studies on the cultivation and domestication of cladocerans. Among them, a euryhaline species of cold water, *D. tibetana* is suitable for the water temperature conditions used during the nursery period of northern marine fish and shrimp. Its low temperature tolerance makes *D. tibetana* preferable to warm-adapted species that need to be raised for cultivation in high-temperature environments. *D. tibetana* has a longer developmental period and lower fecundity than Moinidae and *Daphnia* but higher fecundity than marine copepods and marine zooplankton [3]. Additionally, the amino acid composition of *D. tibetana* can fully meet the essential amino acid needs of most marine and freshwater fish and shrimp, and the contents of certain unsaturated fatty acids in the body are even higher than those of many common species, such as *Moina mongolica*, *Moina* spp., *Tigiopus japonica*, *Brachionus plicatilis*, and *Artemia*.

To date, there have been reports on the influence of *D. tibetana* morphology and structure [4], living habits [5], ecological distribution [3], and classification and evolution [6], and on how environmental factors influence *D. tibetana* population growth and physiological metabolism [7,8]. However, there has been no research on the biology of *D. tibetana* from three lakes in Tibet (Lake Namukacuo, NMKC; Lake Pengcuo, PC; and Lake Zigetangcuo, ZGTC) or on the genetic difference between wild-type and seawater-domesticated *D. tibetana*. This article reports on and compares some biological observations of *D. tibetana* from these three locations that were domesticated indoors to enrich the biological data on *D. tibetana*. This information can be used for in-depth study of indoor seawater domestication and large-scale cultivation of *D. tibetana*.

2. Materials and Methods

2.1. Source and Domestication of Test Animals

Daphniopsis tibetana is an inland saline cladoceran that is widely distributed in saltwater lakes in Tibet and Qinghai, China, including NMKC (31°83′ N, 89°79′ E), PC (31°89′ N, 90°95′ E), and ZTGC (32°00′ N, 90°90′ E) (Figure 1).

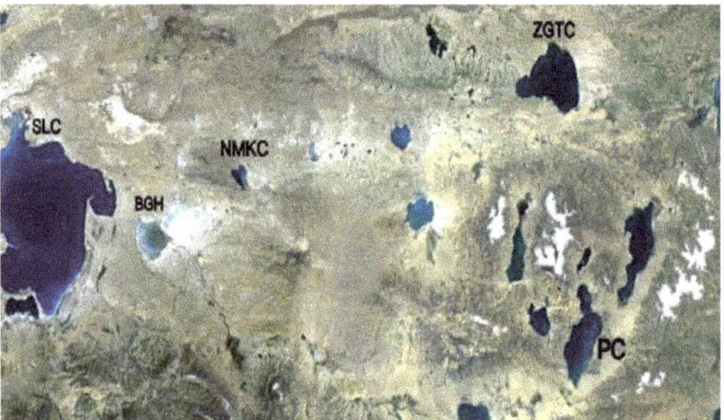

Figure 1. Sampling points of each lake. SLC, Selincuo Lake; BGH, Bangehu Lake.

The *D. tibetana* used in the experiment were collected from NMKC, PC, and ZGTC in Tibet in October 2018. *Daphniopsis tibetana* were brought back to the laboratory and domesticated in diluted seawater with a salinity of 15–16 ppt at 15 ± 0.5 °C and fed with *Chlorella pyrenoidosa*. To avoid the negative impact of individual differences on the experiment, observation of life history started with the larval-stage of *D. tibetana* of the same maternal line. We isolated one gravid mother prior to the experiment and only used its offspring. For the process of domestication, pure water was added to sterilized seawater to adjust the salinity, and the seawater was diluted to 15~16 as culture water. The room temperature was controlled by air conditioning, and the temperature was adjusted to 15 ± 0.5 °C. The Chlorella pyrenoidesa was used as feed with one daily feeding. The culture medium was not replaced during the observation period.

2.2. Experimental Design

Some 30 larval *D. tibetana* cultured under seawater domestication conditions during the same period were selected for experimentation. One larval *D. tibetana* each was placed in a 16-mL test tube. During the experiment, its death, molting, time of first birth, interval between births, and number of births were observed and recorded. The experiment was conducted until the *D. tibetana* died.

2.3. Data Analysis

The population growth parameter was calculated by the following formula:

$$\sum_{x=0}^{\infty} e^{-r_m x} l_x m_x = 1, \; R_0 = \sum l_x m_x, \; \lambda = e^{r_m}, \; T = \ln R_0 \cdot r_m^{-1}$$

In the formula, x is the age period (d), l_x is the survival rate at stage x (%), m_x is the birth rate at stage x, r_m is the intrinsic growth rate (d^{-1}), R_0 is the net reproductive volume (individual), T is the generation period (d), and λ is the weekly growth rate (d^{-1}).

Microsoft Excel 2010 was used to process test data, and IBM SPSS Statistics 23 (IBM Corporation, Armonk, NY, USA) was used for one-way ANOVA and Duncan's multiple comparison test to test for significance and variance homogeneity. The arithmetic mean of replicate groups was taken and expressed as mean $\pm$ standard deviation; $p < 0.05$ indicated significant difference; $p < 0.01$ indicated extremely significant difference.

The growth rate of body length was calculated as follows:

$$\text{LGR} = \frac{L_1 - L_0}{L_0} \times 100\%$$

In the formula, LGR is the growth rate body length (%), L_0 is the initial body length of the *D. tibetana* used in the experiment, and L_1 is the body length at the time of death in the experiment.

2.4. Transcriptome Sequencing

Approximately 400 *D. tibetana* were randomly selected from each treatment group as a biological replicate. Total RNA was extracted from *D. tibetana* using TRIzol (Invitrogen, Carlsbad, CA, USA), and DNase I (TaKara, Dalian, China) was used to remove gene DNA. Using a 2100 Bioanalyzer (Agilent, Santa Clara, CA, USA), the concentration and purity of the extracted RNA were detected by ND-2000 microspectrophotometer (Thermo Scientific, Wilmington, DE, USA) to ensure the integrity of all RNA samples (OD260/OD280 = 1.8–2.2, OD260/OD230 $\geq$ 2.0, RIN $\geq$ 8.5, 28S/18S $\geq$ 1.0) and perform transcriptome sequencing. The mRNA sequencing was conducted using the HiSeq platform, and library construction was performed using the Illumina TruSeq™ RNA sample prep kit method as follows: total RNA was extracted (>1 µg), and mRNA was then enriched, fragmented, and inverted into cDNA; then, adapter ligation and illumina sequencing were performed.

2.5. Differentially Expressed Genes (DEGs)

To explore the differential gene expression of wild-type and domesticated *D. tibetana*, the expression levels of protein-coding genes were calculated by the FPKM method. DEGs were screened, and differential gene expression volcano plots were drawn. Quantitative analysis of gene expression levels was conducted using RSEM (https://deweylab.biostat.wisc.edu/resm/ accessed on 6 June 2019); after obtaining the number of read counts of gene transcripts, DEGseq (http://bioconductor.org/packages/stats/bioc/DESeq/ accessed on 6 June 2019) software was used to analyze gene expression differences between samples. The significance of differential expression was measured by FPKM (fragments per kilobases per million reads) using false discovery rate (FDR) and fold change (FC) as criteria. When a gene exhibited both FDR < 0.05 and |log2FC| > 1, it was considered differentially expressed.

2.6. KEGG Enrichment of DEGs

Functional enrichment analysis of DEGs in different groups was determined using KOBAS (https://kobas.cbi.pku.edu.cn/home.do, accessed on 6 June 2019) leverage of the KEGG database. Genes were classified according to the pathways they participate in or the functions they perform, and the biological processes most relevant to biological phenomena were identified. The Benjamini and Hochberg method was used for multiple test correction, with $p \leq 0.05$ indicating that there was significant enrichment in the GO enrichment function or KEGG pathway.

3. Results and Analysis

3.1. Comparison of Survival Rate and Survival Time of D. tibetana from Different Origins

The survival rate and survival time of *D. tibetana* that originated in each area are shown in Figures 2 and 3. The survival rate of *D. tibetana* from the three areas dropped rapidly after the fifth instar (Figure 2). In general, the survival rate of the NMKC strain was higher than those from the other two areas. However, the NMKC strain had the shortest survival time and failed to complete the seventh instar. This was followed by the PC strain and finally the ZTGC strain, which had the longest survival time and for which there were still survivors after the ninth instar. The longest survival time of each strain was 26 d for NMKC, 35 d for PC35, and 53 d for ZGTC (Figure 3).

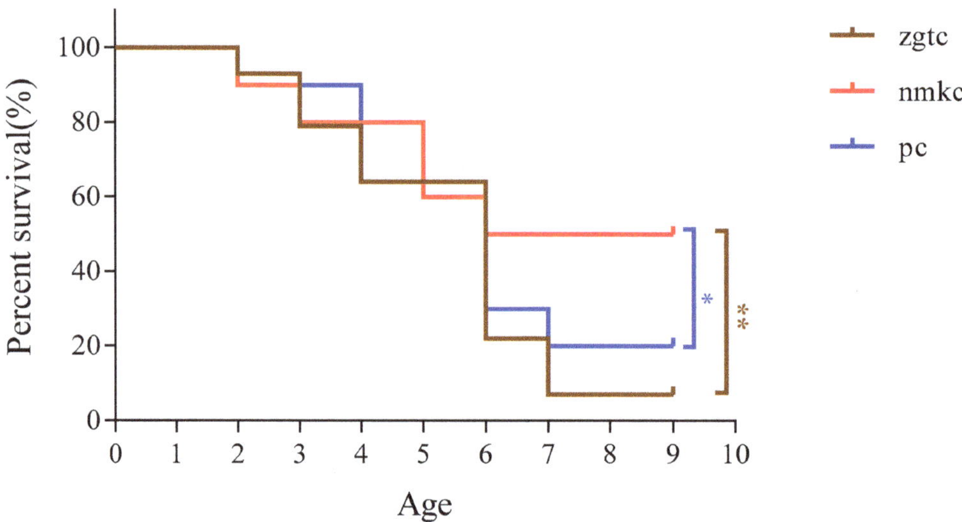

Figure 2. Survival rates of *Daphniopsis tibetana* from different origins.*: Have difference; **: Have a significant difference.

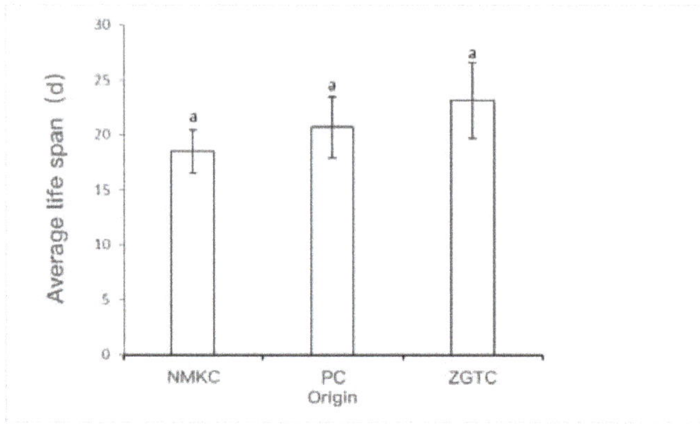

Figure 3. Life span of *Daphniopsis tibetana* from different origins. Means with different lowercase letters are significantly different ($p < 0.05$). a: no significant difference.

3.2. Comparison of Growth and Reproduction of D. tibetana from Different Origins

Growth rates of the NMKC and PC strains were significantly higher than that of the ZGTC strain ($p < 0.05$; Figure 4). The body lengths of the NMKC, PC, and ZGTC strains increased by 1.21 ± 0.91 mm (growth rate, 153.6 ± 12.1%), 1.17 ± 0.13 mm (growth rate, 136.4 ± 16.1%), and 0.70 ± 0.07 mm (growth rate, 86.2 ± 7.6%).

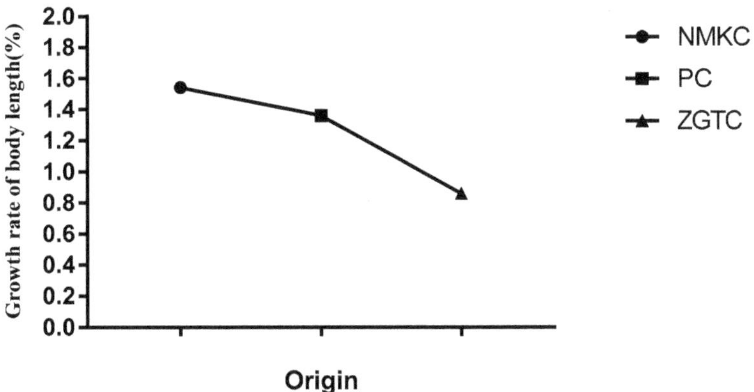

Figure 4. Growth rate of body length of *Daphniopsis tibetana* from different origins. Means with different lowercase letters are significantly different ($p < 0.05$).

Under the conditions of laboratory seawater acclimation (15 ± 1 °C), the NMKC strain had the shortest prenatal development period (19.6 ± 0.25 d), followed by the PC strain (20.25 ± 0.25 d), and the ZGTC strain had the longest prenatal development period (24.5 ± 0.5 d) (Figure 5). The prenatal development period of the ZGTC strain was significantly different from those of the other two strains ($p < 0.05$).

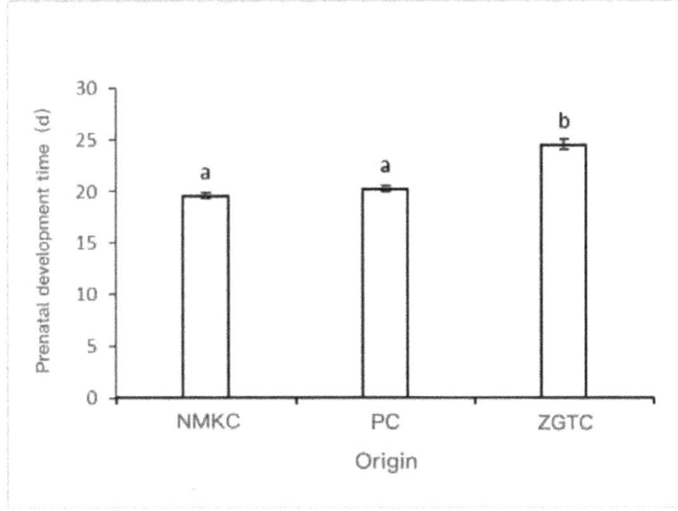

Figure 5. Prenatal development time of *Daphniopsis tibetana* from different origins. Means with different lowercase letters are significantly different ($p < 0.05$). a: no significant difference; b: significant difference.

The average number of offspring per litter for the NMKC and PC strains was higher ($p < 0.05$) than that of the ZGTC strain ($p < 0.05$) (Figure 6). Among them, the NMKC strain had the most offspring per litter (11.5 ± 0.65), followed by the PC strain (9.5 ± 0.57), and then the ZGTC strain (3.0 ± 1.0).

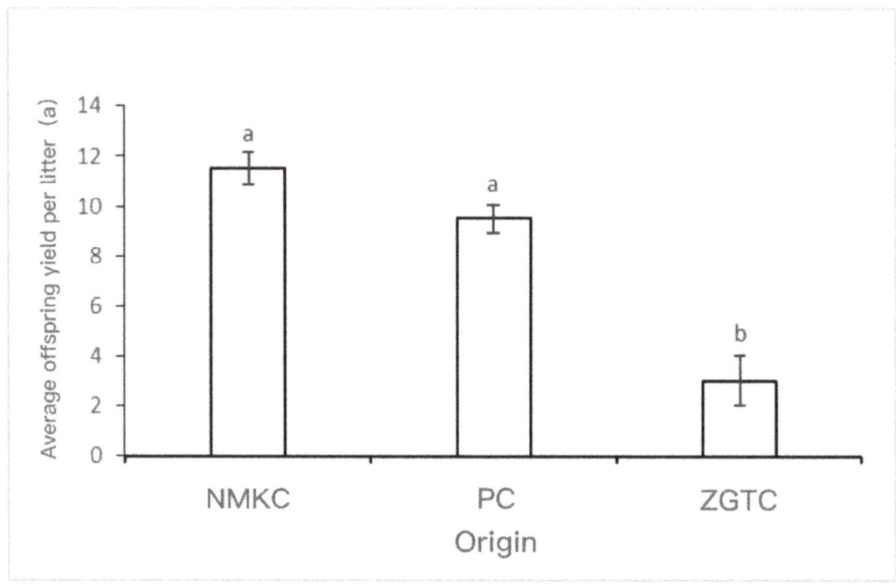

Figure 6. Average number of offspring per litter of *Daphniopsis tibetana* from different origins. Means with different lowercase letters are significantly different ($p < 0.05$). a: no significant difference; b: significant difference.

3.3. Comparison of Population Growth Parameters of D. tibetana from Different Origins

The intrinsic growth rate and net reproductive capacity of the NMKC and PC strains were significantly higher than those of the ZGTC strain ($p < 0.05$), and the generation cycle of ZGTC was significantly longer than those of the other two strains ($p < 0.05$) (Table 1). However, the weekly growth rates of the three strains were not significantly different ($p > 0.05$).

Table 1. Comparison of population growth parameters of *Daphniopsis tibetana* from different origins.

Population Growth Parameter	Origin		
	NMKC	**PC**	**ZGTC**
Intrinsic growth rate (d^{-1})	0.11 ± 0.033 [a]	0.13 ± 0.017 [a]	0.04 ± 0.021 [b]
Weekly growth rate (d^{-1})	1.12 ± 0.036 [a]	1.13 ± 0.019 [a]	1.04 ± 0.022 [b]
Generation cycle (d)	19.40 ± 0.55 [a]	22.37 ± 2.74 [a]	31.01 ± 26.88 [a]
Net reproductive rate (ind.)	9.20 ± 4.00 [a]	19.00 ± 11.83 [a]	3.00 ± 1.41 [a]

Note: a: no significant difference; b: significant difference.

3.4. Embryo Development

The development time and cumulative development time of each instar of each *D. tibetana* strain are shown in Figures 7 and 8. In general, the three *D. tibetana* strains developed to the sixth instar. Among them, the NMKC strain had the shortest survival time; this strain survived after the sixth instar, but soon died and failed to complete the seventh instar. The PC strain developed to the eighth instar. ZGTC had the longest development time but failed to complete the 10th instar. The cumulative development time of the ZGTC strain was longer than those of the other two *D. tibetana* strains.

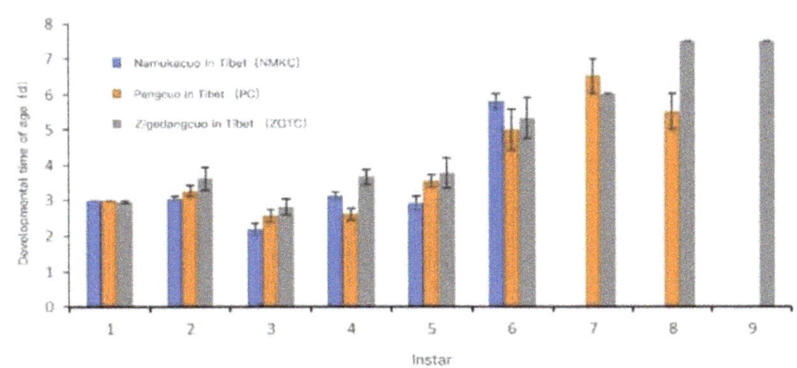

Figure 7. Instar development time of *Daphniopsis tibetana* from different origins.

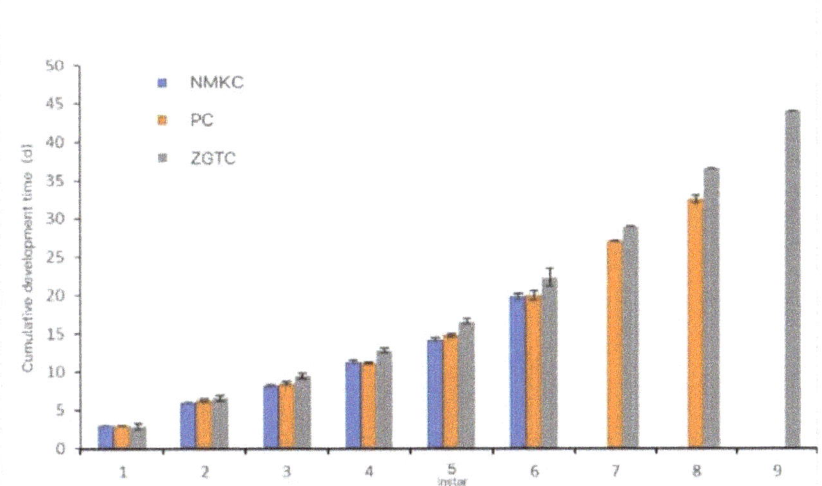

Figure 8. Cumulative development time of *Daphniopsis tibetana* from different origins.

3.5. Transcriptomic Analysis of D. tibetana

Compared with Y, 7252 differentially expressed genes were generated after acclimation with seawater, among which 4146 were up-regulated and 3106 were down-regulated. The results indicated that *D. tibetana* had significant gene expression difference after acclimation in seawater (Figure 9).

Results of GO enrichment analysis performed on the DEGs of wild-type and domesticated *D. tibetana* are shown in Figure 10. The DEGs between wild-type and domesticated *D. tibetana* were mainly enriched in four biological process terms (establishment of localization, transport, single-organism operation, and single-organism localization), one cellular component term (extracellular region), and two molecular function terms (substrate-specific transporter activity and transporter activity).

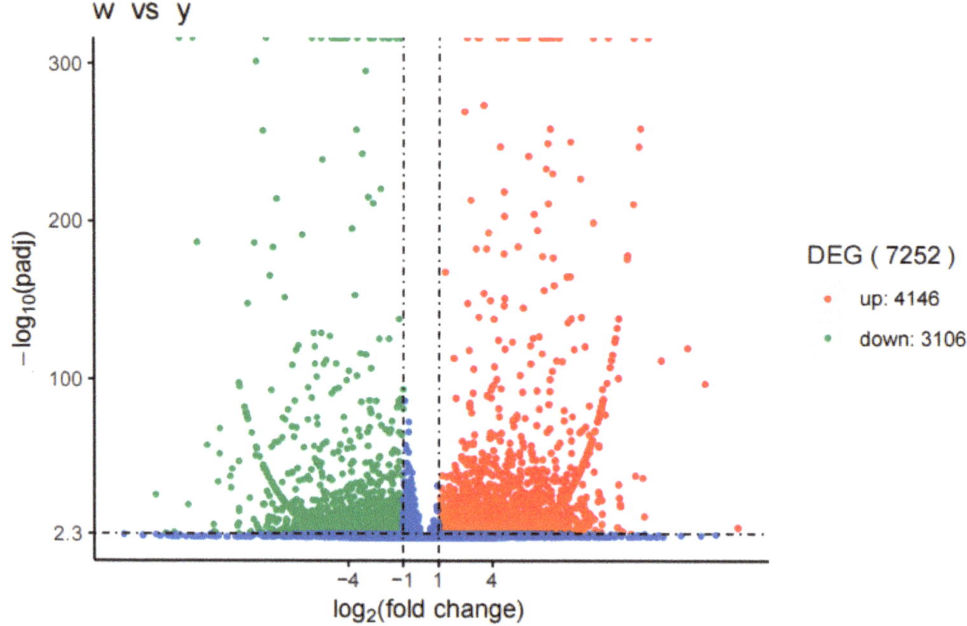

Figure 9. Gene expression differences between wild-type and domesticated *Daphniopsis tibetana*.

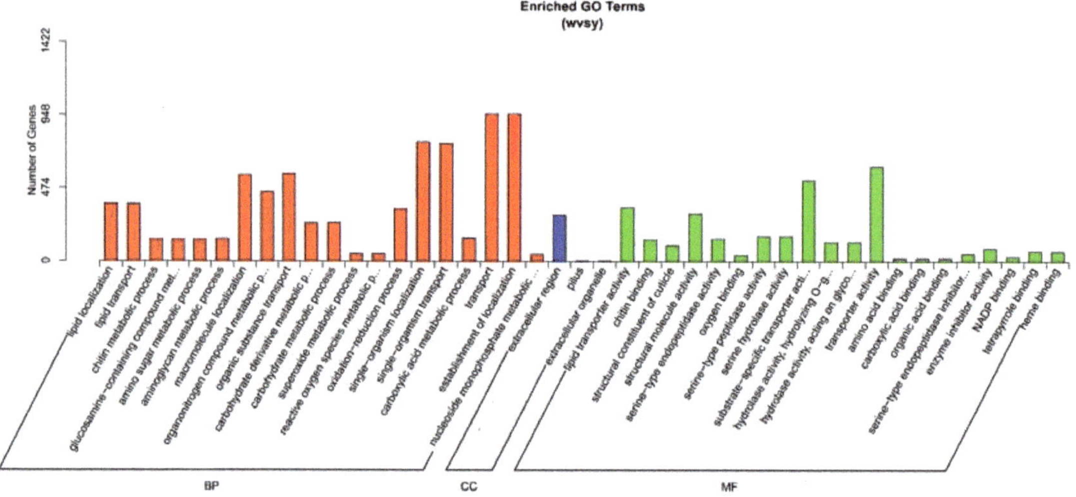

Figure 10. GO enrichment analysis of DEGs between wild-type and domesticated Daphniopsis.

Compared with wild-type *D. tibetana*, domesticated *D. tibetana* had higher enrichment of RNA transport, protein digestion and absorption, and protein processing in endoplasmic reticulum pathways (Figure 11).

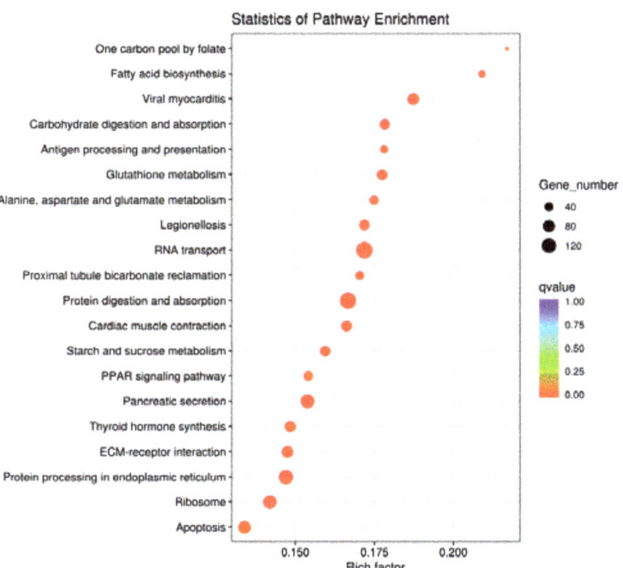

Figure 11. Scatterplot of DEGs between wild-type and domesticated *Daphniopsis tibetana* enriched in the KEGG pathways.

4. Discussion

ZGTC salinity is greater and the water body larger than that of PC, although the ecological environments are similar, and *D. tibetana* is less dense than in NMKC. The salinity of NMKC is between 15–25 ppt, and the composition of salt ions is different from that of other lakes because of its unique geographical location and water environment characteristics; this could be why the *D. tibetana* biomass is greater in NMKC.

A previous study on the salt lakes in northern Tibet revealed that the fish biomass in these salt lakes is low, and *D. tibetana* has become the main food for some water birds in Tibet [9]. Because most of the salt lakes are not connected, the inhabiting activities of birds may be the main reason that *D. tibetana* can be distributed in each salt lake even though the water ecosystem of each salt lake is different and there is a certain amount of geographical isolation. This may be the main reason why *D. tibetana* formed different strains.

Different geographical populations of the same *Daphnia* species must adapt to the specific ecological environment of their habitat; therefore, certain interspecies differences occur. In May and July 2001, Zhao [10] investigated the biological and ecological characteristics of 22 lakes in northern Tibet; the lake salinities ranged between 1 and 390 ppt, and 95 taxa phytoplankton and 42 zooplankton taxa were recorded. Moreover, Na^+ and Mg^{2+} were the main cations in lake water; however, CO_3^{2-} was the dominant anion under low salinity, whereas Cl^- was the dominant anion with increasing salinity. This is consistent with the results of our laboratory's investigation in a few of Tibet's salt lakes in September 2018. Therefore, this experiment used the optimum temperature (15 °C) and salinity (15 ppt) for *D. tibetana* survival and growth to further explore the dynamic changes of *D. tibetana* seasonal populations in three different areas [9].

Under certain environmental conditions, a change in the intrinsic growth rate of a population can reflect small changes in the environment and is an important indicator of the reproductive ability of a species [11,12].This study found that there was no significant difference in the *D. tibetana* intrinsic growth rate, weekly growth rate, and generation cycle between the NMKC and PC strains, but the net reproductive capacity of NMKC was significantly less than that of PC. This is because individuals of the NMKC strain gave birth only once during their entire life cycle, whereas individuals of the PC strain gave birth more than once. However, the ANOVA results for the experimental data of these two strains showed that the average prenatal development period, average reproductive volume per litter, and growth rate of body length were not significant ($p > 0.05$). This finding shows that the *D. tibetana* of NMKC and PC may be the same geographic population. However, on average, the prenatal development period of the NMKC strain was shorter, and the average reproductive capacity per individual was the largest. This may be because the water used in this experiment was closer to the salinity of NMKC and had less impact on this strain. Compared with the other two groups, ZGTC had obvious differences in intrinsic growth rate and net reproductive capacity; this may result from the geographical isolation and salinity changes having important impacts on *D. tibetana* biology.

From the perspective of salinity, the three lakes are all inland salt lakes; however, the populations of these cladocerans in different areas have very different adaptability to salinity domestication [13].The salinity of NMKC and PC are both 16 ppt, whereas that of ZGTC is 21 ppt. Under the experimental conditions, the salinity used was closer to that of NMKC and PC; therefore, compared with the ZGTC strain, the NMKC and PC strains had the characteristics of shorter prenatal development period and larger average reproductive volume per individual. However, because of the dry climate in Tibet, slow changes in salinity during the evaporation and concentration of water also play a natural role in domesticating aquatic organisms.

There is little difference between the pH values of NMKC, PC, and ZGTC (9.54, 9.86, and 10.06, respectively). Moreover, the temperatures of NMKC, PC, and ZGT Care 13 °C, 16.5 °C, and 11.5 °C, respectively, and the control temperature in this experiment (15 ± 0.5 °C) was closer to NMKC and PC. Zhao [14] noted that geographical isolation and salinity changes have important impacts on the genetic diversity of *D. tibetana* from different water bodies. Additionally, Wang [7] found that there were obvious interspecies differences caused by geographical isolation. This study compared the distribution of *D. tibetana* in Tibet with some biological observations of indoor domesticated strains and further confirmed that there are differences in genetic diversity among different geographic populations of *D. tibetana*. However, this difference cannot be attributed simply to geographical isolation. It may be that in Tibet, *D. tibetana* has genetic diversity differences that result from long-term adaptation to different ecological factors. This difference is based mainly on what factors directly or indirectly affect the organisms and can be used to identify differences among different geographic populations.

In addition, 7252 DEGs were identified based on the third-generation transcriptome sequencing data of wild-type and domesticated *D. tibetana* that were analyzed in the laboratory. After *D. tibetana* was moved from the wild to the laboratory, numerous DEGs were generated. Significant enrichment of GO terms revealed that the DEGs are mainly involved in molecular functions, such as substrate-specific transporter activity and transporter activity, and they are mainly located in the cellular components of the extracellular region. Moreover, the majority of DEGs were associated with biological processes and were enriched in the establishment of localization, transport, single-organism operation, and single-organism localization. In the KEGG pathway enrichment analysis of DEGs, the RNA transport pathway, protein digestion and absorption pathway, and protein processing in endoplasmic reticulum pathway were highly enriched. Through these annotations, a large amount of wild-type and domesticated *D. tibetana* transcriptome information, which can more effectively help us understand the genetic characteristics of *D. tibetana* at the molecular level, was obtained. This is of great significance for further exploration of gene function in the future and provides basic data for exploring the functional genes related to *D. tibetana* resistance to environmental stress and studying related physiological functions.

5. Conclusions

Under laboratory domestication at a temperature of 15 ± 0.5 °C and a salinity of 15–16 ppt, the ZGTC strain had the longest life span, but the NMKC and PC strains had significantly higher growth rates of body length than the ZGTC strain ($p < 0.05$).The prenatal development period of the NMKC strain was the shortest (19.6 ± 0.25 d), but the average number of offspring per litter was the largest (11.5 ± 0.65). The intrinsic growth rate and net reproductive capacity of the NMKC and PC strains were significantly higher than those of the ZGTC population (p < 0.05). Three generations of transcriptome sequencing of wild-type D. tibetana after it was moved from the wild to the laboratory were performed in the laboratory, and correlation analysis was performed on the determined DEGs. In total, 7252 DEGs were generated in the comparison between wild-type and domesticated D. tibetana after seawater domestication, of which 4146 were up-regulated and 3106 were down-regulated. After seawater domestication, a series of biological processes and related genes in D. tibetana cells were affected. In GO enrichment analysis, the DEGs were mainly enriched in four biological process terms (establishment of localization, transport, single organism operation, and single organism localization), one cellular component term (extracellular region), and two molecular function terms (substrate-specific transporter activity and transporter activity. In KEGG pathway enrichment analysis, the DEGs were highly enriched in the RNA transport pathway, protein digestion and absorption pathway, and protein processing in the endoplasmic reticulum pathway.

Author Contributions: Conceptualization, W.Z. (Wan Zhang) and J.Z.; methodology, W.Z. (Wen Zhao); software, W.Z. (Wan Zhang); validation, J.W., S.W. and D.Y.; formal analysis, J.Z.; investigation, W.Z. (Wen Zhao); resources, W.Z. (Wen Zhao); data curation, W.Z. (Wen Zhao); writing—original draft preparation, W.Z. (Wan Zhang); writing—review and editing, J.Z.; visualization, W.Z. (Wan Zhang); supervision, W.Z. (Wen Zhao); project administration, J.W.; funding acquisition, W.Z. (Wen Zhao). All authors have read and agreed to the published version of the manuscript.

Funding: The paper was supported by National Key R&D Plan Blue Granary Science and Technology Innovation Project (grant No.2020YFD0900200).

Institutional Review Board Statement: Not applicable.

Informed Consent Statement: Not applicable.

Data Availability Statement: Because we still have a lot of follow-up studies going on, the data is not available.

Conflicts of Interest: The authors declare no conflict of interest.

References

1. He, Z.H.; Zhao, W. Biological resource in inland saline waters in North China. *J. Dalian Fish. Univ.* **2002**, *3*, 157–166. [CrossRef]
2. Jiang, X.Z.; Du, N.S. *Zoology of China (Arthropod Crustacea Freshwater Cladocera)*; Science Press: Beijing, China, 1979; pp. 122–124.
3. Wen, Z.; Qiaohan, W.; Mianping, Z.; Yuanyi, Z.; Hailei, W. A preliminary study on the biology of *Daphniopsis tebitana* Sars. *J. Dalian Fish. Univ.* **2002**, *17*, 209–214. [CrossRef]
4. Zhao, W.; Wang, Q.H. The morphological redescription of *Daphniopsis tebitana* Sars. *J. Dalian Fish. Univ.* **2005**, *20*, 165–173. [CrossRef]
5. Manca, M.; Carnmarano, P.; Spagnuolo, T. Notes on Cladocera and Copepoda from high altitude lakes in the Mount Everest Region (Nepal). *Hydrobiologia* **1994**, *287*, 225–231. [CrossRef]
6. Zhao, W.; Li, R. Molecular phylogeny of four strains of *Daphniopsis tibetana* Sars based on mitochondrial 12S rRNA gene sequences. *J. Dalian Ocean. Univ.* **2012**, *27*, 300–305. [CrossRef]
7. Wang, M.; Zhao, W.; Wei, J.; Wang, S.; Xie, X. Acute effects of UVB radiation on the survival, growth, development, and reproduction of *Daphniopsis tibetana* Sars (Crustacea: Cladocera). *Environ. Sci. Pollut. Res. Int.* **2019**, *26*, 10916–10925. [CrossRef] [PubMed]
8. Zhao, W.; Huo, Y.; Zhang, T.; Wang, S.; Shi, T. Effects of lithium on the survival, growth, and reproduction of *Daphniopsis tibetana* Sars (Crustacea: Cladocera). *Chin. J. Oceanol. Limnol.* **2017**, *35*, 754–762. [CrossRef]
9. Zhao, W.; He, Z.H.; Ren, S.R. *Biology and Technology of Culture and Utilization in Marine Water for Cladocera in Inland Saline Waters*; Science Press: Beijing, China, 2008; p. 163, ISBN 978-7-03-020277-2.
10. Wen, Z.; Mian-Ping, Z.; Xian-Zhong, X.; Xi-Fang, L.; Gan-Lin, G.; Zhi-Hui, H. Biological and ecological features of saline lakes in northern Tibet, China. *Hydrobiologia* **2005**, *541*, 189–203. [CrossRef]
11. Lin, C.S. Theoretical and laboratory studies of animal population dynamics I. The significance and application of laboratory populations and of mathematical models in the studies of animal population dynamics. *Acta Zool. Sin.* **1963**, *15*, 371–381.
12. Krebs, C.J. *Ecology: The Experimental Analysis of Distribution and Abundance*, 4th ed.; Harper Collins College Publishers: New York, NY, USA, 1994; pp. 184–186.

13. He, Z.H.; An, S.S. The adaptability of *Moina rectirostris* to seawater salinity. *Chin. J. Zool.* **1986**, *2*, 25–28. [CrossRef]
14. Zhao, W.; Huo, Y.; Zhang, T.; Wang, S.; Shi, T. Genetic diversity of four population of cladocera *Daphniopsis tebitana*. *J. Dalian Ocean. Univ.* **2011**, *26*, 108–113. [CrossRef]

Article

Effects of *Bacillus pumilus* BP-171 and Carbon Sources on the Growth Performance of Shrimp, Water Quality and Bacterial Community in *Penaeus vannamei* Culture System

Mingyang Wang [1], Yang Liu [1], Kai Luo [1], Tengfei Li [1], Qingbing Liu [2] and Xiangli Tian [1,3,*]

[1] The Key Laboratory of Mariculture, Ocean University of China, Ministry of Education, Qingdao 266003, China
[2] Qingdao Ruizi Group Co., Ltd., Qingdao 266003, China
[3] Function Laboratory for Marine Fisheries Science and Food Production Processes, Qingdao National Laboratory for Marine Science and Technology, Qingdao 266003, China
* Correspondence: xianglitian@ouc.edu.cn

Abstract: A strain of *Bacillus pumilus* BP-171 with the ability of heterotrophic nitrification-aerobic denitrification was isolated from a shrimp culture pond and showed good denitrification ability under laboratory conditions. In order to investigate the effects of strain BP-171 and its combinations with different carbon sources, i.e., poly-3-hydroxybutyrate-co-3-hydroxyvalerate (PHBV) and molasses, on the growth performance of shrimp, water quality and bacterial community in culture system of *Penaeus vannamei*, this experiment was set up. Four experimental groups were designed, i.e., group B applied with a single *B. pumilus* BP-171, the BP added with BP-171 and PHBV, the BM added with BP-171 and molasses, and the control DZ without the probiotic and carbon source. The results showed that the specific growth rate, final body weight, gross weight, feed efficiency rate and survival rate of shrimp in the BP and BM groups were better than those in the control ($p < 0.05$), while the survival rate and gross weight of shrimp in group B were also better than those in the control ($p < 0.05$). Among them, the best growth performance of shrimp was observed in the group BP. The concentrations of ammonia, nitrite, nitrate and total nitrogen were significantly lower in all treatment groups than in the control ($p < 0.05$). The lowest concentrations of ammonia and nitrite were found in group B, while those of nitrate and total nitrogen were found in group BP ($p < 0.05$). The concentrations of dissolved organic carbon and total organic carbon in both BP and BM groups were significantly higher than in group B and the control ($p < 0.05$). Compared to the control, the abundance and diversity of the bacterial community in water did not change with the addition of probiotics and carbon sources. However, altered structure and predicted function, as well as improved stability of the ecological network of the bacterial community, were observed in water. In view of the above, the addition of *B. pumilus* BP-171 and PHBV significantly promoted the growth performance of shrimp, effectively improved water quality, and enhanced the stability of the ecological network of bacterial communities in water, which could have great potential for the application in intensive culture of *P. vannamei*.

Keywords: growth performance; water quality; bacterial community; *Bacillus pumilus* BP-171; molasses; PHBV; *Penaeus vannamei*

Citation: Wang, M.; Liu, Y.; Luo, K.; Li, T.; Liu, Q.; Tian, X. Effects of *Bacillus pumilus* BP-171 and Carbon Sources on the Growth Performance of Shrimp, Water Quality and Bacterial Community in *Penaeus vannamei* Culture System. *Water* **2022**, *14*, 4037. https://doi.org/10.3390/w14244037

Academic Editor: Christophe Piscart

Received: 30 October 2022
Accepted: 8 December 2022
Published: 10 December 2022

Publisher's Note: MDPI stays neutral with regard to jurisdictional claims in published maps and institutional affiliations.

1. Introduction

Penaeus vannamei has become the most dominant species in shrimp farming worldwide [1]. The high demand in the market has led to the expansion of shrimp production and the widespread availability of high stocking density culture patterns. However, with the increasing intensification of shrimp farming, feed residues and shrimp metabolic products led to the accumulation of nitrogen pollutants, including ammonia, nitrite and nitrate nitrogen, in cultured water [2,3]. This problem is of increasing concern in intensive farming practices. The continued increase of nitrogen pollution in culture systems not only leads to the deterioration of water quality in the culture system but also affects the normal physiological functions and immune performance of shrimp and eventually possibly causes frequent diseases in shrimp culture [4–6]. Therefore, it is urgent to develop healthy farming modes and good water quality management technologies for the sustainable development of aquaculture.

Previous studies have shown that the concentration of harmful nitrogen such as ammonia, nitrite and nitrate in culture water could be effectively controlled by adding appropriate probiotics [7–9], ultimately reducing the morbidity of farmed species and increasing the production of farmed animals [10–12]. Probiotics can not only reduce the concentration of hazardous nitrogenous substances accumulated in water through the process of nitrification, denitrification and assimilation but also inhibit the growth of pathogenic microorganisms by competing for physical space and nutrients and secreting bacteriocins and lysozyme [13–15], and thus are widely used in aquaculture.

The addition of a carbon source at an appropriate dosage to the culture water can increase the carbon-to-nitrogen ratio (C/N ratio) and promote the proliferation of heterotrophic bacteria so that ammonia, nitrite and nitrate nitrogen can be removed by the assimilation and denitrification of heterotrophic bacteria [2,16,17]. Molasses is currently one of the most commonly used carbon sources in shrimp aquaculture [18]. However, molasses as an added carbon source is highly soluble and can lead to the rapid growth of heterotrophs and consumption of oxygen, which can easily lead to a rapid decrease in dissolved oxygen, affect the stability of water quality, and eventually threaten the survival of aquatic animals [18,19]. For example, Pérez-Fuentes et al. [18] found that dissolved oxygen decreased significantly from 3.2 mg·L^{-1} to $1–1.5 \text{ mg·L}^{-1}$ when the concentration of added molasses exceeded 0.12 g·L^{-1}, which may lead to mortalities of aquatic animals. PHBV (poly-3-hydroxybutyrate-co-3-hydroxyvalerate) produced by bacteria has many excellent properties, such as thermoplastic, biodegradable, and can exist as a solid material, and has been used as a carbon source for denitrifiers in wastewater treatment systems with good results [20–22]. Compared with conventional carbon sources such as molasses, PHBV is characterized by slow carbon release and easy control, which means the concentration of dissolved organic carbon in water that can be used by heterotrophic bacteria will keep appropriate and stable, so it is a continuous carbon source after application in water and can be used as a biofilm carrier for bacteria [20,22]. However, to our knowledge, the practical application of PHBV as a carbon source in shrimp culture has rarely been reported [3,20].

Biological denitrification is considered one of the most effective, environmentally friendly, and inexpensive biotechnologies for reducing nitrogen levels in aquaculture wastewater [23]. As more heterotrophic nitrifying-aerobic denitrifying bacteria have been isolated, interest in their use for the effective removal of nitrogen accumulated in aquaculture systems has increased in recent years [24–27]. Studies have shown that heterotrophic nitrifying bacteria can convert nitrogen-containing compounds into NH_2OH, $NO_2{}^{-}\text{-N}$ or $NO_3{}^{-}\text{-N}$, etc., by nitrification while using carbon sources for their growth [20,28]. Meanwhile, some bacteria also have simultaneous aerobic denitrification, which can convert $NO_2{}^{-}\text{-N}$ or $NO_3{}^{-}\text{-N}$ into gasses such as NO, N_2O and N_2 [20,28]. Therefore, denitrification treatment with heterotrophic nitrifying-aerobic denitrifying bacteria has the advantages of high economic benefits and environmental friendliness and has gradually become a research focus in recent years [24,26,27]. Unlike the traditional methods of nitrogen removal by autotrophic nitrification and anaerobic denitrification, heterotrophic nitrifying-aerobic denitrifying bacteria can not only avoid the manipulation of separation of aerobic and anaerobic zones but also have the advantage of rapid growth and high denitrification efficiency [24,29]. Heterotrophic nitrifying and aerobic denitrifying microorganisms such as *Halomonas* spp., *Pseudomonas* spp., *Alcaligenes* spp., *Bacillus* spp. and other genera have been isolated successively [26,27]. However, the relevant research was conducted only on the laboratory scale, with few reports on the pilot scale and above, and most of them are based on biofortification, i.e., aerobic denitrifying microorganisms are exogenously added to the bioreactor in the form of microbial agents to improve the denitrification efficiency of the reactor [24,25,29]. A strain of *Bacillus pumilus* BP-171 with the ability of heterotrophic nitrification-aerobic denitrification was isolated from a shrimp culture pond and showed good denitrification ability under laboratory conditions [29]. In addition, Li et al. [20] found that BP-171 significantly reduced nitrite concentration when added to a *P. vannamei* culture system with two other probiotic strains and PHBV simultaneously.

In this study, a shrimp culture experiment was set up to determine the possible effects of different treatments by adding a single BP-171, the combination of BP-171 and PHBV, and the combination of BP-171 and molasses on shrimp growth performance, water quality, and water microbiota in a *P. vannamei* culture system to understand the mechanisms of action of *B. pumilus* BP-171 and provide necessary information for its potential application in shrimp culture practice.

2. Materials and Methods

2.1. Experimental Animals

The juvenile *P. vannamei* were purchased from Qingdao Zhengda Agricultural Development Co., Ltd. (Qingdao, China). Before the culture experiment, shrimp were allowed to acclimate under experimental conditions for two weeks. During acclimation, the water temperature was controlled at 25 ± 0.5 °C and salinity at 29.0 ± 1.0‰. Water was exchanged once daily at a 10% exchange rate and continuously aerated. *P. vannamei* was fed three times a day (7:00, 12:00 and 18:00). At the end of acclimation, healthy *P. vannamei* of similar size were used for the culture experiment. The feed was obtained from Guangdong Yuehai Feeds Group; the main components and nutrient contents of the feed are shown in Supplementary Table S1.

2.2. Experimental Strain and Carbon Sources

Bacillus pumilus BP-171 was obtained from the Microbial Culture Collection, Lab of Aquaculture Ecology, Ocean University of China, a heterotrophic nitrifying-aerobic denitrifying strain [29]. PHBV was purchased from Ningbo Tianan Biological Material Co., Ltd. (Ningbo, China). It is white and cylindrical, with a height of 4 mm and an inner diameter of 1 mm. PHBV was activated in seawater with sufficient aeration for 10 d before the experiment [22]. Molasses (23.7% of total organic carbon content) was purchased from Jinan Pengduo Trading Co., Ltd. (Jinan, China).

2.3. Experimental Design

The experiments were performed in 12 white polyethylene tanks with a volume of 500 L. Four experimental groups were designed, i.e., the probiotic group B applied with a single *B. pumilus* BP-171, the group BP added with the strain BP-171 and PHBV, the group BM with the strain BP-171 and molasses, and the control group DZ without the probiotic and carbon source. Three replicates were set up for each treatment group, and each replicate was randomly stocked with 70 shrimp, and the average weight of the shrimp was 6.06 ± 0.02 g.

The viable bacteria of *B. pumilus* BP-171 were regularly added to the water every seven days. The final concentration of probiotic bacteria in the water of each treatment group was designed as 1×10^7 cfu·L^{-1}.

PHBV particles were placed in a PVC pipe with an inner diameter of 10 cm and a height of 35 cm, and the ends of the PVC pipe were covered with suitable sieves to prevent the PHBV particles from leaking. For aeration, an air stone was placed in the PVC pipe to allow the continuous release of the carbon source into the water with water currents. After assembly, the entire device containing 500 g of PHBV was placed in the corresponding tanks [20].

The molasses was applied with reference to the formula of Avnimelech [30] as follows:

$$\Delta N = (Feed \times N\%) \times \%N \text{ excretion}$$

$$\Delta CH = \Delta N \times [C/N] \text{ mic}/(\%C \times E)$$

ΔN is the amount of nitrogen required to produce new bacteria. Feed is the amount of feed fed, and N% is the percentage of nitrogen in the feed, %N excretion is the percentage of feed nitrogen converted to ammonia in the culture system and is approximately 50%. ΔCH is the amount of molasses added. [C/N] mic is the C/N ratio of the heterotrophic bacteria themselves, %C is the carbon content of the added carbohydrate, and E is the efficiency of assimilation by heterotrophic bacteria, approximately 0.4.

An appropriate amount of molasses was diluted with seawater and poured evenly into the experimental tanks twice daily.

2.4. Experimental Management

The strain of *B. pumilus* BP-171 was inoculated into 2216E liquid medium and cultured at 160 r/min, (28.0 $\pm$ 1.0) °C to logarithmic phase, and the concentration of viable bacteria in the fermentation broth was 1×10^9 cfu·mL^{-1}.

During the experimental period, *P. vannamei* was fed 5% of the total weight of shrimp three times daily (7:00, 12:00, and 18:00). Uneaten feed particles and feces were collected for 1 h after feeding, dried at 60 °C, and weighed. Using a portable dissolved oxygen meter (YSI 550A, Fisher Scientific, Hanover Park, IL, USA), a salinity meter (YSI EC300A, Fisher Scientific, Hanover Park, IL, USA), and a pH meter (YSI pH100A, Fisher Scientific, Hanover Park, IL, USA), temperature (25–28 °C), salinity (28–31‰), pH (7.8–8.0), and dissolved oxygen (>5 mg·L^{-1}) were measured daily.

The feeding trial was conducted in workshop 16 of Qingdao Ruizi Group Co. (Qingdao, China) and lasted for 30 days.

2.5. Sample Collection and Measurement
2.5.1. Growth Performance of Shrimp

The shrimp were counted and weighed for each treatment group at the beginning and end of the experiment, respectively. The survival rate, feed efficiency rate, and specific growth rate of *P. vannamei* were calculated as follows.

$$\text{Survival rate (SR)} = (N_t/N_0) \times 100\%;$$

$$\text{Feed efficiency rate (FER)} = (W_t - W_0)/W_f \times 100\%;$$

$$\text{Specific growth rate (SGR)} = (\ln W_t - \ln W_0)/T \times 100\%.$$

N_t is the number of alive shrimp on the day the feeding trial ended, and N_0 is the number of shrimp put in the tank when the feeding trial started. W_t is the final wet weight of shrimp when the feeding trial ended, and W_0 is the initial wet weight when the feeding trial started. T represented the days from the start to the end of the experiment.

2.5.2. Water Quality Parameters

The water sample of 500 mL was collected every seven days. The parameters of ammonia, nitrite, nitrate, total nitrogen, soluble reactive phosphate, and total phosphorus were determined using an automatic chemical analyzer (Clever Chem 380G, DeChem-Tech. GmbH, Germany) according to the instructions. Water samples for DOC (Dissolved organic carbon) and TOC (Total organic carbon) were analyzed by a multi-2100s TOC analyzer (Analytik Jena). Besides, the average values at different time points of the above parameter were calculated to compare the difference of corresponding parameters among different treatments.

2.5.3. DNA Extraction, Amplification, Purification, and Sequencing

Bacterial samples were collected on Day 30 when the feeding trial ended. A 1L water sample was filtered through a filter membrane with a pore size of 0.22 µm, then the bacterial samples were stored in a −80 °C refrigerator. Total genomic DNA was extracted from the water sample using the E.Z.N.A.® Water DNA Kit (Omega, GA, USA), and PCR amplification was performed using primers 338F (5′-ACTCCTACGGGAGGCAGCA-3′)

and 806R (5′-GGACTACHVGGGTWTCTAAT-3′) specific for the V3 and V4 regions of the 16S rRNA gene. PCR products were then recovered to generate sequencing libraries, and the constructed libraries were sequenced at high throughput using the Illumina HiSeq platform. Raw reads were processed by splicing, filtering, and removal of chimeras to obtain effective reads. The sequences were clustered to obtain operational taxonomic units (OTUs) using UPARSE software (version 7.0) [31] with sequenced reads at a 97.0% similarity level. OTUs were taxonomically annotated using the Silva database (http://www.arb-silva.de/ (accessed on 1 July 2022)).

2.6. Statistical Analysis

Mothur software (version 1.30.2) was used to analyze the diversity of sample sequences, including alpha diversity such as ACE index, Chao1 index, Shannon index, Simpson index, and Good-coverage as well as beta diversity such as PCA (Principal Component Analysis), PCoA (Principal Co-ordinates analysis), and PLS-DA (Partial Least Squares Discriminant Analysis). Chao index and ACE index were used to estimate the number of OTU, the total number of species, reflecting the species richness of α diversity in a community, but the algorithms are different. Both the Shannon index and Simpson index were used to estimate the α diversity of the bacterial community in samples. They consider not only the richness of species in the community but also the evenness of species. However, the algorithms of the two are different. In addition, the higher the Shannon value is, the higher α-diversity is. However, the higher the Simpson value is, the lower α-diversity is. Analysis and visualization of OTU-based Venn diagrams were performed using the VennDiagram package in R (v3.3.1). Based on the species composition of each treatment group at each taxonomic level, bar graphs were generated using the ggplot2 package in R (v.3.3.1), which can be used to visualize the dominant species of each group at a given taxonomic level and the relative abundance of each dominant species. Using the stats package in R (v.3.3.1) and the scipy package in Python, hypothesis tests were performed among species in the different groups based on the species abundance data of bacterial community using the one-way test ANOVA or the Kruskal-Wallis H test to evaluate the significance level of differences in species abundance and obtain species with significant differences between groups.

RDA analysis (redundancy analysis) was performed using the vegan package in R (v.3.3.1), and the significance of RDA analysis was determined by permutest analysis similar to ANOVA. Spearman correlation coefficients between environmental factors and selected species were calculated using the vegan package in R (v.3.3.1) for correlation analysis, and Spearman correlation significance tests were performed using the corrplot package in R (v.3.3.1). Ecological networks were created based on CoNet software in Cytoscape (v.3.8.2) and visualized using Cytoscape (Faust et al., 2016). The KEGG Module database was used to link bacterial taxa to gene sets with particular metabolic capacities and other phenotypic traits. The Shapiro-Wilk test was used to test the data for normal distribution ($p > 0.05$), and Levene's test was used to test for chi-square ($p > 0.05$). One-way analysis of variance (ANOVA) and Duncan's multiple comparison method in IBM SPSS Statistics 24.0 software were used to analyze the significance of differences between groups. The Kruskal-Wallis test was used for analysis when the data were not normally distributed or when there was unequal overall variance. $p < 0.05$ indicated significant differences.

3. Results

3.1. Growth Performance of Shrimp

The growth performance of shrimp is shown in Table 1. The survival rate of shrimp in groups B, BP, and BM was significantly higher than that in the control ($p < 0.05$), and there was no significant difference among groups B, BP, and BM ($p > 0.05$). The final body weight and specific growth rate (SGR) of shrimp were significantly higher in groups BP and BM than in groups B and the control ($p < 0.05$), and they were highest in group BP ($p < 0.05$). The gross weight of shrimp in the BP group was the highest and significantly higher than that of the other groups ($p < 0.05$), while that of groups B and BM was significantly higher than that of the control group ($p < 0.05$). The feed efficiency rate of shrimp in the BP group was significantly higher than that in other groups ($p < 0.05$), while no significant difference was found between other groups and the control ($p > 0.05$).

Table 1. Gross weight, survival rate, specific growth rate, and the feed efficiency rate of *P. vannamei* (Mean $\pm$ S.E.).

Treatment	B	BP	BM	DZ
Initial body weight (g)	6.05 ± 0.05	6.07 ± 0.07	6.03 ± 0.06	6.12 ± 0.01
Final body weight (g)	13.64 ± 0.63 [a]	17.23 ± 0.18 [c]	15.12 ± 0.13 [b]	12.92 ± 0.53 [a]
Gross weight (g)	504.23 ± 7.74 [b]	637.78 ± 14.89 [c]	526.13 ± 16.88 [b]	478.15 ± 12.56 [a]
Survival rate (%)	70.00 ± 0.06 [b]	88.57 ± 0.01 [b]	82.74 ± 0.03 [b]	64.31 ± 0.03 [a]
SGR (%)	2.70 ± 0.13 [a]	3.47 ± 0.07 [c]	3.07 ± 0.06 [b]	2.48 ± 0.14 [a]
FER (%)	51.36 ± 6.96 [a]	70.27 ± 2.65 [b]	55.77 ± 3.50 [a]	45.34 ± 1.94 [a]

Note: The group B, a single *B. pumilus* BP-171 was added; the group BP, *B. pumilus* BP-171 and PHBV were added; the group BM, *B. pumilus* BP-171 and molasses were added; the group DZ, the control without any probiotics and carbon sources addition. Data are expressed as mean $\pm$ standard error, $n = 3$. Values in the same row with different superscripts are significantly different among treatments ($p < 0.05$).

3.2. Water Quality Parameters

3.2.1. Changes in Ammonia, Nitrite, Nitrate, and Total Nitrogen

The average values of the temperature, dissolved oxygen, and pH in the water changed from 25.6 ± 0.3 °C, 7.21 ± 0.51 mg/L, and 7.46 ± 0.21 °C to 25.9 ± 0.3 mg/L, 7.43 ± 0.17 and 7.62 ± 0.19 during the period of the experiment, and no significant differences were found among the groups ($p > 0.05$). The salinity of the water was (30.0 ± 1.0) ‰ during the feeding experiment.

The concentrations and changes of nitrogen in water during the feeding experiment are shown in Figure 1. The concentration of total ammonia nitrogen (TAN) in groups B, BP, and BM was significantly lower than that in the control ($p < 0.05$), and the concentration in group BM was significantly higher than that in groups B and BP. The lowest concentration occurred in the B group and was reduced by 70.22% compared with the control ($p < 0.05$). The concentrations of TAN in the control increased during the experimental period (Figure 1A). In contrast, the TAN concentration in the BM group peaked at about Day 18, whereas the concentration in the B and BP groups leveled off after Day 6 and was significantly lower than that in the control group during the experiment ($p < 0.05$).

Similarly, the concentration of nitrite nitrogen in the B, BP, and BM groups was significantly lower than that in the control ($p < 0.05$). The concentration in the B group was significantly lower than that in the other groups ($p < 0.05$) and reduced by 76.88% compared to the control. In addition, the nitrite-nitrogen concentration in the control group increased from Day 1 to Day 24, stabilized after Day 24, and was significantly higher than all treatment groups (Figure 1B). In comparison, the concentration in the treatment groups leveled off after Day 6 and was significantly lower than that in the control during the experiment ($p < 0.05$).

The concentration of nitrate nitrogen in groups B, BP, and BM was significantly lower than that in the control group ($p < 0.05$). The concentration in the BP group was significantly lower than that in the other groups ($p < 0.05$) and reduced by 26.24% compared to the control. Moreover, the nitrate-nitrogen concentration in water showed an increasing trend in all treatment groups until Day 18 and stabilized after Day 18 in groups B and DZ (Figure 1C). However, concentration in groups BP and BM continued to increase until the end of the experiment. From Day 18 to 30, the nitrate-nitrogen concentration was significantly higher in the control group than in groups B and BP.

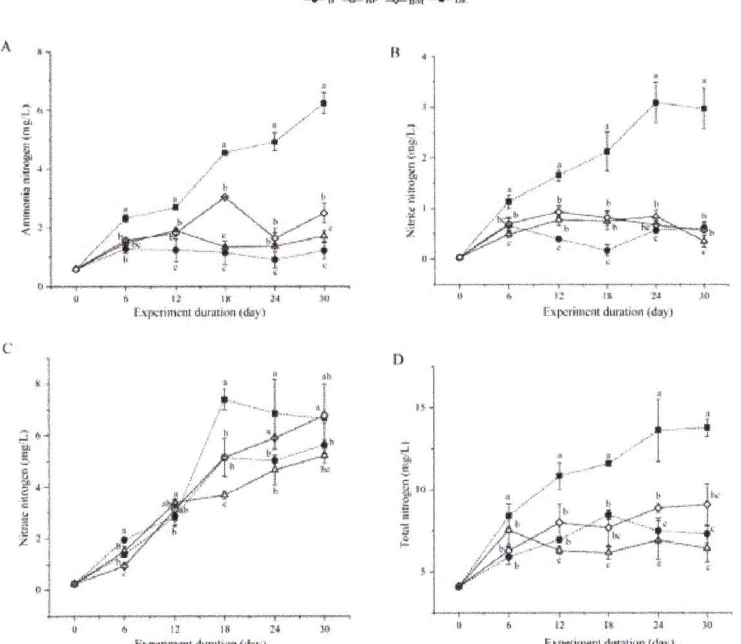

Figure 1. Changes of Ammonia nitrogen (**A**), Nitrite nitrogen (**B**), Nitrate nitrogen (**C**), and Total nitrogen (**D**) in the water of different groups. Note: group B, a single *B. pumilus* BP-171, was added; the group BP, *B. pumilus* BP-171, and PHBV were added; the group BM, *B. pumilus* BP-171, and molasses were added; the group DZ, the control without any probiotics and carbon sources addition. Values with different superscripts on the same day are significantly different among treatments in each figure ($p < 0.05$).

The concentration of total nitrogen was significantly lower in groups B, BP, and BM than in the control group ($p < 0.05$). The concentration in group BP was significantly lower than in the other groups ($p < 0.05$) and was reduced by 40.02% compared to the control. The concentration of total nitrogen in the control group showed an increasing trend (Figure 1D). The values in the treatment groups leveled off between Day 6 and 30 and were significantly lower than that in the control group ($p < 0.05$).

3.2.2. Changes in Soluble Reactive Phosphorus (SRP) and Total Phosphorus (TP)

The concentrations of SRP and TP are presented in Figure 2. There was no significant difference in SRP and TP concentrations among all the groups during the experiment, with an overall increasing trend.

3.2.3. Changes of Dissolved Organic Carbon (DOC) and Total Organic Carbon (TOC)

The concentrations and changes of organic carbon are presented in Figure 3, respectively. The concentrations of DOC and TOC in the groups without the addition of carbon sources (DZ and B) were significantly lower than those in the groups with the addition of carbon sources (BP and BM) ($p < 0.05$). The DOC and TOC concentrations in the DZ and B groups showed a trend of stabilization, while the concentrations in the culture system with carbon source addition increased until Day 18. After Day 18, the concentrations of DOC and TOC in the BP group increased slowly and stabilized gradually, while the concentrations in the BM group decreased rapidly. The average concentrations of DOC and TOC between the B group and the control group showed no significant difference ($p > 0.05$).

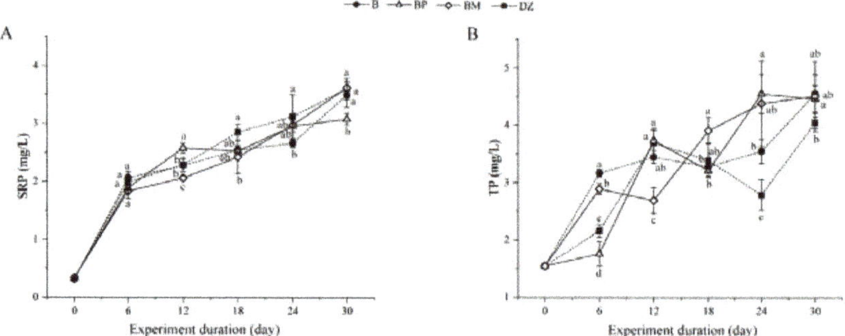

Figure 2. Changes of SPR (**A**) and TP (**B**) in the water of different groups. Note: group B, a single *B. pumilus* BP-171, was added; the group BP, *B. pumilus* BP-171, and PHBV were added; the group BM, *B. pumilus* BP-171, and molasses were added; the group DZ, the control without any probiotics and carbon sources addition. Values with different superscripts on the same day are significantly different among treatments in each figure ($p < 0.05$).

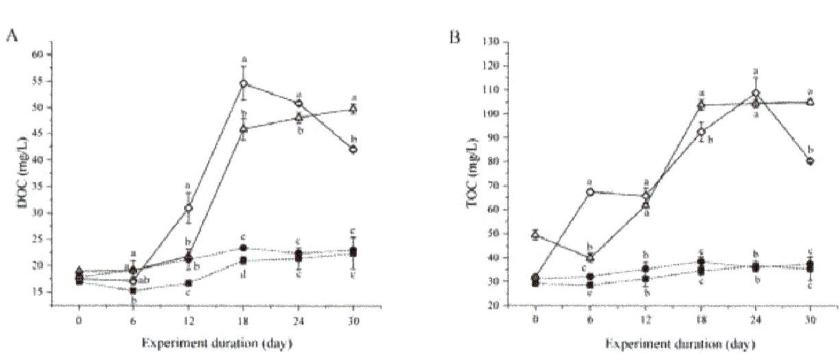

Figure 3. Changes of DOC (**A**) and TOC (**B**) in the water of different groups. Note: group B, a single *B. pumilus* BP-171, was added; the group BP, *B. pumilus* BP-171, and PHBV were added; the group BM, *B. pumilus* BP-171, and molasses were added; the group DZ, the control without any probiotics and carbon sources addition. Values with different superscripts on the same day are significantly different among treatments in each figure ($p < 0.05$).

3.3. The Microbial Diversity, Community Structure, and Function

3.3.1. Microbial Diversity

The bacterial diversity indices are shown in Table 2. The coverage indices of all groups were above 0.99, indicating that the sequencing data were adequate to represent the bacterial community structure. The Shannon, Simpson, Chao, and Ace indices of bacterial communities of all groups did not show significant differences, indicating that the addition of probiotics and carbon sources did not change the abundance and diversity of bacterial communities in the culture system.

As shown in Figure 4, the similarity analysis among groups using partial least squares discriminant analysis (PLS-DA) showed that the samples of group BP were separated from those of groups BM, DZ, and B along the COMP1 axis. Meanwhile, the samples of group B were separated from those of groups BM and DZ along the COMP2 axis. In addition, the samples within the same group clustered together while the samples from different groups moved away from each other.

Table 2. α-diversity indices of bacterial communities in shrimp culture systems.

Groups	α-Diversity Index				Coverage
	Shannon	Simpson	Ace	Chao	
B	3.23 ± 0.69	0.10 ± 0.07	438.93 ± 97.23	440.16 ± 87.69	0.998
BP	2.74 ± 0.69	0.18 ± 0.11	443.16 ± 134.35	435.90 ± 122.11	0.997
BM	3.56 ± 0.46	0.06 ± 0.02	459.62 ± 30.45	447.41 ± 31.13	0.998
DZ	3.36 ± 0.08	0.07 ± 0.01	411.01 ± 29.19	370.11 ± 19.18	0.998

Note: group B, a single *B. pumilus* BP-171, was added; the group BP, *B. pumilus* BP-171, and PHBV were added; the group BM, *B. pumilus* BP-171, and molasses were added; the group DZ, the control without any probiotics and carbon sources addition. Data are expressed as mean ± standard error, n = 3.

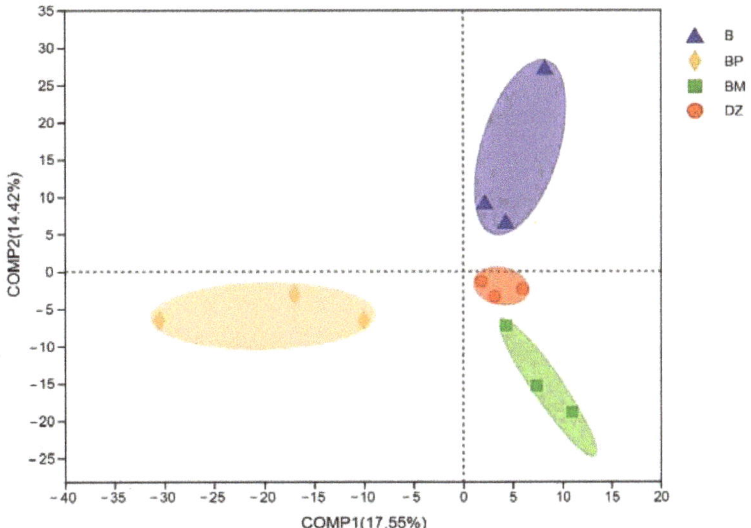

Figure 4. PLS-DA analysis based on OTU level of different *P. vannamei* culture systems. Note: group B, a single *B. pumilus* BP-171, was added; the group BP, *B. pumilus* BP−171, and PHBV were added; the group BM, *B. pumilus* BP-171, and molasses were added; the group DZ, the control without any probiotics and carbon sources addition.

3.3.2. Bacterial Community Composition

There were 501, 605, 626, and 565 OTUs in the DZ, B, BP, and BM groups, respectively (Figure 5). In addition, the unique OTUs in the DZ, B, BP, and BM groups were 34, 86, 144, and 66, respectively. The group with the lowest number of specific OTUs was the control group, while the most were found in the BP group.

The predominant phyla were *Proteobacteria*, *Bacteroidetes*, *Actinobacteria*, and *Verrucomicrobia* (Figure 6A). The relative abundance of the phylum *Bacteroidetes* was significantly lower in the BP group than in the control group ($p < 0.05$). However, the relative abundance of the phylum *Proteobacteria* was distinctly higher compared with the control ($p < 0.05$). In addition, the relative abundance of the phylum *Verrucomicrobia* in the B group was observably higher than that in the other groups ($p < 0.05$).

Additionally, the classes *Alphaproteobacteria*, *Flavobacteriia*, *Actinobacteria*, and *Verrucomicrobiae* dominated each group (Figure 6B). The relative abundance of *Flavobacteria* was significantly lower in the BP group than in the other groups ($p < 0.05$). Compared to the control, the relative abundance of *Alphaproteobacteria* was significantly increased in the BP group ($p < 0.05$). Besides, the relative abundance of *Verrucomicrobiae* was remarkably higher in the B group than in the other groups ($p < 0.05$).

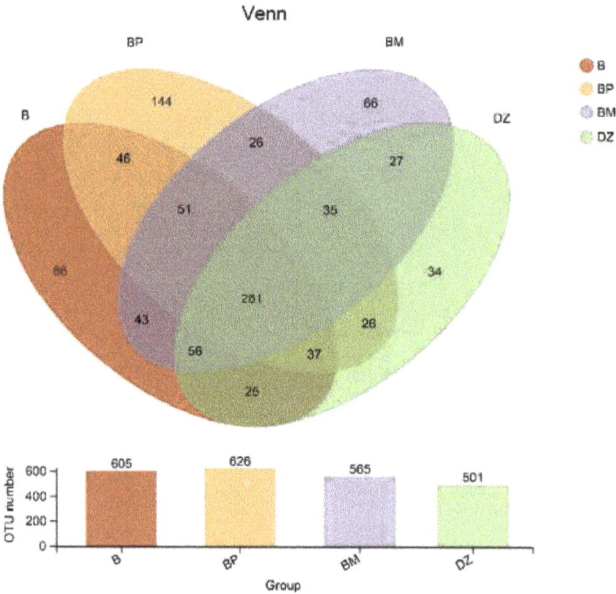

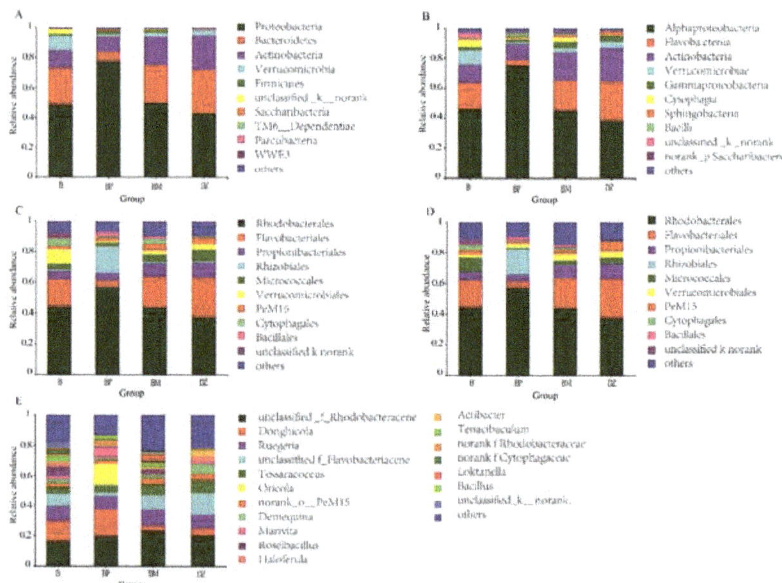

Figure 5. Venn diagram based on OTU level for different *P. vannamei* culture systems. Note: group B, a single *B. pumilus* BP-171, was added; the group BP, *B. pumilus* BP-171, and PHBV were added; the group BM, *B. pumilus* BP-171, and molasses were added; the group DZ, the control without any probiotics and carbon sources addition.

Figure 6. Bacterial composition at different levels of phylum (**A**), class (**B**), order (**C**), family (**D**), and genus level (**E**) in different culture systems. Note: group B, a single *B. pumilus* BP-171, was added; the group BP, *B. pumilus* BP-171, and PHBV were added; the group BM, *B. pumilus* BP-171, and molasses were added; the group DZ, the control without any probiotics and carbon sources addition.

At the order level, *Rhodobacterales, Flavobacteriales, Propionibacteriales, Rhizobiales,* and *Verrucomicrobiales* were the dominant bacterial taxa in each treatment (Figure 6C). The relative abundance of *Flavobacteriales* was significantly lower in the BP group than in the other groups ($p < 0.05$). In contrast, the abundance of *Rhizobiales* and *Rhodobacterales*, which belong to the bacterial taxonomic class *Alphaproteobacteria* of the phylum *Proteobacteria*, was significantly increased in the BP group compared with the control ($p < 0.05$). In addition, the relative abundance of *Verrucomicrobiales* was prominently higher in the B group than in the other groups ($p < 0.05$).

At the family level, *Rhodobacteraceae, Flavobacteriaceae, Propionibacteriaceae, Phyllobacteriaceae,* and *Verrucomicrobiaceae* were relatively abundant and dominant in each treatment (Figure 6D). The relative abundance of *Flavobacteriaceae* in the BP group was notably lower than that in the other groups ($p < 0.05$), while the abundance of *Phyllobacteriaceae* and *Rhodobacteraceae* in the BP group was significantly higher than that in the control group ($p < 0.05$). Furthermore, *Verrucomicrobiaceae* in the B group was more abundant than in the other groups ($p < 0.05$).

At the genus level, the unclassified *Rhodobacteraceae, Donghicola, Ruegeria, Tessaracoccus, Oricola,* and *Marivita* dominated the B, BP, and BM groups (Figure 6E). As depicted in Figure 7, the relative abundance of *Oricola* spp., *Donghicola* spp., and *Marivita* spp. was significantly higher in the BP group than in the other groups ($p < 0.05$). Compared with the control, the relative abundance of *Bacillus* spp. in groups B, BP, and BM was significantly increased, and the relative abundance of *Bacillus* spp. in the BP group was significantly higher than that in groups B and BM ($p < 0.05$).

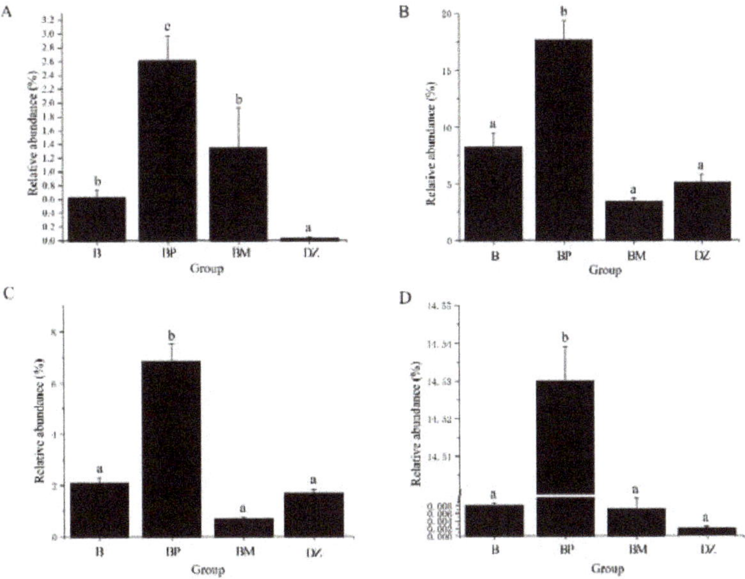

Figure 7. The relative abundance of *Bacillus* spp. (**A**), *Donghicola* spp. (**B**), *Marivita* spp. (**C**) and *Oricola* spp. (**D**) in different groups. Note: group B, a single *B. pumilus* BP-171, was added; the group BP, *B. pumilus* BP-171, and PHBV were added; the group BM, *B. pumilus* BP-171, and molasses were added; the group DZ, the control without any probiotics and carbon sources addition. Values with different superscripts are significantly different among treatments in each figure ($p < 0.05$).

3.3.3. The Correlation of Water Quality Indexes and Microbial Community Structure

Redundancy analysis (RDA) revealed the effects of environmental variables (TAN, NO_2^--N, NO_3^--N, TN, SRP, TP, DOC, and TOC) on microbial community structure. It can be seen that TAN, NO_2^--N, NO_3^--N, and TN had strong positive impacts on the distribution of flora at the phylum level on the first typical axis, and the above four environmental factors had synergistic effects on community structure (Figure 8A). In contrast, DOC and TOC had strong negative effects on the distribution of flora at the phylum level on the first typical axis, and the two environmental factors had synergistic effects on community structure. In addition, nitrate nitrogen was the most influential variable that had a significant effect on bacterial community structure ($r^2 = 0.538$, $p = 0.043$).

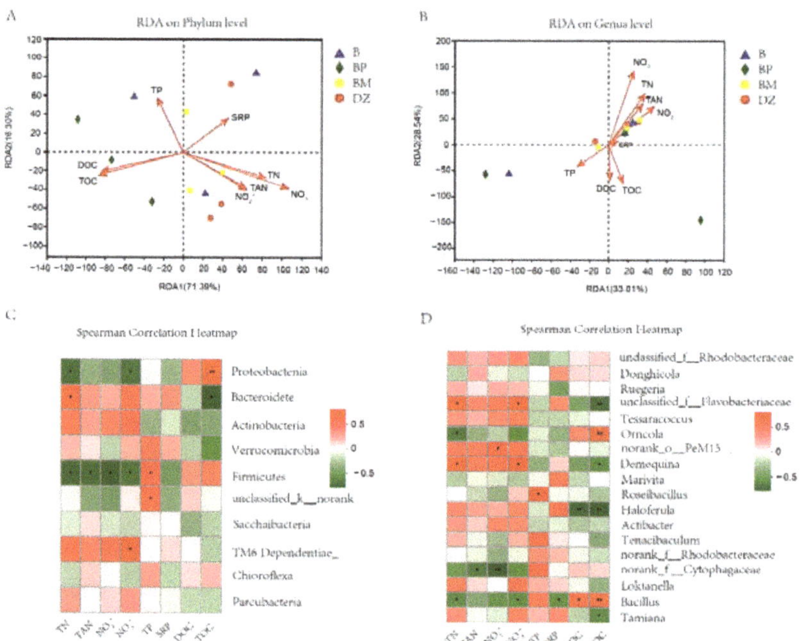

Figure 8. The Redundancy Analysis on phylum (**A**) and genus (**B**) level and the correlation heatmap analysis between bacterial taxa and water quality parameters on phylum (**C**) and genus (**D**). Note: group B, a single *B. pumilus* BP-171, was added; the group BP, *B. pumilus* BP-171, and PHBV were added; the group BM, *B. pumilus* BP-171, and molasses were added; the group DZ, the control without any probiotics and carbon sources addition.

Similar to the phylum level, TAN, NO_2^--N, NO_3^--N, and TN had strong positive effects on the distribution of flora at the genus level on the second typical axis, and the four environmental factors had synergistic effects on community structure (Figure 8B). In addition, DOC and TOC had strong negative effects on the distribution of flora at the genus level on the second typical axis, and the two environmental factors also had synergistic effects on community structure. Moreover, nitrate nitrogen had the greatest effect on bacterial community structure at the genus level ($r^2 = 0.409$, $p = 0.038$).

The correlation between environmental variables and bacteria was explored by calculating Spearman coefficients of water quality factors and bacterial taxa at phylum and genus levels, respectively. At the phylum level, *Proteobacteria* showed a significant positive correlation with TOC ($p < 0.05$), while it showed a significant negative correlation with nitrate nitrogen and total nitrogen, respectively ($p < 0.05$) (Figure 8C). In addition, the phylum *Firmicutes* was significantly negatively correlated with ammonia nitrogen, nitrite nitrogen, nitrate nitrogen, and total nitrogen, respectively ($p < 0.05$). Besides, the phylum Bacteroidetes had a significant negative correlation with TOC ($p < 0.05$) but had a significant positive correlation with total nitrogen ($p < 0.05$).

At the genus level, *Bacillus* spp. showed a significant positive correlation ($p < 0.05$) with DOC and TOC but had a significant negative correlation with nitrate nitrogen and total nitrogen ($p < 0.05$) (Figure 8D). Additionally, *Oricola* spp. was positively correlated with TOC ($p < 0.05$) while negatively correlated with total nitrogen ($p < 0.05$). The *unclassified_f__Flavobacteriaceae* spp. and *Demequina* spp. both had a significant negative correlation with TOC, while they had a significant positive correlation with nitrate nitrogen and total nitrogen ($p < 0.05$).

3.3.4. Ecological Network Analysis

Ecological network analysis showed that there were more nodes in groups B, BP, and BM than in the control group (Table 3). However, no significant difference was found in negative or positive interactions among different groups. There were more functional modules in the groups to which carbon sources were added (BP and BM) than in the control and B groups, with the highest number of modules in the BP group (Figure 9). The number of functional modules in the B, BP, BM and DZ groups was 7, 26, 16, and 7, respectively.

Table 3. Topological properties of the networks.

	B	BP	BM	DZ
Nodes	328	323	334	285
Edges	1270	1056	1256	1223
Positive relationship	800	616	753	733
Negative relationship	470	440	503	490
negative interactions/positive interaction ratio	58.75%	71.43%	66.80%	66.85%
Module number	7	26	16	7

Note: group B, a single *B. pumilus* BP-171, was added; the group BP, *B. pumilus* BP-171, and PHBV were added; the group BM, *B. pumilus* BP-171, and molasses were added; the group DZ, the control without any probiotics and carbon sources addition. Data are expressed as mean $\pm$ standard error, n = 3.

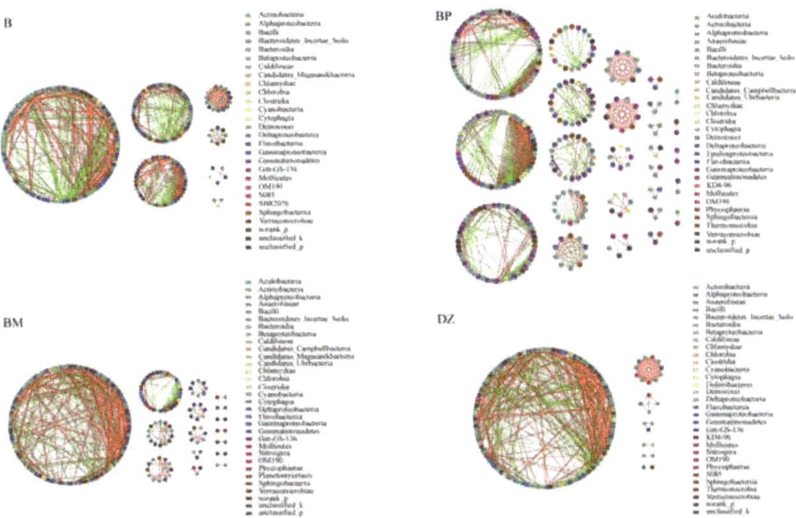

Figure 9. Ecological Network based on OTUs of the bacterial community in the different culture systems. Note: In the ecological network diagram, different nodes represent bacteria from different OTUs, and the line between two nodes indicates that there is some interaction between bacteria from two OTUs, and the red line represents a positive relationship between bacteria from two OTUs, the green line represents a negative relationship. The group B, a single *B. pumilus* BP-171 was added; the group BP, *B. pumilus* BP-171, and PHBV were added; the group BM, *B. pumilus* BP-171, and molasses were added; the group DZ, the control without any probiotics and carbon sources addition.

3.3.5. Predictive Functions of Microbiota in the Water

As shown in Figure 10, compared to the control, eight functional modules (Cationic antimicrobial peptide resistance, photosystem II, coenzyme M biosynthesis, bacilysin biosynthesis, kanosamine biosynthesis, cytochrome aa3-600 menaquinol oxidase, photosystem I and lysine biosynthesis) were significantly improved in the B group ($p < 0.05$). In the BP group, 12 functional modules (ketone body biosynthesis, pentose phosphate pathway, ethylmalonyl pathway, entner-Doudoroff pathway, hydroxypropionate-hydroxybutylate cycle, urea cycle, cobalamin biosynthesis, D-Galacturonate degradation, purine degradation, tyrosine biosynthesis, and molybdenum cofactor biosynthesis and nicotinate degradation) were significantly higher than the control ($p < 0.05$). However, 19 functional modules (tetrahydrobiopterin biosynthesis, C4-dicarboxylic acid cycle, nodulation, N-glycan precursor biosynthesis, tetrahydrofolate biosynthesis, threonine biosynthesis, biotin biosynthesis, biotin biosynthesis of BioW pathway, methionine degradation, assimilatory sulfate reduction, abscisic acid biosynthesis, tetrahydrofolate biosynthesis, ascorbate biosynthesis, dTDP-L-rhamnose biosynthesis, pyrimidine ribonucleotide biosynthesis, and coenzyme A biosynthesis) were significantly decreased ($p < 0.05$). In the BM group, three functional modules (coenzyme M biosynthesis, N-glycan biosynthesis, and assimilatory nitrate reduction) were enriched ($p < 0.05$), whereas one functional module (D-Glucuronate degradation) was significantly decreased ($p < 0.05$).

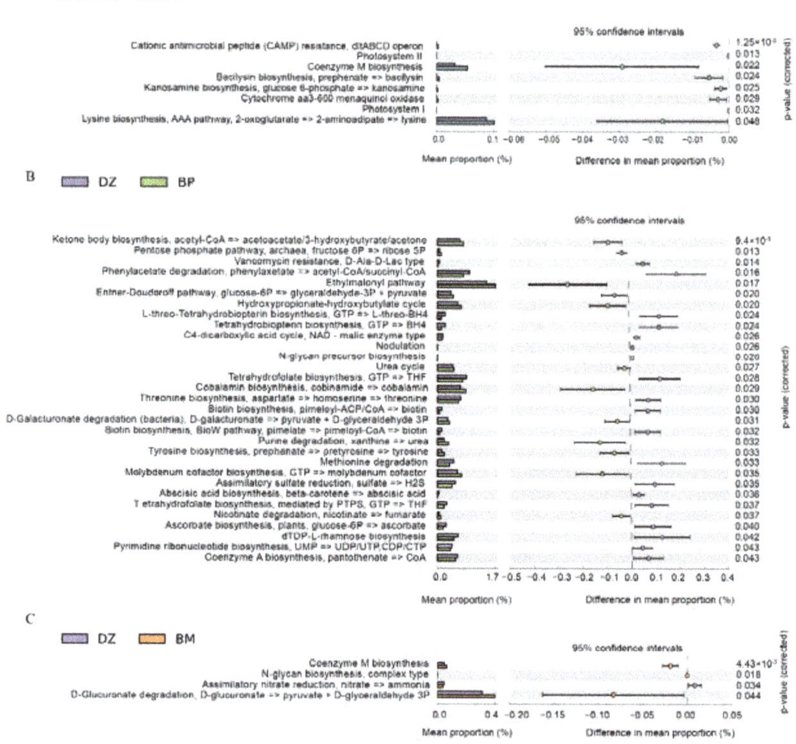

Figure 10. Differences of predicted functions based on the KEGG Module database using STAMP. Note: Only data with significant differences ($p < 0.05$) between groups are shown. (**A**) DZ and B group, (**B**) DZ and BP group, (**C**) DZ and BM group. The group B, a single *B. pumilus* BP-171 was added; the group BP, *B. pumilus* BP-171, and PHBV were added; the group BM, *B. pumilus* BP-171, and molasses were added; the group DZ, the control without any probiotics and carbon sources addition.

4. Discussion

4.1. Effects of Addition of B. pumilus 171 and Carbon Sources on the Growth Performance of Shrimp

Many previous studies have shown that the addition of probiotics to culture water can promote the growth of aquatic animals by improving FER and regulating the balance of aquatic flora, and reducing toxic substances such as ammonia and nitrite [8,9,32,33]. The results from the present study were consistent with the above findings. Higher survival rate and gross weight of shrimp were observed in the group with the single addition of *B. pumilus* BP-171 compared to the control. In addition to probiotics, carbon sources with the property of improving growth performance were also observed [34–39]. In this study, the shrimp in the group to which both *B. pumilus* BP-171 and carbon sources were added showed better growth performance than the group to which only one probiotic was added, and the group using PHBV as a carbon source showed the best performance. The above results suggest that *B. pumilus* BP-171 promoted the growth performance of shrimp, and a suitable carbon source such as PHBV could further enhance the growth-promoting function of probiotics.

4.2. Effects of the Addition of B. pumilus 171 and Carbon Sources on the Water Quality

Some probiotics, such as *Bacillus* spp. could be used to improve water quality by reducing the concentration of ammonia and nitrite in the culture system [7–9]. For example, Lee et al. [40] found that total NH_4^+ concentration was significantly lower when *Bacillus* spp. were added to the culture system. Decreased nitrite concentration was observed when *B. subtilis* FY99-01 was used in the culture system of *P. vannamei* [32]. Barman et al. [23] also found that *B. cereus* PB45 could consume nitrite in the culture pond effluent. The addition of carbon sources could also promote the removal of nitrogen from water by increasing the C/N ratio [2,16,17]. However, to our knowledge, the effects of combining probiotics and carbon sources on aquaculture water quality have rarely been reported. In this study, *B. pumilus* BP-171, a heterotrophic nitrifying-aerobic denitrifying strain isolated from shrimp ponds [29], was tested in a shrimp culture system. This strain can not only convert ammonia nitrogen into bacterial biomass by heterotrophic assimilation but also convert nitrite and nitrate nitrogen into gaseous nitrogen by denitrification [29]. This study showed that the concentration of ammonia nitrogen was reduced by more than 60% in the B and BP groups compared to the control, while the concentration of nitrite nitrogen was reduced by more than 69% in the B and BP groups. The removal rates of ammonia and nitrite nitrogen in the B group reached 70.22% and 76.88%, respectively. The average concentrations of nitrate nitrogen in the B and BP groups were reduced by more than 17%, while the concentrations of total nitrogen in the B and BP groups were reduced by more than 35%. The nitrate and total nitrogen removal rates in group BP were 26.24% and 40.02%, respectively. In conclusion, the addition of *B. pumilus* BP-171 alone could reduce the concentrations of ammonia and nitrite in the culture system, while the simultaneous addition of *B. pumilus* BP-171 and PHBV could reduce the concentrations of nitrate and total nitrogen. In addition, PHBV was better than molasses both as a solid carbon source and as a biofilm carrier when used together with *B. pumilus* BP-171.

4.3. Effects of Addition of B. pumilus 171 and Carbon Sources on the Microbial Diversity and Microbiota Compositions in Water

Bacillus spp. is a type of common probiotic used as a water quality improver in aquaculture systems [7–9,41]. BP-171 is a strain of heterotrophic nitrifying-aerobic denitrifying bacteria isolated from shrimp environments with high nitrogen removal capacity [29]. In this study, a single *B. pumilus* BP-171 and various combinations of *B. pumilus* BP-171 with PHBV and molasses were added to the shrimp culture system. Interestingly, no significant difference in the Ace, Chao, Shannon, and Simpson indices of microbiota was observed between the groups in this study, which was in agreement with the results of Kokkuar et al. [3]. However, the addition of *B. pumilus* BP-171 and carbon sources altered the microbiota composition in the water. The number of OTUs and unique OTUs in the B, BP, and BM groups were all higher than that in the control, and that in the BP group was the highest. In addition, the microbial composition at various taxonomic levels differed distinctly in the different groups.

The abundance of the phylum *Verrucomicrobia*, the class *Verrucomicrobiae*, the order *Verrucomicrobiales*, and the family *Verrucomicrobiaceae* was significantly higher in group B, to which only *B. pumilus* BP-171 was added, than in the other groups. Although the role of *Verrucomicrobia* in aquaculture has rarely been reported, some studies have shown that they were widely distributed in drinking water, freshwater lakes, and marine sediments [42]. In addition, some *Verrucomicrobia* taxa isolated from seawater have been shown to be strictly aerobic chemoheterotrophs that use mono- or disaccharides as carbon and energy sources and can convert nitrate nitrogen to nitrite nitrogen [43,44]. Besides, some *Verrucomicrobia* taxa, which can utilize a variety of organic and inorganic gas molecules such as methane, carbon dioxide, ammonia, and nitrogen gas, were involved in the natural carbon and nitrogen cycles [45]. Thus, *Verrucomicrobia* might involve in the conversion of nitrogen in the culture system, leading to the low concentration of ammonia and nitrite nitrogen in the B group.

The relative abundance of the phylum *Proteobacteria* in the BP group was significantly increased compared with the control when *B. pumilus* BP-171 and PHBV were added simultaneously. Previous studies have shown that the phylum *Proteobacteria* was widely distributed in various regions of the marine environments [46], and many microorganisms involved in nitrogen removal belong to this phylum, including nitrifying and denitrifying bacteria [47]. In this study, it was demonstrated that the abundance of phylum *Proteobacteria* had a negative correlation with the concentration of nitrate and total nitrogen, which might be one origin of a lower concentration of nitrate and total nitrogen in the BP group. In addition, the relative abundance of the class *Alphaproteobacteria* in the BP group reached 75.01%, which was strikingly higher than in other groups. *Alphaproteobacteria* have been shown to have excellent denitrification ability [47,48]. Moreover, the abundance of several dominant bacteria of various taxonomic levels belonging to the class *Alphaproteobacteria* was significantly higher in the BP group than in the control. For example, the order *Rhodobacterales* and *Rhizobiales*, the family *Erythrobacteriaceae* and *Rhizobacteriaceae*, and the genus *Donghicola*, *Oricola*, and *Marivita* as well as *Bacillus*, whose relative abundance was significantly higher in the BP group than in the control. *Rhodobacterales* is considered to be the most abundant denitrifying bacterium widely distributed in the environment [49]. Hu et al. [50] found that the family *Rhodobacteraceae*, as one of the core taxa in shrimp culture ponds, removed nitrite nitrogen from the system mainly by denitrification. Besides, the genus taxa of *Donghicola* and *Marivita*, which were isolated from seawater and are both Gram-negative and aerobic, belong to the class *Rhodobacteraceae*, but their role in bacterial communities has hardly been studied [51]. The order *Rhizobiales*, a type of heterotrophic bacteria with denitrification character, was found to be the second most abundant functional bacterium in ammonia-oxidizing anaerobic systems with a relative abundance of 18.2% [52]. The family *Phyllobacteriaceae* was a group of aerobic bacteria that can utilize various forms of nitrogen for reproduction and was found in marine environments [53,54]. In addition, Zheng et al. [55] identified the most abundant transporter proteins involved in the transport and uptake of carbohydrates from a strain of *Oricola* sp. based on metagenomic and metaproteomic analysis. Among them, three proteins involved in ammonia assimilation and a large number of genes involved in the uptake and metabolism of inorganic nitrogen were also observed in this strain [55]. These results suggest that *Oricola* sp. might be able to utilize carbon sources in the environment and participate in the conversion and removal of nitrogen.

Many studies have shown that *Bacillus* plays an important role in nitrogen cycling via nitrification [56] and denitrification [57]. *B. pumilus* BP-171 was periodically added into different shrimp culture systems in this study. Although an increase in the relative abundance of *Bacillus* spp. compared to other taxa in the microbial community was not observed, the relative abundance of *Bacillus* spp. in groups B, BP, and BM was significantly higher than in the control. Of course, due to methodological limitations, it could not be determined whether the *Bacillus* spp. was the strain BP-171. The results of the present study showed that the microbial composition shifted distinctly at different taxonomic levels when *B. pumilus* BP-171 and different combinations of the strain BP-171 with PHBV and molasses were added. It was also found that the relative abundance of *Oricola* spp. was positively correlated with TOC concentration, while the relative abundance of *Bacillus* spp. was positively correlated with the concentration of TOC and DOC, indicating that the addition of carbon source promoted the proliferation of *Oricola* spp. and *Bacillus* spp. Furthermore, the relative abundance of *Bacillus* spp. showed a significant negative correlation between the concentration of nitrate nitrogen and total nitrogen, while the relative abundance of *Oricola* spp. showed a significant negative correlation with total nitrogen concentration, suggesting an increase in the abundance of *Bacillus* spp. and *Oricola* spp. promoted the conversion and removal of nitrogen. The above correlations between the relative abundance of bacteria, the concentration of TOC and DOC, and the concentration of nitrate and total nitrogen might partially explain the higher removal rate of nitrate nitrogen and total nitrogen in the BP group.

4.4. Effects of Addition of B. pumilus 171 and Carbon Sources on the Ecological Network and Function of the Microbial Community

The complicated ecological network consisted of negative interactions and positive interactions of interspecies in the bacterial community, which sustained the stability of the bacterial ecosystem in water [58,59]. In general, the cooperative network involving mutualism or synergy bacteria can be efficient but not stable [60]. Negative interactions such as competition can weaken the efficiency of the cooperating network but enhance its stability [60]. In this study, a higher ratio of negative to positive interactions was observed in the ecological network of the BP group, suggesting that the addition of PHBV might strengthen the stability of the shrimp culture ecosystem. Moreover, each module was considered a functional unit, performing an identifiable task [58,61]. In the present study, the largest number of modules was observed in the BP group, which indicated that the addition of PHBV could alter the bacteria in the water to perform more biological functions.

The functional modules in the KEGG Module database represent cellular and organismal level functions, and these modules generally contain various molecular level functions stored in the KO (KEGG Orthology) database [62]. The bacterial community in water can affect the growth of aquatic animals in various ways, such as the inhibition of pathogenic bacteria and the secretion of nutrients [9,11,13,15]. The antimicrobial effect of organic acids has been demonstrated [63–65]. Hydroxybutyrate can exert its inhibitory effect against pathogenic *Vibrio* bacteria [66,67]. In this study, the function prediction analysis showed that 8, 12, and 3 functional modules were significantly enhanced in groups B, BP, and BM, respectively. The hydroxypropionate-hydroxybutylate cycle, ethylmalonyl pathway, and the metabolic activity of organic acids, such as fumarate, were significantly enhanced in the BP group. In addition, previous studies have shown that urea dissolved free amino acids, as well as inorganic nitrogen together sustained the nitrogen demand of bacteria for growth in natural water [68,69]. In the present study, the urea cycle, metabolic activities of nutrient substances such as tyrosine, pyruvate, glyceraldehyde-3P, ribose 5P, and cobalamin, as well as molybdenum cofactor were also remarkably enhanced. Just as previous research has shown that many invertebrates like shrimps have demonstrated the ability to take up a variety of organic compounds, including amino acids, even against the concentration gradient [70–74]. Therefore, the overall promotion of numerous metabolic functions of the microbial community in the water might be partially responsible for the improvement in shrimp growth performance.

5. Conclusions

In summary, probiotics and various combinations of probiotics with different carbon resources had differential impacts on the growth performance of shrimp, water quality, and bacterial community in the *P. vannamei* culture system. The addition of BP-171 and carbon sources could promote the growth of shrimp to varying degrees and improve the yield of farmed shrimp, with the best in the group of simultaneous addition of BP-171 and PHBV. The single addition of BP-171 could effectively reduce the concentration of ammonia and nitrite nitrogen in the culture system, and the simultaneous addition of BP-171 and PHBV could effectively improve the removal rates of nitrate and total nitrogen. In addition, the addition of BP-171 and carbon sources did not change the abundance and diversity of the bacterial community in the shrimp culture system but altered the structure and function of the bacterial community and enhanced the stability of the community's ecological network.

Supplementary Materials: The following supporting information can be downloaded at: https://www.mdpi.com/article/10.3390/w14244037/s1, Table S1: Composition of the basic feed; Figure S1: Analysis of rarefaction curves.

Author Contributions: Conceptualization, methodology, and writing—original draft preparation, X.T. and M.W.; investigation, Y.L., K.L. and Q.L.; formal analysis, M.W. and T.L.; visualization, Y.L. and K.L.; data curation and project administration, M.W. and T.L.; writing—review and editing, validation, resources, supervision and funding acquisition, X.T. All authors have read and agreed to the published version of the manuscript.

Funding: This study was funded by the National Key Research and Development Program of China (2020YFD0900201, 2019YFD0900403), and the Joint Fund of Think Tank for Biomanufacturing Industry of Qingdao (QDSWZK202111).

Data Availability Statement: The data from this study are available from the corresponding author upon reasonable request.

Acknowledgments: We thank all the students whom participated the field works and laboratory analysis.

Conflicts of Interest: The authors declare no conflict of interest.

References

1. Goh, J.X.H.; Tan, L.T.; Law, J.W.; Ser, H.; Khaw, K.; Letchumanan, V.; Lee, L.; Goh, B. Harnessing the Potentialities of Probiotics, Prebiotics, Synbiotics, Paraprobiotics, and Postbiotics for Shrimp Farming. *Rev. Aquac.* **2022**, *14*, 1478–1557. [CrossRef]
2. Robles-Porchas, G.R.; Gollas-Galván, T.; Martínez-Porchas, M.; Martínez-Cordova, L.R.; Miranda-Baeza, A.; Vargas-Albores, F. The Nitrification Process for Nitrogen Removal in Biofloc System Aquaculture. *Rev. Aquac.* **2020**, *12*, 2228–2249. [CrossRef]
3. Kokkuar, N.; Li, L.; Srisapoome, P.; Dong, S.; Tian, X. Application of Biodegradable Polymers as Carbon Sources in Ex Situ Biofloc Systems: Water Quality and Shift of Microbial Community. *Aquac. Res.* **2021**, *52*, 3570–3579. [CrossRef]
4. Hlordzi, V.; Kuebutornye, F.K.A.; Afriyie, G.; Abarike, E.D.; Lu, Y.; Chi, S.; Anokyewaa, M.A. The Use of Bacillus Species in Maintenance of Water Quality in Aquaculture: A Review. *Aquac. Rep.* **2020**, *18*, 100503. [CrossRef]
5. Lieke, T.; Meinelt, T.; Hoseinifar, S.H.; Pan, B.; Straus, D.L.; Steinberg, C.E.W. Sustainable Aquaculture Requires Environmental-friendly Treatment Strategies for Fish Diseases. *Rev. Aquac.* **2020**, *12*, 943–965. [CrossRef]
6. Zokaeifar, H.; Babaei, N.; Saad, C.R.; Kamarudin, M.S.; Sijam, K.; Balcazar, J.L. Administration of *Bacillus subtilis* Strains in the Rearing Water Enhances the Water Quality, Growth Performance, Immune Response, and Resistance against Vibrio Harveyi Infection in Juvenile White Shrimp, *Litopenaeus vannamei*. *Fish Shellfish Immunol.* **2014**, *36*, 68–74. [CrossRef]
7. Devaraja, T.; Banerjee, S.; Yusoff, F.; Shariff, M.; Khatoon, H. A holistic approach for selection of *Bacillus* spp. as a bioremediator for shrimp postlarvae culture. *Turk. J. Biol.* **2013**, *37*, 92–100. [CrossRef]
8. Kuebutornye, F.K.A.; Abarike, E.D.; Lu, Y. A Review on the Application of Bacillus as Probiotics in Aquaculture. *Fish Shellfish Immunol.* **2019**, *87*, 820–828. [CrossRef]
9. Soltani, M.; Ghosh, K.; Hoseinifar, S.H.; Kumar, V.; Lymbery, A.J.; Roy, S.; Ringø, E. Genus *Bacillus*, Promising Probiotics in Aquaculture: Aquatic Animal Origin, Bio-Active Components, Bioremediation and Efficacy in Fish and Shellfish. *Rev. Fish. Sci. Aquac.* **2019**, *27*, 331–379. [CrossRef]
10. Crab, R.; Defoirdt, T.; Bossier, P.; Verstraete, W. Biofloc Technology in Aquaculture: Beneficial Effects and Future Challenges. *Aquaculture* **2012**, *356–357*, 351–356. [CrossRef]
11. Abbaszadeh, A.; Keyvanshokooh, S.; Yavari, V.; Naderi, M. Proteome Modifications of Pacific White Shrimp (*Litopenaeus vannamei*) Muscle under Biofloc System. *Aquac. Nutr.* **2019**, *25*, 358–366. [CrossRef]

12. Gao, F.; Liao, S.; Liu, S.; Bai, H.; Wang, A.; Ye, J. The Combination Use of *Candida tropicalis* HH8 and *Pseudomonas stutzeri* LZX301 on Nitrogen Removal, Biofloc Formation and Microbial Communities in Aquaculture. *Aquaculture* **2019**, *500*, 50–56. [CrossRef]

13. El-Saadony, M.T.; Alagawany, M.; Patra, A.K.; Kar, I.; Tiwari, R.; Dawood, M.A.O.; Dhama, K.; Abdel-Latif, H.M.R. The Functionality of Probiotics in Aquaculture: An Overview. *Fish Shellfish Immunol.* **2021**, *117*, 36–52. [CrossRef] [PubMed]

14. Duan, Y.; Zhang, Y.; Dong, H.; Wang, Y.; Zhang, J. Effect of the Dietary Probiotic Clostridium Butyricum on Growth, Intestine Antioxidant Capacity and Resistance to High Temperature Stress in Kuruma Shrimp *Marsupenaeus japonicus*. *J. Therm. Biol.* **2017**, *66*, 93–100. [CrossRef] [PubMed]

15. Dash, P.; Tandel, R.S.; Bhat, R.A.H.; Mallik, S.; Pandey, N.N.; Singh, A.K.; Sarma, D. The Addition of Probiotic Bacteria to Microbial Floc: Water Quality, Growth, Non-Specific Immune Response and Disease Resistance of Cyprinus Carpio in Mid-Himalayan Altitude. *Aquaculture* **2018**, *495*, 961–969. [CrossRef]

16. Dauda, A.B. Biofloc Technology: A Review on the Microbial Interactions, Operational Parameters and Implications to Disease and Health Management of Cultured Aquatic Animals. *Rev. Aquac.* **2020**, *12*, 1193–1210. [CrossRef]

17. Ray, A.J.; Lotz, J.M. Comparing a Chemoautotrophic-Based Biofloc System and Three Heterotrophic-Based Systems Receiving Different Carbohydrate Sources. *Aquac. Eng.* **2014**, *63*, 54–61. [CrossRef]

18. Pérez-Fuentes, J.A.; Hernández-Vergara, M.P.; Pérez-Rostro, C.I.; Fogel, I. C: N Ratios Affect Nitrogen Removal and Production of Nile Tilapia *Oreochromis niloticus* Raised in a Biofloc System under High Density Cultivation. *Aquaculture* **2016**, *452*, 247–251. [CrossRef]

19. De Schryver, P.; Crab, R.; Defoirdt, T.; Boon, N.; Verstraete, W. The Basics of Bio-Flocs Technology: The Added Value for Aquaculture. *Aquaculture* **2008**, *277*, 125–137. [CrossRef]

20. Li, Y.; Li, T.; Tian, X.; Luo, K.; Wang, L.; Zhang, S.; Wei, C.; Liu, Y. Effects of probiotics and polyhydroxybutyrate on growth and non-specific immunity of *Litopenaeus vannamei* and its culturing water quality. *Period. Ocean Univ. China* **2021**, *51*, 22–31. [CrossRef]

21. Xu, Z.; Song, L.; Dai, X.; Chai, X. PHBV Polymer Supported Denitrification System Efficiently Treated High Nitrate Concentration Wastewater: Denitrification Performance, Microbial Community Structure Evolution and Key Denitrifying Bacteria. *Chemosphere* **2018**, *197*, 96–104. [CrossRef]

22. Xu, Z.; Dai, X.; Chai, X. Biological Denitrification Using PHBV Polymer as Solid Carbon Source and Biofilm Carrier. *Biochem. Eng. J.* **2019**, *146*, 186–193. [CrossRef]

23. Barman, P.; Kati, A.; Mandal, A.K.; Bandyopadhyay, P.; Mohapatra, P.K.D. Biopotentiality of *Bacillus cereus* PB45 for Nitrogenous Waste Detoxification in Ex Situ Model. *Aquac. Int.* **2017**, *25*, 1167–1183. [CrossRef]

24. Sun, X.; Li, Q.; Zhang, Y.; Liu, H.; Zhao, J.; Qu, K. Phylogenetic analysis and nitrogen removal characteristics of a heterotrophic nitrifying-aerobic denitrifying bacteria strain from marine environment. *Acta Microbiol. Sin.* **2012**, *52*, 687–695. [CrossRef]

25. Zhao, K.; Tian, X.; Li, H.; Dong, S.; Jiang, W. Characterization of a Novel Marine Origin Aerobic Nitrifying–Denitrifying Bacterium Isolated from Shrimp Culture Ponds. *Aquac. Res.* **2019**, *50*, 1770–1781. [CrossRef]

26. Xie, F.; Thiri, M.; Wang, H. Simultaneous Heterotrophic Nitrification and Aerobic Denitrification by a Novel Isolated *Pseudomonas mendocina* X49. *Bioresour. Technol.* **2021**, *319*, 124198. [CrossRef]

27. Lei, X.; Jia, Y.; Chen, Y.; Hu, Y. Simultaneous Nitrification and Denitrification without Nitrite Accumulation by a Novel Isolated *Ochrobactrum Anthropic* LJ81. *Bioresour. Technol.* **2019**, *272*, 442–450. [CrossRef]

28. Chen, P.; Li, J.; Li, Q.X.; Wang, Y.; Li, S.; Ren, T.; Wang, L. Simultaneous Heterotrophic Nitrification and Aerobic Denitrification by Bacterium *Rhodococcus* Sp. CPZ24. *Bioresour. Technol.* **2012**, *116*, 266–270. [CrossRef]

29. Song, J.; Zhao, K.; Tian, X.; Xie, Y.; He, Y.; Dong, S. Effects of different environmental factors and carbon and nitrogen sources on the nitrogen removing performance of *Bacillus pumilus* BP-171. *Period. Ocean. Univ. China* **2019**, *49*, 34–42. [CrossRef]

30. Avnimelech, Y. Carbon/Nitrogen Ratio as a Control Element in Aquaculture Systems. *Aquaculture* **1999**, *176*, 227–235. [CrossRef]

31. Edgar, R.C. UPARSE: Highly Accurate OTU Sequences from Microbial Amplicon Reads. *Nat. Methods* **2013**, *10*, 996–998. [CrossRef] [PubMed]

32. Wu, D.X.; Zhao, S.M.; Peng, N.; Xu, C.P.; Wang, J.; Liang, Y.X. Effects of a Probiotic (*Bacillus subtilis* FY99-01) on the Bacterial Community Structure and Composition of Shrimp (*Litopenaeus vannamei*, Boone) Culture Water Assessed by Denaturing Gradient Gel Electrophoresis and High-Throughput Sequencing. *Aquac. Res.* **2016**, *47*, 857–869. [CrossRef]

33. Moriarty, D.J.W. Control of Luminous Vibrio Species in Penaeid Aquaculture Ponds. *Aquaculture* **1998**, *164*, 351–358. [CrossRef]

34. Azim, M.E.; Little, D.C. The Biofloc Technology (BFT) in Indoor Tanks: Water Quality, Biofloc Composition, and Growth and Welfare of Nile Tilapia (Oreochromis Niloticus). *Aquaculture* **2008**, *283*, 29–35. [CrossRef]

35. Megahed, M.E.; Mohamed, K. Sustainable Growth of Shrimp Aquaculture Through Biofloc Production as Alternative to Fishmeal in Shrimp Feeds. *J. Agric. Sci.* **2014**, *6*, p176. [CrossRef]

36. Khanjani, M.H.; Sajjadi, M.M.; Alizadeh, M.; Sourinejad, I. Nursery Performance of Pacific White Shrimp (*Litopenaeus vannamei* Boone, 1931) Cultivated in a Biofloc System: The Effect of Adding Different Carbon Sources. *Aquac. Res.* **2017**, *48*, 1491–1501. [CrossRef]

37. Crab, R.; Avnimelech, Y.; Defoirdt, T.; Bossier, P.; Verstraete, W. Nitrogen Removal Techniques in Aquaculture for a Sustainable Production. *Aquaculture* **2007**, *270*, 1–14. [CrossRef]

38. Emerenciano, M.; Cuzon, G.; Arévalo, M.; Miquelajauregui, M.M.; Gaxiola, G. Effect of Short-Term Fresh Food Supplementation on Reproductive Performance, Biochemical Composition, and Fatty Acid Profile of *Litopenaeus vannamei* (Boone) Reared under Biofloc Conditions. *Aquac. Int.* **2013**, *21*, 987–1007. [CrossRef]

39. Ahmad, I.; Babitha Rani, A.M.; Verma, A.K.; Maqsood, M. Biofloc Technology: An Emerging Avenue in Aquatic Animal Healthcare and Nutrition. *Aquac. Int.* **2017**, *25*, 1215–1226. [CrossRef]

40. Lee, C.; Kim, S.; Shin, J.; Kim, M.-G.; Gunathilaka, B.E.; Kim, S.H.; Kim, J.E.; Ji, S.-C.; Han, J.E.; Lee, K.-J. Dietary Supplementations of Bacillus Probiotic Improve Digestibility, Growth Performance, Innate Immunity, and Water Ammonia Level for Pacific White Shrimp, *Litopenaeus vannamei*. *Aquac. Int.* **2021**, *29*, 2463–2475. [CrossRef]

41. Abdel-Latif, H.M.R.; Yilmaz, E.; Dawood, M.A.O.; Ringø, E.; Ahmadifar, E.; Yilmaz, S. Shrimp Vibriosis and Possible Control Measures Using Probiotics, Postbiotics, Prebiotics, and Synbiotics: A Review. *Aquaculture* **2022**, *551*, 737951. [CrossRef]

42. Op den Camp, H.J.M.; Islam, T.; Stott, M.B.; Harhangi, H.R.; Hynes, A.; Schouten, S.; Jetten, M.S.M.; Birkeland, N.-K.; Pol, A.; Dunfield, P.F. Environmental, Genomic and Taxonomic Perspectives on Methanotrophic *Verrucomicrobia*: Perspectives on Methanotrophic *Verrucomicrobia*. *Environ. Microbiol. Rep.* **2009**, *1*, 293–306. [CrossRef]

43. Yoon, J.; Matsuo, Y.; Adachi, K.; Nozawa, M.; Matsuda, S.; Kasai, H.; Yokota, A. Description of *Persicirhabdus sediminis* Gen. Nov., Sp. Nov., *Roseibacillus ishigakijimensis* Gen. Nov., Sp. Nov., *Roseibacillus ponti* Sp. Nov., *Roseibacillus persicicus* Sp. Nov., *Luteolibacter pohnpeiensis* Gen. Nov., Sp. Nov. and *Luteolibacter algae* Sp. Nov., Six Marine Members of the Phylum "*Verrucomicrobia*", and Emended Descriptions of the Class *Verrucomicrobiae*, the Order *Verrucomicrobiales* and the Family *Verrucomicrobiaceae*. *Int. J. Syst. Evol. Microbiol.* **2008**, *58*, 998–1007. [CrossRef] [PubMed]

44. Szuróczki, S.; Abbaszade, G.; Szabó, A.; Bóka, K.; Schumann, P.; Tóth, E. *Phragmitibacter flavus* Gen. Nov., Sp. Nov. a New Member of the Family *Verrucomicrobiaceae*. *Int. J. Syst. Evol. Microbiol.* **2020**, *70*, 2108–2114. [CrossRef] [PubMed]

45. Schmitz, R.A.; Peeters, S.H.; Versantvoort, W.; Picone, N.; Pol, A.; Jetten, M.S.M.; Op den Camp, H.J.M. Verrucomicrobial Methanotrophs: Ecophysiology of Metabolically Versatile Acidophiles. *FEMS Microbiol. Rev.* **2021**, *45*, fuab007. [CrossRef] [PubMed]

46. Arrigo, K.R. Marine Microorganisms and Global Nutrient Cycles. *Nature* **2005**, *437*, 349–355. [CrossRef]

47. Gao, Y.; Wang, X.; Li, J.; Lee, C.T.; Ong, P.Y.; Zhang, Z.; Li, C. Effect of Aquaculture Salinity on Nitrification and Microbial Community in Moving Bed Bioreactors with Immobilized Microbial Granules. *Bioresour. Technol.* **2020**, *297*, 122427. [CrossRef]

48. Wyman, M.; Hodgson, S.; Bird, C. Denitrifying Alphaproteobacteria from the Arabian Sea That Express *NosZ*, the Gene Encoding Nitrous Oxide Reductase, in Oxic and Suboxic Waters. *Appl. Environ. Microbiol.* **2013**, *79*, 2670–2681. [CrossRef]

49. Wang, Y.; Qi, L.; Huang, R.; Wang, F.; Wang, Z.; Gao, M. Characterization of Denitrifying Community for Application in Reducing Nitrogen: A Comparison of NirK and NirS Gene Diversity and Abundance. *Appl. Biochem. Biotechnol.* **2020**, *192*, 22–41. [CrossRef]

50. Hu, D.; Wang, L.; Zhao, R.; Zeng, J.; Shao, Z. Core Microbiome Involved in Nitrite Removal in Shrimp Culture Ponds. *Aquac. Res.* **2022**, *53*, 1663–1675. [CrossRef]

51. Hameed, A.; Shahina, M.; Lin, S.-Y.; Nakayan, P.; Liu, Y.-C.; Lai, W.-A.; Hsu, Y.-H. *Youngimonas vesicularis* Gen. Nov., Sp. Nov., of the Family *Rhodobacteraceae*, Isolated from Surface Seawater, Reclassification of *Donghicola xiamenensis* Tan et al. 2009 as *Pseudodonghicola xiamenensis* Gen. Nov., Comb. Nov. and Emended Description of the Genus *Donghicola yoon* et al. 2007. *Int. J. Syst. Evol. Microbiol.* **2014**, *64*, 2729–2737. [CrossRef] [PubMed]

52. Islam Chowdhury, M.M.; Nakhla, G. Anammox Enrichment: Impact of Sludge Retention Time on Nitrogen Removal. *Environ. Technol.* **2021**, *43*, 4426–4437. [CrossRef] [PubMed]

53. Brailo, M.; Schreier, H.J.; McDonald, R.; Maršić-Lučić, J.; Gavrilović, A.; Pećarević, M.; Jug-Dujaković, J. Bacterial Community Analysis of Marine Recirculating Aquaculture System Bioreactors for Complete Nitrogen Removal Established from a Commercial Inoculum. *Aquaculture* **2019**, *503*, 198–206. [CrossRef] [PubMed]

54. Mergaert, J.; Swings, J. *Phyllobacterium*. In *Bergey's Manual of Systematics of Archaea and Bacteria*; Whitman, W.B., Rainey, F., Kämpfer, P., Trujillo, M., Chun, J., DeVos, P., Hedlund, B., Dedysh, S., Eds.; Wiley: Hoboken, NJ, USA, 2015; pp. 1–7. ISBN 978-1-118-96060-8.

55. Zheng, Q.; Wang, Y.; Lu, J.; Lin, W.; Chen, F.; Jiao, N. Metagenomic and Metaproteomic Insights into Photoautotrophic and Heterotrophic Interactions in a *Synechococcus* Culture. *mBio* **2020**, *11*, e03261-19. [CrossRef]

56. Rout, P.R.; Bhunia, P.; Dash, R.R. Simultaneous Removal of Nitrogen and Phosphorous from Domestic Wastewater Using *Bacillus cereus* GS-5 Strain Exhibiting Heterotrophic Nitrification, Aerobic Denitrification and Denitrifying Phosphorous Removal. *Bioresour. Technol.* **2017**, *244*, 484–495. [CrossRef]

57. Verbaendert, I.; Boon, N.; De Vos, P.; Heylen, K. Denitrification Is a Common Feature among Members of the Genus *Bacillus*. *Syst. Appl. Microbiol.* **2011**, *34*, 385–391. [CrossRef]

58. Deng, Y.; Jiang, Y.-H.; Yang, Y.; He, Z.; Luo, F.; Zhou, J. Molecular Ecological Network Analyses. *BMC Bioinform.* **2012**, *13*, 113. [CrossRef]

59. Venturelli, O.S.; Carr, A.V.; Fisher, G.; Hsu, R.H.; Lau, R.; Bowen, B.P.; Hromada, S.; Northen, T.; Arkin, A.P. Deciphering Microbial Interactions in Synthetic Human Gut Microbiome Communities. *Mol. Syst. Biol.* **2018**, *14*, e8157. [CrossRef]

60. Coyte, K.Z.; Schluter, J.; Foster, K.R. The Ecology of the Microbiome: Networks, Competition, and Stability. *Science* **2015**, *350*, 663–666. [CrossRef]

61. Liu, L.; Wang, M.; Wei, C.; Liu, Y.; Pan, M.; Wang, S.; Cui, L.; Tian, X. Effects of Dietary Poly-β-Hydroxybutyrate Supplementation on the Growth, Non-Specific Immunity, and Intestinal Microbiota of the Sea Cucumber *Apostichopus japonicus*. *Front. Mar. Sci.* **2022**, *9*, 855938. [CrossRef]
62. Kanehisa, M. Enzyme Annotation and Metabolic Reconstruction Using KEGG. In *Protein Function Prediction*; Kihara, D., Ed.; Methods in Molecular Biology; Springer: New York, NY, USA, 2017; Volume 1611, pp. 135–145. ISBN 978-1-4939-7013-1.
63. Peh, E.; Kittler, S.; Reich, F.; Kehrenberg, C. Antimicrobial Activity of Organic Acids against Campylobacter Spp. and Development of Combinations—A Synergistic Effect? *PLoS ONE* **2020**, *15*, e0239312. [CrossRef]
64. Yu, H.; Huang, G.H.; Zhang, X.D.; Li, Y. Inhibitory Effects of Organic Acids on Bacteria Growth During Food Waste Composting. *Compos. Sci. Util.* **2010**, *18*, 55–63. [CrossRef]
65. Nieto-Peñalver, C.G.; Savino, M.J.; Bertini, E.V.; Sánchez, L.A.; de Figueroa, L.I.C. Gluconic Acid Produced by Gluconacetobacter Diazotrophicus Pal5 Possesses Antimicrobial Properties. *Res. Microbiol.* **2014**, *165*, 549–558. [CrossRef] [PubMed]
66. Defoirdt, T.; Mai Anh, N.T.; De Schryver, P. Virulence-Inhibitory Activity of the Degradation Product 3-Hydroxybutyrate Explains the Protective Effect of Poly-β-Hydroxybutyrate against the Major Aquaculture Pathogen *Vibrio campbellii*. *Sci. Rep.* **2018**, *8*, 7245. [CrossRef] [PubMed]
67. Fukami, K.; Takagi, F.; Sonoda, K.; Okamoto, H.; Kaneno, D.; Horikawa, T.; Takita, M. Effects of the Monomeric Components of Poly-Hydroxybutyrate-Co-Hydroxyhexanoate on the Growth of *Vibrio penaeicida* In Vitro and on the Survival of Infected Kuruma Shrimp (*Marsupenaeus japonicus*). *Animals* **2021**, *11*, 567. [CrossRef]
68. Cho, B.; Park, M.; Shim, J.; Azam, F. Significance of Bacteria in Urea Dynamics in Coastal Surface Waters. *Mar. Ecol. Prog. Ser.* **1996**, *142*, 19–26. [CrossRef]
69. Jørgensen, N. Uptake of Urea by Estuarine Bacteria. *Aquat. Microb. Ecol.* **2006**, *42*, 227–242. [CrossRef]
70. Siebers, D. Bacterial—Invertebrate Interactions in Uptake of Dissolved Organic Matter. *Am Zool* **1982**, *22*, 723–733. [CrossRef]
71. Preston, R.L. Transport of Amino Acids by Marine Invertebrates. *J. Exp. Zool.* **1993**, *265*, 410–421. [CrossRef]
72. Moss, S.M.; Pruder, G.D. Characterization of Organic Particles Associated with Rapid Growth in Juvenile White Shrimp, *Penaeus vannamei* Boone, Reared under Intensive Culture Conditions. *J. Exp. Mar. Biol. Ecol.* **1995**, *187*, 175–191. [CrossRef]
73. Vogt, G. Synthesis of Digestive Enzymes, Food Processing, and Nutrient Absorption in Decapod Crustaceans: A Comparison to the Mammalian Model of Digestion. *Zoology* **2021**, *147*, 125945. [CrossRef] [PubMed]
74. Li, X.; Han, T.; Zheng, S.; Wu, G. Nutrition and Functions of Amino Acids in Aquatic Crustaceans. *Amino Acids Nutr. Health* **2021**, *1285*, 169–198. [CrossRef]

Article

Effects of Long-Term High Carbonate Alkalinity Stress on the Ovarian Development in *Exopalaemon carinicauda*

Xiuhong Zhang [1,2,†], Jiajia Wang [1,2,†], Chengwei Wang [1,2], Wenyang Li [1,2], Qianqian Ge [3], Zhen Qin [1,2], Jian Li [1,2] and Jitao Li [1,2,*]

[1] Key Laboratory of Sustainable Development of Marine Fisheries, Ministry of Agriculture and Rural Affairs, Yellow Sea Fisheries Research Institute, Chinese Academy of Fishery Sciences, Qingdao 266071, China
[2] Laboratory for Marine Fisheries Science and Food Production Processes, Pilot National Laboratory for Marine Science and Technology, Qingdao 266071, China
[3] Pilot National Laboratory for Marine Science and Technology, Qingdao 266237, China
* Correspondence: lijt@ysfri.ac.cn; Tel.: +86-532-85826690
† These authors have contributed equally to this work.

Abstract: Saline–alkaline water limits the growth and survival of aquatic animals due to its high carbonate alkalinity, high pH, and various ion imbalances. The ridgetail white prawn *Exopalaemon carinicauda* is strongly adaptable to the saline–alkaline water, making it an excellent candidate species for large-scale aquaculture in saline–alkaline areas. To explore the effect of long-term high carbonate alkalinity stress on ovarian development in *E. carinicauda* for assisting the development of saline–alkaline aquaculture, we performed ovary histology analysis and RNA–sequencing of the eyestalk and ovary in order to compare the transcriptomic responses of individuals in high carbonate alkalinity (8 mmol/L) with a control group (2 mmol/L) for 60 days. It was found that high carbonate alkalinity stress resulted in a loose arrangement of oogonia and a small number of surrounding follicular cells. A total of 1102 differentially expressed genes (DEGs) in ovary tissue were identified under high carbonate alkalinity stress, and the 18 important DEGs were associated with ovarian development. The majority of the DEGs were enriched in ECM–receptor interaction, Folate biosynthesis, the FoxO signaling pathway, insect hormone biosynthesis, and lysosome, which were involved in the ovarian development of *E. carinicauda*. A total of 468 DEGs were identified in eyestalk tissue under high carbonate alkalinity stress, and the 13 important DEGs were associated with ovarian development. KEGG enrichment analysis found that ECM–receptor interaction, folate biosynthesis, lysosome, metabolic pathways, and retinol metabolism may be involved in the ovarian development under high carbonate alkalinity stress. Our results provide new insights and reveal the genes and pathways involved in the ovarian development of *E. carinicauda* under long-term high carbonate alkalinity stress.

Keywords: carbonate alkalinity stress; *Exopalaemon carinicauda*; reproduction; ovary; eyestalk; transcriptome

Citation: Zhang, X.; Wang, J.; Wang, C.; Li, W.; Ge, Q.; Qin, Z.; Li, J.; Li, J. Effects of Long-Term High Carbonate Alkalinity Stress on the Ovarian Development in *Exopalaemon carinicauda*. *Water* **2022**, *14*, 3690. https://doi.org/10.3390/w14223690

Academic Editor: Heiko L. Schoenfuss

Received: 17 October 2022
Accepted: 12 November 2022
Published: 15 November 2022

Publisher's Note: MDPI stays neutral with regard to jurisdictional claims in published maps and institutional affiliations.

1. Introduction

Saline–alkaline water, as the third kind of water, differs from seawater and freshwater and accounts for a large proportion of the world's water resources [1,2]. There are about 46 million hectares of saline–alkaline water areas, which are widely distributed in the northwest, northeast, and north of China, involving 19 provinces, cities, and autonomous regions [3–5]. Saline–alkaline water has a poor buffering capacity, which is characterized by high pH, high carbonate alkalinity, and various types of ion imbalances [1]. Excessive carbonate alkalinity in water substantially affects the development, survival, and reproduction of organisms. Several fishes can adapt to high alkalinity, such as *Leuciscus waleckii* [6], *Gymnocypris przewalskii* [7], and tilapia [8], which provide excellent breeding organisms for the development and utilization of carbonate alkalinity waters for aquaculture.

The ridgetail white prawn *E. carinicauda* is one of the most important commercial shrimps [9], which is widely distributed in the coastal areas of the Yellow Sea and the Bohai Sea in China [10]. Due to its multiple advantages, such as fast growth, high reproductive performance, and good disease resistance [11], the breeding area of *E. carinicauda* in China has expanded in recent years, which may account for one third of the total output of mixed culture ponds [12,13]. Recently in China, *E. carinicauda* has been successfully bred and cultured in saline–alkaline ponds at Dongying city, Shandong province (with salinity of 5–8 and carbonate alkalinity of 1.4–8.0 mmol/L) and Cangzhou city, Hebei province (with salinity of 10–20 and carbonate alkalinity of 3.5–13.0 mmol/L), suggesting that *E. carinicauda* have a high tolerance to saline–alkaline stress [13,14]. *E. carinicauda* shows strong tolerance to salinity and high carbonate alkalinity, and can carry eggs during culture. Therefore, it is a potential species suitable for large-scale culture in saline–alkaline waters. However, as most studies focused on changes in the physiology and gill transcriptomics in *E. carinicauda*, the effects of long-term carbonate alkalinity stress on the reproductive mechanisms in *E. carinicauda* still remains unknown.

Carbonate alkalinity has been considered to be the main stress source affecting the survival, growth, and reproductive traits of aquatic animals in saline–alkaline water [15–17]. High carbonate alkalinity can alter normal metabolism, osmotic pressure regulation capacity, and antioxidant capacity [18]. Yao et al. [3] reported that when the carbonate alkalinity exceeded 15.7 mmol/L, the survival rate of medaka (*Oryzias latipes*) decreased, and morphological abnormalities such as embryo coagulation, embryonic development stagnation, and hatching failure were present. Previous studies have shown that the growth, development, and reproduction indexes in *Moina mongolica* Daday are optimal when the alkalinity is 2.05–4.58 mmol/L. However, when the alkalinity is 6.43–8.98 mmol/L, all indicators showed a downward trend [19]. Xu et al. [20] found that survival rate was not affected by water with a salinity < 3.2 and alkalinity < 14.32 mmol/L in a 72 h embryo tolerance experiment with *Barbus capito*. Water with salinity < 5.1 and alkalinity < 14.32 mmol/L did not affect the 96 h survival of larvae. Liu et al. [21] also found the growth and reproduction of *E. carinicauda* was not affected by low carbonate alkalinity. *E. carinicauda* is better adapted to environments with high carbonate alkalinity, as it adjusts immune enzyme activities. However, research on the influence of carbonate alkalinity in aquatic organisms mainly focuses on survival and growth. Few reports address carbonate alkalinity influences on the mechanisms involved in gonadal development. Therefore, we urgently need to understand the reproduction of aquatic organisms in saline–alkaline waters to aid aquaculture development, especially breeding in saline–alkaline water.

In recent years, transcriptome sequencing technology has been widely used to study the differential expression and molecular pathways of genes under specific environmental stresses [22,23]. For example, RNA–seq compares the transcriptomic responses of *Litopenaeus vannamei* under salinity stress [24]. Li et al. [25] used transcriptome sequencing to reveal the genes and pathways related to salt stress in *Eriocheir sinensis*. However, research on the effect of carbonate alkalinity on aquatic organisms focuses on osmoregulation, with relatively few studies involved in ovarian development. This is the first study involving the transcriptome of the *E. carinicauda* ovary and eyestalk being sequenced by RNA–seq technology. We analyzed the transcriptome data of the eyestalks and ovaries under high carbonate alkalinity stress in order to identify the genes and pathways involved in ovarian development. This study helps to clarify the ovarian development mechanisms in *E. carinicauda* when adapting to high carbonate alkalinity.

2. Materials and Methods

2.1. Sample Collection

Adult female shrimps (body length 2.67 ± 0.20 cm, body weight 0.12 ± 0.08 g) were collected from Haichen Aquaculture Co., Ltd., in Rizhao, Shandong province, China. The shrimps were domesticated in the environment (25 °C) for two weeks. The experiment was performed in 200 L PVC tanks. One hundred and eighty shrimps were divided into two groups randomly, including the high carbonate alkalinity group (carbonate alkalinity 8 ± 0.5 mmol/L, salinity 25 ppt, temperature $26 \pm 0.5\,^{\circ}$C, pH 8.1 ± 0.3, dissolved oxygen 7.5 ± 0.5 mg L^{-1}) and the control group (carbonate alkalinity 2 ± 0.5 mmol/L, salinity 25 ppt, temperature $26 \pm 0.5\,^{\circ}$C, pH 8.8 ± 0.5, dissolved oxygen 7.5 ± 0.5 mg L^{-1}). The experimental design included three replicates of 30 shrimps in each group. The shrimp were fed 3–5% of their body weight twice daily (8:00 and 18:00) during the experimental period. The water was aerated and 30% of the water was changed daily (with the new sea-water adjusted to maintain the original carbonate alkalinity).

After 60 days, thirty–six female shrimps (6 individuals × 3 three replicates × 2 groups) were randomly sampled for Illumina (San Diego, CA, USA) RNA–seq. The ovary and eyestalk samples were obtained and rapidly frozen in liquid nitrogen, then stored at $-80\,^{\circ}$C until RNA isolation, respectively. The ovary samples in the high carbonate alkalinity group are labeled HC_O and the eyestalk samples are labeled HC_E. The ovary samples in the control group are labeled CG_O and the eyestalk samples are labeled CG_E. Six female shrimps (2 individuals × 3 three replicates × 2 groups) were sacrificed for the tissue slices.

2.2. Ovary Histology

The ovaries were fixed in 4% paraformaldehyde for 24 h before washing with $1 \times$ PBS and dehydrated using a graded ethanol series (80% ethanol for 1 h, followed by 95% ethanol for 1 h, followed by 100% ethanol for 1 h). Transparency was improved using xylene (pure ethanol: xylene (1:1) for 1 h, then xylene for 1 h). Samples were infiltrated with paraffin (xylene: paraffin (1:1) at 62 °C for 1 h, then paraffin at 62 °C for 2 h) and processed for paraffin embedding. Sections were cut to 6 µm before staining with hematoxylin and eosin. Samples were scanned using a microscope slide scanner (Pannoramic MIDI, 3DHISTECH Ltd., Budapest, Hungary).

2.3. RNA Isolation, Library Construction and Illumine Sequencing

TRIzol® reagent (Invitrogen, San Diego, CA, USA) extracted the total RNA by following the manufacturer's instructions. The total RNA was treated with DNase I to implement DNA digestion and obtain pure RNA products. Finally, the RNA purity was examined using NanoPhotometer® spectrophotometer NanoDrop 2000 (IMPLEN, Westlake Village, CA, USA). The RNA integrity and concentration was determined using Agilent 2100/4200 system software (Agilent Technologies, Santa Clara, CA, USA). Next, equal amounts of RNA from the same group of different individuals were pooled in order to construct the library. Next–generation sequencing library preparations were constructed according to the manufacturer's instructions (NEBNext® Ultra™ RNA Library Prep Kit for Illumina® (NEB, Ipswich, MA, USA)). After the mRNA library successfully passed the quality inspection, PE150 sequencing was performed using the Illumina NovaSeq 6000 platform (Thermo, Waltham, MA, USA).

2.4. Basic Analysis of Sequencing Data

We removed any technical sequences or their fragments (including adapters, polymerase chain reaction (PCR) primers), and any low quality sequences (with a base quality < 20). We achieved this by filtering the data (in the fastq format) using Trimmomatic (v0.30) to provide only high–quality, clean data in our analysis. Firstly, the whole genome sequence of *E. carinicauda* assembled by our research group was taken as the reference genome (the data have not been published yet). Next, Hisat2 (v2.0.1) was indexed to the reference genome sequence. Finally, we mapped the clean reads to the Silva database to remove the rRNA. All subsequent date analyses were based on clean data without rRNA.

2.5. Differential Expression Genes (DEGs) Analysis and Enrichment Analysis

The DESeq2 and edgeR [26] methods were used to perform the differential expression analysis. The influence of gene length and sequencing volume on the calculated gene expression level was eliminated by the fragments per kilobase per million reads (FPKM) method. The calculated gene expression level can be directly used to compare the expression variations between different genes. We used FPKM to normalize the data. The Benjamini and Hochberg's approach controlled the false discovery rate. $|\log_2 FC| \geq 1$ and p-values < 0.05 were set to detect any significant DEGs. Gene Ontology (GO) and Kyoto Encyclopedia of Genes and Genome (KEGG) enrichment analyses of differentially expressed gene sets were implemented by the topG and KOBAS packages 3.0 [27], respectively.

2.6. RNA-Seq Data Validation by Real–Time Quantitative PCR

Ten DEGs were randomly selected for quantitative reverse transcription–PCR (qRT–PCR) analysis to validate the differential expression of mRNAs. The qRT–PCR assay was performed using SYBR Green PCR Master Mix (Life Technologies, MASS, Waltham, MA, USA) in the 7500 fast Real–Time PCR system (Applied Biosystems, Foster, CA, USA) according to the manufacturer's agreement. The 18S rRNA of *E. carinicauda* was used as the internal reference [28]. All primer sequences and 18S rRNA sequences are listed in Table S1. The relative expression of target genes was calculated with $2^{-\Delta\Delta CT}$ methods. One–way ANOVA method and Duncan's test in the statistical software SPSS 22.0 (SPSS, Chicago, IL, USA) were used for statistical analysis. The results are presented as the mean ± standard error, and differences in gene expression were considered statistically significant at $p < 0.05$.

3. Results

3.1. Histopathology of Ovary

Normal reproduction characteristics were observed in the histological sections of the ovary. The *E. carinicauda* ovary was in Phase I. The oogonia were oval and proliferative, and the nuclei were round. Most cells had one nucleolus, while a few had two nucleoli. The nucleolus stained the deepest color. The cells of the oogonia were closely arranged. A single layer of follicle cells was closely arranged around the oogonia (Figure 1a). After 60 days of high carbonate alkalinity stress, most of the cells in the ovarian cavity were previtellogenic follicles, and the nuclei were large and original, accounting for about half of the cell volume (Figure 1b).

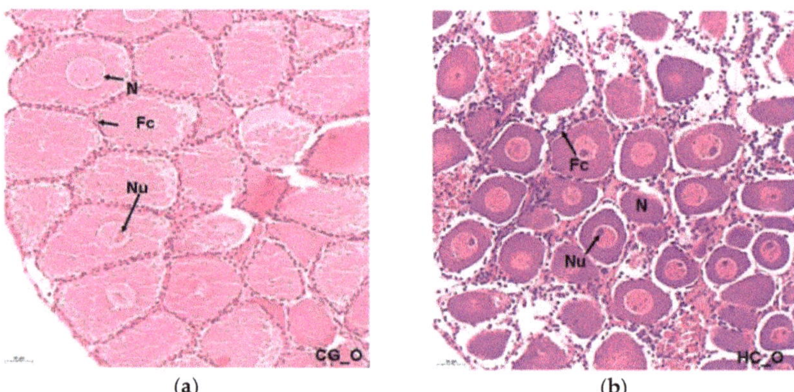

(a)　　　　　(b)

Figure 1. Histological sections of the ovarian status of *E. carinicauda*: (**a**) CG_O group; (**b**) HC_O group. Bar: 50 μm. N: nucleus; Nu: nucleolus; Fc: follicular cell.

3.2. Summary of the RNA–Sequencing Data

Six mRNA libraries ($n = 6$) were constructed from the ovary and eyestalk, respectively, to identify the effects of high carbonate alkalinity on the underlying molecular signaling pathways in *E. carinicauda* ovarian development. The raw data was submitted to NCBI with accession numbers PRJNA881755 and PRJNA881756. A total of 273,545,646 clean reads were obtained from ovaries, with the Q30 (%) varying from 92.63–94.20%. Of these reads, 70.91–72.74% mapped to the reference genome of *E. carinicauda*. A total of 264,648,418 clean reads were obtained from eyestalks, with the Q30 (%) varying from 91.72–92.67%. Of these reads, 79.84–83.02% mapped to the reference genome of *E. carinicauda* (Table S2).

3.3. Differential Gene Expression (DEGs) Analysis

A total of 1102 significant DEGs were identified between the HC_O and CG_O groups. Compared to the control group, 758 up−regulated genes and 344 down−regulated genes were expressed in the ovarian tissue of the HC group (Figure 2a). Compared with the CG_E group, 468 DEG (139 up−regulated and 329 down−regulated) were identified in the HC_E group (Figure 2b). These results suggest that carbonate alkalinity has a significant effect on transcription in the ovary and eyestalk.

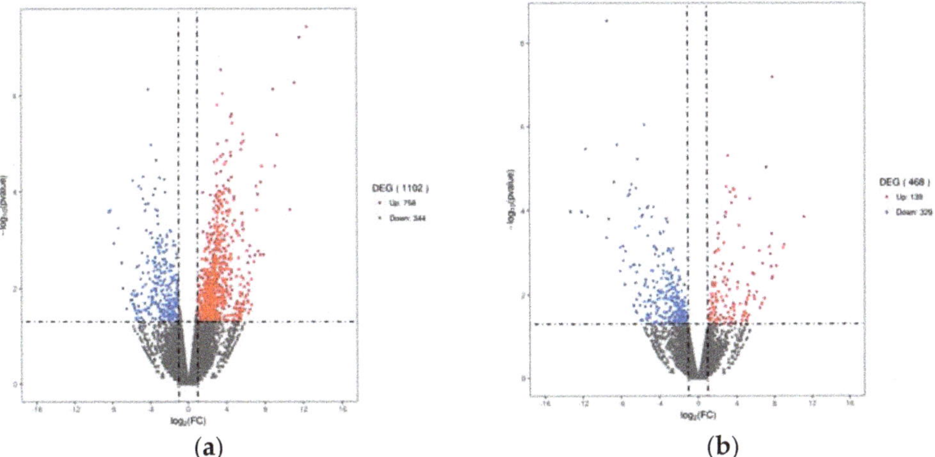

(a)   (b)

Figure 2. The number of differentially expressed genes between different groups. (**a**) HC_O vs. CG_O group; (**b**) HC_E vs. CG_E group. Significantly up−regulated and down−regulated genes are indicated in red and blue, respectively, and those not significantly different are in gray.

3.4. Gene Ontology (GO)Analysis of Significant DEGs

The DEGs of the comparison of HC_O and CG_O were classified into biological process (BP), cellular component (CC), and molecular function (MF). Compared with the CG_O group, up−regulated genes in the HC_O group had enriched steroid metabolic process, sterol homeostasis, molting cycle, follicle cell microvillus organization, steroid biosynthetic process, sterol transport, and estrogen secretion. Conversely, the down−regulated genes in the HC_O group had an enriched positive regulation of ovulation, regulation of ovulation, sterol transport, ovulation, and sterol homeostasis (Table 1).

The DEGs of the comparison of HC_E and CG_E were also classified into biological process (BP), cellular component (CC), and molecular function (MF). Compared with the CG_E group, the up−regulated genes in the HC_E group were enriched in the positive regulation of sterol transport, regulation of sterol transport, embryonic morphogenesis, embryonic organ morphogenesis, embryo development, entry into reproductive diapause, and sterol metabolic process. The down−regulated genes in the HC_E group were enriched in sterol transport, regulation of metaphase/anaphase transition of meiosis I, ecdysteroid metabolic process, sterol homeostasis, maternal placenta development, metaphase/anaphase transition of meiosis I, and the molting cycle (Table 2).

Table 1. The GO term of related to ovarian development in the HC_O vs. CG_O group.

Term	Significant	Up/Down	ID	p Value
Steroid metabolic process	31	up	0008202	5.70×10^{-7}
Sterol homeostasis	11	up	0055092	9.50×10^{-5}
Molting cycle	22	up	0042303	0.00016
Follicle cell microvillus organization	3	up	0032529	0.00031
Retinoid metabolic process	12	up	0001523	0.00052
Steroid biosynthetic process	16	up	0006694	0.00076
Sterol transport	11	up	0015918	0.00486
Response to estrogen	9	up	0043627	0.00875
Estrogen secretion	2	up	0035937	0.01058
Positive regulation of ovulation	3	down	0060279	0.00099
Regulation of ovulation	3	down	0060278	0.00176
Sterol transport	6	down	0015918	0.02138
Ovulation	3	down	0030728	0.03153
Sterol homeostasis	4	down	0055092	0.04769

Table 2. The GO term of related to ovarian development in the HC_E vs. CG_E group.

Term	Significant	Up/Down	ID	p Value
Positive regulation of sterol transport	2	up	0032373	0.00482
Regulation of sterol transport	2	up	0032371	0.01030
Embryonic morphogenesis	10	up	0048598	0.02561
Embryonic organ morphogenesis	5	up	0048562	0.02572
Embryo development	16	up	0009790	0.02894
Entry into reproductive diapause	1	up	0055116	0.04250
Sterol metabolic process	5	up	0008202	0.02501
Sterol transport	7	down	0015918	0.00226
Regulation of metaphase/anaphase transition of meiosis I	1	down	1905186	0.01765
Ecdysteroid metabolic process	3	down	0045455	0.02708
Sterol homeostasis	4	down	0055092	0.02759
Maternal placenta development	2	down	0001893	0.02798
Metaphase/anaphase transition of meiosis I	1	down	1905186	0.01765
Molting cycle	8	down	0042303	0.03756

3.5. Kyoto Encyclopedia of Genes and Genomes (KEGG) Analysis

Comparing the DEGs to the KEGG database of pathway enrichment, the potential functions of the significant DEGs were analyzed in order to further understand the ovarian development of *E. carinicauda* under high carbonate alkalinity stress.

For high carbonate alkalinity stress treatment, 24 KEGG pathways (19 up−regulated and 5 down−regulated pathways) were significantly enriched in the HC_O group (Figure 3). Among these KEGG pathways, some were associated with ovarian development, such as ECM−receptor interaction, folate biosynthesis, FoxO signaling pathway, insect hormone biosynthesis, lysosome, metabolic pathways and neuroactive ligand−receptor interaction.

In the comparison of HC_E and CG_E group, 17 KEGG pathways (5 up−regulated and 12 down−regulated pathways) were significantly enriched (Figure 4). Concurrently, the ECM-receptor interaction, folate biosynthesis, lysosome, metabolic pathways, and retinol metabolism were associated with ovarian development.

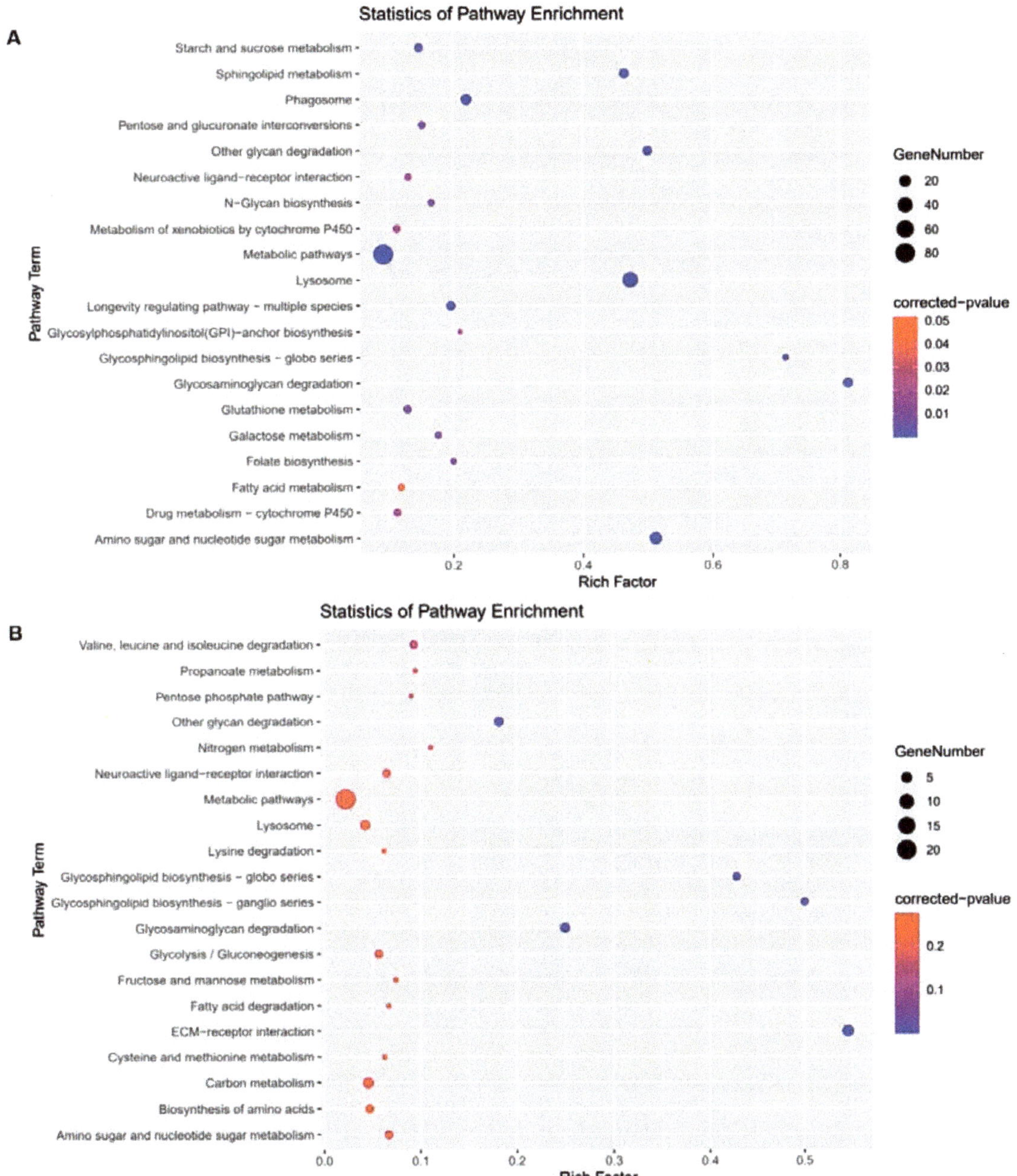

Figure 3. The KEGG significant enrichment pathway in HC_O vs. CG_O group. (**A**) Up−regulation pathways in HC_O vs. CG_O group; (**B**) down−regulation pathways in HC_O vs. CG_O group.

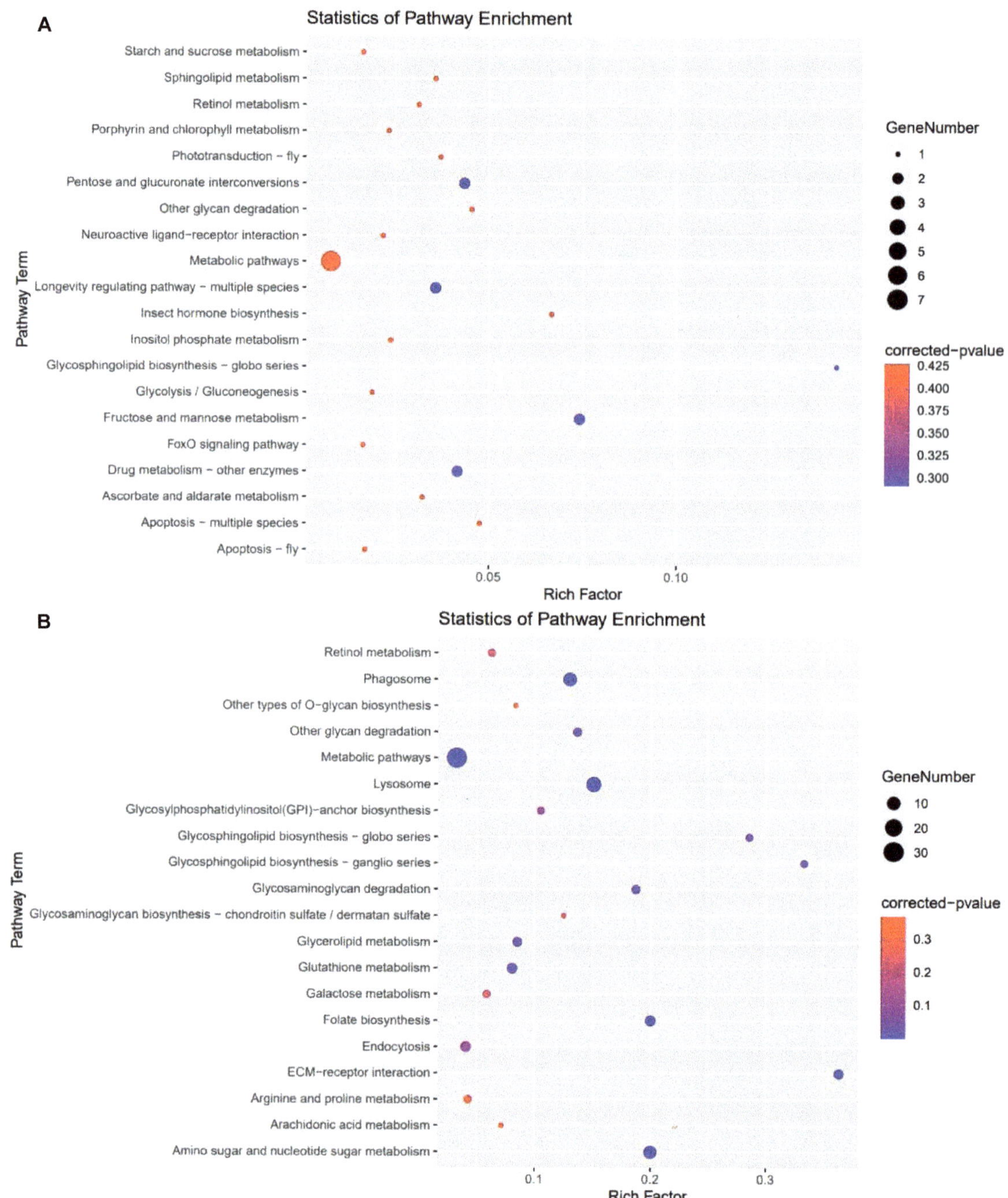

Figure 4. The KEGG significant enrichment pathway in HC_E vs. CG_E group. (**A**) up−regulation pathways in HC_E vs. CG_E group; (**B**) down−regulation pathways in HC_E vs. CG_E group.

3.6. DEGs Involved in Ovarian Development

In the ovary, differential genes involved in ovarian development were selected high carbonate alkalinity stress, such as the G protein−coupled receptor, vitelline membrane outer layer protein 1 (VMO−1), retinol dehydrogenase 11, ecdysone−induced protein 78, ecdysteroid regulated−like protein, voltage−dependent calcium channel gamma−7, methyl farnesoate epoxidase, and estradiol 17−beta−dehydrogenase. A clustering heatmap of these results is provided (Figure 5).

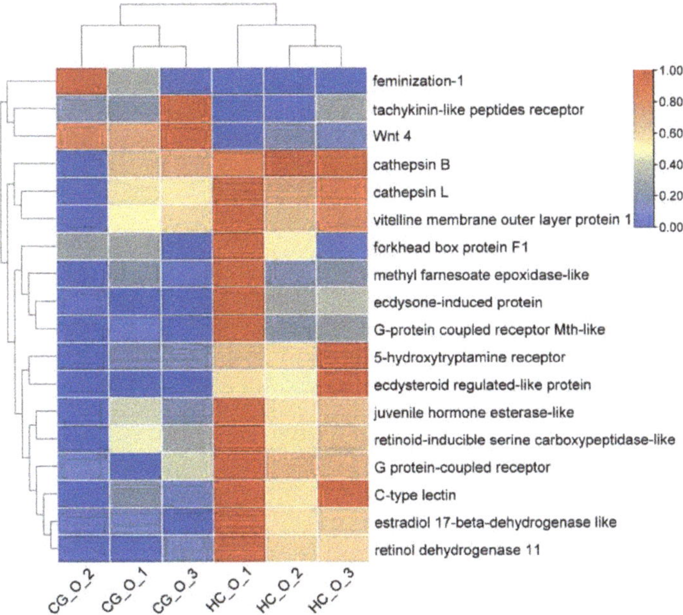

Figure 5. The heatmap of DEGs related to ovarian development in the HC_O vs. CG_O group. The columns and rows indicate individuals and genes, respectively. The color scale represents FPKM after standard normalization, the same as below.

In the eyestalk, the participation of differential genes in ovarian development was observed, and genes were selected which showed high carbonate alkalinity stress. Some DEGs related to ovarian development were screened, such as the vitelline membrane outer layer protein 1, pigment−dispersing hormone 1, insulin−like growth factor binding protein, insulin receptor−related protein, NF−kappa B inhibitor alpha, neuronal acetylcholine receptor, neuroligin−2, and DNA replication complex GINS. A clustering heatmap of these results is provided (Figure 6).

3.7. The Validation of DEGs by qRT−PCR

In order to further validate the reliability of the DEGs identified by RNA−Seq, we randomly selected 10 DEGs from two comparisons: HC_O vs. CG_O and HC_E vs. CG_E. The qRT−PCR results were consistent with those of RNA−seq, indicating that the RNA−Seq data was accurate (Figure 7).

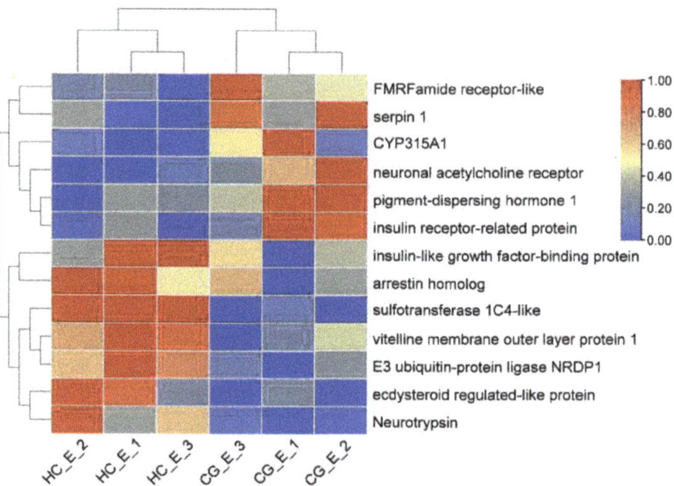

Figure 6. The heatmap of DEGs related to ovarian development in the HC_E vs. CG_E group.

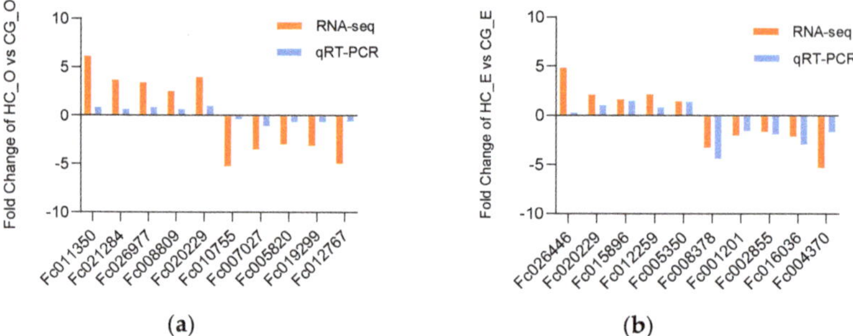

(**a**)   (**b**)

Figure 7. The results were verified by qRT−PCR. (**a**) Relative fold change of DEGs between qRT−PCR and RNA−seq results in HC_O vs. CG_O group. (**b**) Relative fold change of DEGs between qRT-PCR and RNA−seq results in HC_E vs. CG_E group. Relative expression levels from the RNA−seq results were calculated as log2FC values. Fc011350 ecdysteroid regulated; Fc021284 legumain; Fc026977 lysosome−associated membrane glycoprotein 1; Fc008809 baculoviral IAP repeat-containing; Fc020229 N−acetylated−alpha−linked acidic dipeptidase; Fc010755 high−affinity choline transporter 1; Fc007027 Wnt 4; Fc019299 tachykinin−like peptides receptor; Fc005820 reelin 3; Fc012767 facilitated trehalose transporter; Fc026446 neuroligin 2; Fc015896 insulin receptor−related; Fc012259 serpin 1; Fc005350 UDP−glucosyltransferase 2; Fc008378 crustacyanin−C1; Fc001201 legumain; Fc002855 cathepsin L; Fc016036 multidrug resistance−associated protein; Fc004370 macrophage mannose receptor 1.

4. Discussion

Ovarian maturation is an elaborate and complex molecular process, and a large number of genes need to be regulated to ensure the normal development of the oocytes. In this study, we found that 8 mmol/L carbonate alkalinity stress retarded *E. carinicauda* ovarian development, even in non−developing females [21]. The ovaries and eyestalks are important organs in crustacean ovarian development. Eyestalk ablation is commonly used to induce ovarian maturation in shrimp farming [29]. Although the mechanism by which eyestalk ablation leads to ovarian maturation remains inconclusive, some genes expressed in the eyestalk regulate ovarian development. The histological study investigates the influence of high carbonate alkalinity on ovary and oocyte development. We used high-throughput sequencing of ovaries and eyestalks for the first time to understand the ovarian development mechanisms in *E. carinicauda* when in high carbonate alkalinity osmotic stress.

KEGG pathway enrichment can identify major biochemical metabolic and signal transduction pathways involved in genes. In the HC_O group, the insect hormone biosynthesis pathway was significantly enriched, with methyl farnesoate epoxidase, which can catalyze methyl farnesoate into juvenile hormone (JH) III, differentially up−regulated. Methyl farnesoate is a non−epoxidized form of insect juvenile hormone III, which is secreted by the mandibular organs of crustaceans and has a significant stimulating effect on Vg synthesis in various Decapoda species [30–32]. Although JH III plays a very important role in regulating life processes such as growth, molting, and reproduction, it is considered to be an important endogenous hormone in insects and other arthropods [33,34]. Up-regulated methylfarnesoate epoxidase may lead to a decrease in methylfarnesoate synthesis. However, in *Macrobrachium rosenbergii* [35] and *Sagmariasus verreauxi* [36], metabolic enzymes convert methyl farnesoate into JH, which supports the view that JH is also an active hormone in crustaceans. Furthermore, methyl farnesoate epoxidase can directly convert farnesoic acid into JH III acid to form JH III. Therefore, the up−regulation of methyl farnesoate epoxidase promotes the synthesis of JH, thus affecting ovarian development in crustaceans.

Lysosomes were the most significant pathways, with a larger number of DEGs varying between the HC_O and CG_O groups as well as between the HC_E and CG_E groups. Lysosomes are important for intracellular trafficking, metabolic signaling, lipid metabolism, and immune responses [37]. Lysosomes are involved in the preparation of free cholesterol for steroidogenesis and degradation of steroidogenesis regulators, as well as follicle rupture during ovulation in the ovaries of vertebrates [38]. As the central digestive organ of cells, various macromolecules are sent to lysosomes for degradation. Vitellogenin is an important precursor of egg yolk in nearly all oviparous animals [39]. Lysosomes play an important role in the degradation of vitellogenin, which is internalized by endocytosis [40]. Lysosomes are related to the hydrolysis of vitellogenin and energy demand during *Macrobrachium nipponense* ovarian maturation [41]. Lysosomal enzymes, especially cathepsin B and L, are associated with ovarian development in crustaceans [42,43]. We also found that cathepsin B and L in the ovaries of *E. carinicauda* were significantly up−regulated in response to carbonate alkalinity stress, which was consistent with the above results. In addition, we found that the lysosomal pathway was significantly up−regulated in the ovaries and significantly down−regulated in the eyestalks. Therefore, we speculate that this change in lysosomes is closely related to the ovarian development of *E. carinicauda* under carbonate alkalinity stress. However, the specific mechanism is unknown.

In this study, significant differential genes provide evidence of response to high carbonate alkalinity stress in the ovary and eyestalk, which are the target tissues that regulate ovarian development. As the largest membrane receptor family in eukaryotes, G protein coupled receptors are involved in regulating many key physiological and biochemical processes, including sexual maturation and reproduction [44,45]. There are many important factors controlling the development and maturation of shrimp ovaries, such as neurotransmitters, hormones, and their receptors [46,47], which mainly bind and activate G protein coupled receptors on the cell surface [48] and initiate multiple downstream cascades [49]. In this study, G protein−coupled receptor and Mth were significantly up−regulated in the HC_O group, which indicated that the expression of G protein−coupled receptor and Mth may regulate the ovarian development of *E. carinicauda* under high carbonate alkalinity stress.

As a glycoprotein bound to the cell surface, lectin specifically recognizes carbohydrates [50] and plays an important role in the innate immunity of invertebrates [51,52]. Qin et al. found that pmcl1 plays an important role in the immune response to pathogen infection and ammonia nitrogen stress [53]. Tateno et al. suggests that lectins in fish oocytes may prevent polysperm fertilization, regulate carbohydrate metabolism, participate in the formation of the fertilization shell after binding with glycoproteins, determine the source of disease, and have antibacterial effects [50]. Therefore, we speculate that C−type lectins may perform similar functions in *E. carinicauda*, but the specific role requires further research. Additionally, lectin and vitellin VMO1/2 are closely bound to ovomucin to form the basic skeleton of the outer membrane of vitellin [54]. The VMO−1 protein is essential for the formation of the outer membrane in chicken eggs [55], and the adhesion between yolk membranes may be related to the VMO−2 protein in quail (*Coturnix japonica*) eggs [56]. VMO−1 is an important protein in the development of oocytes [57], and its main function is to prevent mixing between yolk and protein in crayfish [58]. In this study, the expression of VMO−1 and C−type lectin was up−regulated in the HC_O group, suggesting that VMO−1 and lectin affects ovarian development in *E. carinicauda* under osmotic stress.

The crustacean eyestalk is known to regulate reproduction, molting, and energy metabolism [59,60]. Removing the eyestalk can induce ovarian maturation and oviposition in many crustaceans [61,62]. Eyestalk−derived neuropeptides regulate vitellogenesis in crustaceans. Pigment−dispersing hormone (PDH) is involved in the regulation of ovarian maturation in crustaceans [63]. The PDH may participate in vitellogenesis according to their spatiotemporal expression patterns, which maintained a high level from the pre−vitellogenesis stage and decreased significantly in the mature stage in *Scylla paramamosain* [64,65]. Wei et al. provided evidence of the inductive effect of PDH on oocyte meiotic maturation in *E. sinensis* [66]. In this study, the PDH was up−regulated in the HC_E group, suggesting that it may participate in the ovarian development of *E. carinicauda* under high carbonate alkalinity stress.

Retinol and its derivatives play key roles in the meiosis of mammalian fetal ovarian germ cells [67], follicular development [68], ovarian steroidogenesis [69], and oocyte maturation [70]. The retinol dehydrogenase (RDH) is a member of the short−chain dehydrogenase/reductase (SDR) superfamily, which includes three RDHs, RDH11, RDH12, and RDH13 in transcriptome sequences. The expression level of RDH13 in a vitellogenesis ovary of zebrafish was significantly higher than that in a non−vitellogenesis ovary [71]. In addition, knockout of RDH11 resulted in decreased transcription of vitellogenin and vitellogenin receptors in *Procambarus clarkii* [72], suggesting that RDH11 might play an important role in the synthesis and conveyance of vitellogenin in crustaceans. Our previous study revealed that RDH11 is critical for ovarian development in *E. carinicauda* [73]. In this study, RDH11, associated with ovarian development in *E. carinicauda*, was significantly expressed under high carbonate alkalinity stress.

5. Conclusions

In the present study, the transcriptome analysis of the ovary and eyestalk of *E. carinicauda* under high carbonate alkalinity stress, as well as the effects of high carbonate alkalinity on ovarian development-related genes and signaling pathways, are described for the first time. Eighteen and thirteen DEGs in ovary and eyestalk tissue were identified, respectively. The key genes were identified to be involved in the folate biosynthesis, insect hormone biosynthesis, lysosome, and retinol metabolism, which play an essential role in the response of the ovaries and eyestalks of *E. carinicauda* to high carbonate alkalinity stress. This study provided new insights into the ovarian development of *E. carinicauda* under high carbonate alkalinity stress, which could be useful for saline–alkaline water aquaculture and related studies on the reproduction of crustaceans.

Supplementary Materials: The following supporting information can be downloaded at: https://www.mdpi.com/article/10.3390/w14223690/s1, Table S1: Primers of RT−PCR designed for validation experiment of DEGs; Table S2: Number of high-throughput clean reads and mapped clean reads generated from *E. carinicauda* ovary and eyestalk mRNA library.

Author Contributions: X.Z., J.W. and J.L. (Jitao Li) conceived and designed the research; X.Z., J.W., C.W., W.L., Q.G. and Z.Q. performed the experiments and analyzed the data; J.L. (Jian Li) provided research ideas for the experiments. All authors have read and agreed to the published version of the manuscript.

Funding: This research was funded by National Key R&D Program of China (2019YFD0900404-03), National Natural Science Foundation of China (32072974), China Agriculture Research System of MOF and MARA (CARS-48) and Central Public-Interest Scientific Institution Basal Research Fund, CAFS (2020TD46).

Institutional Review Board Statement: No human subjects were included in this study. The animal study was reviewed and approved by the Animal Ethics Committee of Yellow Sea Research Institute, CAFS (ID Number: YSFRI–2022033).

Informed Consent Statement: Not applicable.

Data Availability Statement: The datasets presented in this study can be found in online repositories. The names of the repository/repositories and accession number(s) can be found below: Center for Biotechnology Information (NCBI) with the accession number PRJNA881755 and PRJNA881756.

Conflicts of Interest: The authors declare no conflict of interest.

References

1. Sharma, B.R.; Minhas, P.S. Strategies for managing saline/alkali waters for sustainable agricultural production in South Asia. *Agric. Water Manag.* **2005**, *78*, 136–151. [CrossRef]
2. Wang, Y.; Yang, C.; Liu, G.; Jing, J. Development of a cDNA microarray to identify gene expression of *Puccinellia tenuiflora* under saline-alkali stress. *Plant Physiol. Bioch.* **2007**, *45*, 567–576. [CrossRef] [PubMed]
3. Yao, Z.L.; Lai, Q.F.; Zhou, K.; Rizalita, R.E.; Wang, H. Developmental biology of medaka fish (*Oryzias latipes*) exposed to alkalinity stress. *J. Appl. Ichthyol.* **2010**, *26*, 397–402. [CrossRef]
4. Fan, Z.; Wu, D.; Zhang, Y.Y.; Li, J.; Xu, Q.Y.; Wang, L.S. Carbonate alkalinity and dietary protein levels affected growth performance, intestinal immune responses and intestinal microflora in Songpu mirror carp (*Cyprinus carpio* Songpu). *Aquaculture* **2021**, *545*, 737135. [CrossRef]
5. Lin, T.; Lai, Q.; Yao, Z.; Lu, J.; Zhou, K.; Wang, H. Combined effects of carbonate alkalinity and pH on survival, growth and haemocyte parameters of the Venus clam *Cyclina sinensis*. *Fish Shellfish Immun.* **2013**, *35*, 525–531. [CrossRef]
6. Xu, J.; Ji, P.; Wang, B.; Zhao, L.; Wang, J.; Zhao, Z.; Zhang, Y.; Li, J.; Xu, P.; Sun, X. Transcriptome sequencing and analysis of wild amur ide (*Leuciscus waleckii*) inhabiting an extreme alkaline–saline lake reveals insights into stress adaptation. *PLoS ONE* **2013**, *8*, e59703. [CrossRef]
7. Li, H.; Lai, Q.; Yao, Z.; Liu, Y.; Sun, Z. Ammonia excretion and blood gas variation in naked carp (*Gymnocypris przewalskii*) exposed to acute hypoxia and high alkalinity. *Fish Physiol. Biochem.* **2020**, *46*, 1981–1990. [CrossRef]
8. Su, H.; Ma, D.; Zhu, H.; Liu, Z.; Gao, F. Transcriptomic response to three osmotic stresses in gills of hybrid tilapia (*Oreochromis mossambicus* female × *O. urolepis hornorum* male). *BMC Genom.* **2020**, *21*, 110. [CrossRef]
9. Li, J.T.; Li, J.; Duan, Y.F.; Chen, P.; Liu, P. The roles of heat shock proteins 70 and 90 in *Exopalaemon carinicauda* after WSSV and *Vibrio anguillarum* challenges. *J. Ocean Univ. China* **2018**, *17*, 399–406. [CrossRef]
10. Feng, Y.Y.; Zhai, Q.Q.; Wang, J.J.; Li, J.T.; Li, J. Comparison of florfenicol pharmacokinetics in *Exopalaemon carinicauda* at different temperatures and administration routes. *J. Vet. Pharmacol. Ther.* **2019**, *42*, 230–238. [CrossRef]
11. Li, J.T.; Li, J.; Chen, P.; Liu, P.; He, Y.Y. Transcriptome analysis of eyestalk and hemocytes in the ridgetail white prawn *Exopalaemon carinicauda*: Assembly, annotation and marker discovery. *Mol. Biol. Rep.* **2015**, *42*, 135–147. [CrossRef] [PubMed]
12. Li, J.T.; Ma, P.; Ping, L.; Chen, P.; Jian, L. The roles of Na^+/K^+-ATPase α-subunit gene from the ridgetail white prawn *Exopalaemon carinicauda* in response to salinity stresses. *Fish Shellfish Immun.* **2015**, *42*, 264–271. [CrossRef] [PubMed]
13. Ge, Q.Q.; Li, J.; Wang, J.J.; Li, Z.D.; Li, J.T. Characterization, functional analysis, and expression levels of three carbonic anhydrases in response to pH and saline–alkaline stresses in the ridgetail white prawn *Exopalaemon carinicauda*. *Cell Stress Chaperon.* **2019**, *24*, 503–515. [CrossRef] [PubMed]
14. Chang, Z.Q.; Neori, A.; He, Y.Y.; Li, J.T.; Li, J. Development and current state of seawater shrimp farming, with an emphasis on integrated multi–trophic pond aquaculture farms, in China–a review. *Rev. Aquacult.* **2020**, *4*, 2544–2558. [CrossRef]
15. González-Vera, C.; Brown, J.H. Effects of alkalinity and total hardness on growth and survival of postlarvae freshwater prawns, *Macrobrachium rosenbergii* (De Man 1879). *Aquaculture* **2017**, *473*, 521–527. [CrossRef]
16. Yao, Z.L.; Wang, H.; Chen, L.; Zhou, K.; Ying, C.Q.; Lai, Q.F. Transcriptomic profiles of Japanese medaka (*Oryzias latipes*) in response to alkalinity stress. *Genet. Mol. Res.* **2012**, *11*, 2200–2246. [CrossRef]
17. Mcfarland, K.; Donaghy, L.; Volety, A.K. Effect of acute salinity changes on hemolymph osmolality and clearance rate of the non-native mussel, *Perna viridis*, and the native oyster, *Crassostrea virginica*, in Southwest Florida. *Aquat. Invasions* **2013**, *8*, 299–310. [CrossRef]
18. Sun, Y.C.; Han, S.C.; Yao, M.Z.; Liu, H.B.; Wang, Y.M. Exploring the metabolic biomarkers and pathway changes in crucian under carbonate alkalinity exposure using high–throughput metabolomics analysis based on UPLC–ESI–QTOF–MS. *RSC Adv.* **2020**, *10*, 1552–1571. [CrossRef]
19. Zhao, W.; Wang, C.; Zhang, L.; Wei, J.; Yang, W.; Teng, L. Effects of alkalinity and pH on the survival, growth and neonate production of two strains of *Moina mongolica* Daday. *Acta Ecol. Sin.* **2009**, *2*, 589–598, (In Chinese with English Abstract).

20. Xu, W.; Geng, L.W.; Li, C.T.; Jin, G.X.; Liu, X.Y. The artificial propagation, embryonic development and saline–alkali tolerant experiment of *Barbus capito*. *J. Fish. China* **2011**, *2*, 255–260, (In Chinese with English Abstract).

21. Liu, F.; Li, J.; Li, J.T.; Ge, Q.Q.; Ge, H.X.; Shen, M.M. Effects of carbonate alkalinity stress on the survival, growth, reproduction, and immune enzyme activities of *Exopalaemon carinicauda*. *J. Fish. Sci. China* **2016**, *5*, 1137–1147, (In Chinese with English Abstract).

22. Dugas, D.V. Functional annotation of the transcriptome of *Sorghum bicolor* in response to osmotic stress and abscisic acid. *BMC Genom.* **2011**, *12*, 514. [CrossRef] [PubMed]

23. Teets, N.M.; Peyton, J.T.; Colinet, H.; Renault, D.; Denlinger, D.L. Gene expression changes governing extreme dehydration tolerance in an Antarctic insect. *Proc. Natl. Acad. Sci. USA* **2013**, *109*, 20744–20749. [CrossRef] [PubMed]

24. Chen, K.; Li, E.C.; Li, T.Y.; Xu, C.; Wang, X.D.; Lin, H.Z.; Qin, J.G.; Chen, L.Q. Transcriptome and molecular pathway analysis of the hepatopancreas in the pacific white shrimp *Litopenaeus vannamei* under chronic low–salinity stress. *PLoS ONE* **2015**, *10*, e131503. [CrossRef] [PubMed]

25. Li, E.; Wang, S.; Li, C.; Wang, X.; Chen, K.; Chen, L. Transcriptome sequencing revealed the genes and pathways involved in salinity stress of Chinese mitten crab, *Eriocheir sinensis*. *Physiol. Genom.* **2014**, *46*, 177–190. [CrossRef]

26. Robinson, M.D.; Mccarthy, D.J.; Smyth, G.K. edgeR: A Bioconductor package for differential expression analysis of digital gene expression data. *Biogeosciences* **2010**, *26*, 139–140. [CrossRef]

27. Bu, D.H.; Luo, H.T.; Huo, P.P.; Wang, Z.H.; Zhang, S.; He, Z.H.; Wu, Y.; Zhao, L.H.; Liu, J.J.; Guo, J.C.; et al. KOBAS–i: Intelligent prioritization and exploratory visualization of biological functions for gene enrichment analysis. *Nucleic Acids Res.* **2021**, *49*, 317–325. [CrossRef]

28. Duan, Y.F.; Liu, P.; Li, J.T.; Li, J.; Gao, B.Q. cDNA cloning, characterization and expression analysis of peroxiredoxin 5 gene in the ridgetail white prawn *Exopalaemon carinicauda*. *Mol. Biol. Rep.* **2013**, *40*, 6569–6577. [CrossRef]

29. Sittikankaew, K.; Pootakham, W.; Sonthirod, C.; Sangsrakru, D.; Yoocha, T.; Khudet, J.; Nookaew, I.; Uawisetwathana, U.; Rungrassamee, W.; Karoonuthaisiri, N. Transcriptome analyses reveal the synergistic effects of feeding and eyestalk ablation on ovarian maturation in black tiger shrimp. *Sci. Rep.* **2020**, *10*, 3229.

30. Mak, A.S.C.; Choi, C.L.; Tiu, S.H.K.; Hui, J.H.L.; He, J.G.; Tobe, S.S.; Chan, S.M. Vitellogenesis in the red crab *Charybdis feriatus*: Hepatopancreas–specific expression and farnesoic acid stimulation of vitellogenin gene expression. *Mol. Reprod. Dev.* **2005**, *70*, 288–300. [CrossRef]

31. Yang, Y.; Ye, H.H.; Huang, H.Y.; Jin, Z.X.; Li, S.J. Cloning, expression and functional analysis of farnesoic acid O-methyltransferase (FAMeT) in the mud crab, *Scylla paramamosain*. *Mar. Freshw. Behav. Phy.* **2012**, *45*, 209–222. [CrossRef]

32. Miyakawa, H.; Toyota, K.; Sumiya, E.; Iguchi, T. Comparison of JH signaling in insects and crustaceans. *Curr. Opin. Insect Sci.* **2014**, *1*, 81–87. [CrossRef] [PubMed]

33. Luo, W.; Veeran, S.; Wang, J.; Li, S.; Liu, S.N. Dual roles of juvenile hormone signaling during early oogenesis in *Drosophila*. *Insect Sci.* **2019**, *27*, 665–674. [CrossRef]

34. Suang, S.; Hiruma, K.; Yu, K.; Manaboon, M. Diapause hormone directly stimulates the prothoracic glands of diapause larvae under juvenile hormone regulation in the bamboo borer, *Omphisa fuscidentalis* Hampson. *Arch. Insect Biochem.* **2019**, *102*, e21603. [CrossRef] [PubMed]

35. Tomer, V.; Rivka, M.; Aflalo, E.D.; Vered, C.C.; Simy, W.; Omri, S.; Amir, S.; Yu, J. Post–embryonic transcriptomes of the prawn *Macrobrachium rosenbergii*: Multigenic succession through metamorphosis. *PLoS ONE* **2013**, *8*, e55322.

36. Ventura, T.; Fitzgibbon, Q.P.; Battaglene, S.C.; Elizur, A. Redefining metamorphosis in spiny lobsters: Molecular analysis of the phyllosoma to puerulus transition in *Sagmariasus verreauxi*. *Sci. Rep.* **2015**, *5*, 13537. [CrossRef]

37. Lamming, D.W.; Bar-Peled, L. Lysosome: The metabolic signaling hub. *Traffic* **2019**, *20*, 27–38. [CrossRef]

38. Li, Y.; Wang, Z.; Andersen, C.L.; Ye, X. Functions of lysosomes in mammalian female reproductive system. *Reprodu. Dev. Med.* **2020**, *4*, 109–122. [CrossRef]

39. Harwood, G.; Amdam, G. Vitellogenin in the honey bee midgut. *Apidologie* **2021**, *52*, 837–847. [CrossRef]

40. Carnevali, O.; Cionna, C.; Tosti, L.; Lubzens, E.; Maradonna, F. Role of cathepsins in ovarian follicle growth and maturation. *Gen. Comp. Endocr.* **2006**, *146*, 195–203. [CrossRef]

41. Zhang, Y.N.; Jiang, S.F.; Qiao, H.; Xiong, Y.W.; Fu, H.T.; Zhang, W.Y.; Gong, Y.S.; Jin, S.B.; Wu, Y. Transcriptome analysis of five ovarian stages reveals gonad maturation in female updates *Macrobrachium nipponense*. *BMC Genom.* **2021**, *22*, 510. [CrossRef] [PubMed]

42. Luo, B.Y.; Qian, H.L.; Jiang, H.C.; Xiong, X.Y.; Ye, B.Q.; Liu, X.; Guo, Z.Q.; Ma, K.Y. Transcriptional changes revealed water acidification leads to the immune response and ovary maturation delay in the Chinese mitten crab *Eriocheir sinensis*. *Comp. Biochem. Phys. D* **2021**, *39*, 100868. [CrossRef] [PubMed]

43. Wu, P.; Qi, D.; Chen, L.Q.; Zhang, H.; Zhang, X.W.; Qin, J.G.; Hu, S.N. Gene discovery from an ovary cDNA library of oriental river prawn *Macrobrachium nipponense* by ESTs annotation. *Comp. Biochem. Phys. D* **2009**, *4*, 111–120. [CrossRef] [PubMed]

44. Nguyen, T.V.; Rotllant, G.E.; Cummins, S.F.; Elizur, A.; Ventura, T. Insights into sexual maturation and reproduction in the norway lobster (*Nephrops norvegicus*) via in silico prediction and characterization of neuropeptides and G protein–coupled receptors. *Front. Endocrinol.* **2018**, *9*, 430. [CrossRef]

45. Flaherty, P.; Radhakrishnan, M.L.; Dinh, T.; Rebres, R.A.; Roach, T.I.; Jordan, M.I.; Arkin, A.P. A dual receptor crosstalk model of G protein coupled signal transduction. *PLoS Comput. Biol.* **2008**, *4*, e1000185. [CrossRef]

46. Subramoniam, T. Mechanisms and control of vitellogenesis in crustaceans. *Fish. Sci.* **2011**, *77*, 1–21. [CrossRef]

47. Crown, A.; Clifton, D.K.; Steiner, R.A. Neuropeptide signaling in the integration of metabolism and reproduction. *Neuroendocrinology* **2016**, *86*, 175–182. [CrossRef]

48. Sharabi, O.; Manor, R.; Weil, S.; Aflalo, E.D.; Lezer, Y.; Levy, T.; Aizen, J.; Ventura, T.; Mather, P.B.; Khalaila, I. Identification and characterization of an insulin–like receptor involved in crustacean reproduction. *Endocrinology* **2016**, *157*, 928–941. [CrossRef]

49. Christie, A.E.; Stemmler, E.A.; Dickinson, P.S. Crustacean neuropeptides. *Cell. Mol. Life Sci.* **2010**, *67*, 4135–4169. [CrossRef]

50. Tateno, H.; Saneyoshi, A.; Ogawa, T.; Muramoto, K.; Kamiya, H.; Saneyoshi, M. Isolation and characterization of rhamnose-binding lectins from eggs of steelhead trout (*Oncorhynchus mykiss*) homologous to low density lipoprotein receptor superfamily. *J. Biol. Chem.* **1998**, *273*, 19190–19197. [CrossRef]

51. Dodd, R.B.; Kurt, D. Lectin-like proteins in model organisms: Implications for evolution of carbohydrate–binding activity. *Glycobiology* **2001**, *11*, 71–79. [CrossRef] [PubMed]

52. Luo, Z.; Zhang, J.Q.; Li, F.H.; Liu, C.Z.; Xiang, J.H. Cloning of a novel C–type lectin LvLec2 from the shrimp *Litopenaeus vannamei* and its immune response to different challenges. *Mar. Sci.* **2010**, *34*, 103–110.

53. Qin, Y.; Jiang, S.; Huang, J.; Zhou, F.; Yang, Q.; Jiang, S.; Yang, L. C–type lectin response to bacterial infection and ammonia nitrogen stress in tiger shrimp (*Penaeus monodon*). *Fish Shellfish Immun.* **2019**, *90*, 188–198. [CrossRef]

54. Kido, S.; Morimoto, A.; Kim, F.; Doi, Y. Isolation of a novel protein from the outer layer of the vitelline membrane. *Biochem. J.* **1992**, *286*, 17–22. [CrossRef] [PubMed]

55. Back, J.F.; Bain, J.M.; Vadehra, D.V.; Burley, R.W. Proteins of the outer layer of the vitelline membrane of hen's eggs. *Biochim. Biophys. Acta.* **1982**, *705*, 12–19. [CrossRef]

56. Rahman, M.A.; Akihiko, M.; Atsushi, I.; Norio, Y. VMO–II mediates the binding of the chalaziferous layer with the vitelline membrane in quail eggs. *J. Poult. Sci.* **2009**, *46*, 240–248. [CrossRef]

57. Gismondi, E.; Thome, J.P.; Urien, N.; Uher, E.; Baiwir, D.; Mazzucchelli, G.; De Pauw, E.; Fechner, L.C.; Lebrun, J.D. Ecotoxicoproteomic assessment of the functional alterations caused by chronic metallic exposures in gammarids. *Environ. Pollut.* **2017**, *225*, 428–438. [CrossRef]

58. Sricharoen, S.; Kim, J.J.; Tunkijjanukij, S.; Soderhall, I. Exocytosis and proteomic analysis of the vesicle content of granular hemocytes from a crayfish. *Dev. Comp. Immunol.* **2005**, *29*, 1017–1031. [CrossRef]

59. Philip, B.; Abigail, E.; Richard, W.; Cummins, S.F.; Wayne, K. Gene expression profiling of the cephalothorax and eyestalk in penaeus monodon during ovarian maturation. *Int. J. Biol. Sci.* **2012**, *8*, 328–343.

60. Jung, H.T.; Lyons, R.E.; Hurwood, D.A.; Mather, P. Genes and growth performance in crustacean species: A review of relevant genomic studies in crustaceans and other taxa. *Rev. Aquacult.* **2013**, *5*, 77–110. [CrossRef]

61. Koshio, S.; Teshima, S.I.; Kanazawa, A. Effects of unilateral eyestalk ablation and feeding frequencies on growth, survival, and body compositions of juvenile freshwater prawn *Macrobrachium rosenbergii*. *Nippon Suisan Gakk.* **1992**, *58*, 1419–1425. [CrossRef]

62. Qiao, H.; Xiong, Y.; Zhang, W.; Fu, H.; Jiang, S.; Sun, S.; Bai, H.; Jin, S.; Gong, Y. Characterization, expression, and function analysis of gonad–inhibiting hormone in Oriental River prawn, *Macrobrachium nipponense* and its induced expression by temperature. *Comp. Biochem. Phys. A* **2015**, *185*, 1–8. [CrossRef] [PubMed]

63. Rotllant, G.; Nguyen, T.V.; Aizen, J.; Suwansa-ard, S.; Ventura, T. Toward the identification of female gonad–stimulating factors in crustaceans. *Hydrobiologia* **2018**, *825*, 91–119. [CrossRef]

64. Huang, X.; Ye, H.; Huang, H.; Yu, K.; Huang, Y. Two beta-pigment–dispersing hormone (β–PDH) isoforms in the mud crab, *Scylla paramamosain*: Implication for regulation of ovarian maturation and a photoperiod-related daily rhythmicity. *Anim. Reprod. Sci.* **2014**, *150*, 139–147. [CrossRef] [PubMed]

65. Bao, C.; Yang, Y.; Huang, H.; Ye, H. Neuropeptides in the cerebral ganglia of the mud crab, *Scylla paramamosain*: Transcriptomic analysis and expression profiles during vitellogenesis. *Sci. Rep.* **2015**, *5*, 17055. [CrossRef] [PubMed]

66. Wei, L.L.; Chen, T.T.; Luo, B.Y.; Qiu, G.F. Evidences for Red Pigment Concentrating Hormone (RPCH) and Beta-Pigment Dispersing Hormone (beta–PDH) Inducing Oocyte Meiotic Maturation in the Chinese Mitten Crab, *Eriocheir sinensis*. *Front. Endocrinol.* **2021**, *12*, 802768. [CrossRef]

67. Jiang, Y.W.; Li, C.J.; Chen, L.; Wang, F.G.; Zhou, X. Potential role of retinoids in ovarian physiology and pathogenesis of polycystic ovary syndrome. *Clin. Chim. Acta* **2017**, *469*, 87–93. [CrossRef]

68. Kawai, T.; Yanaka, N.; Richards, J.S.; Shimada, M. De novo–synthesized retinoic acid in ovarian antral follicles enhances FSH–mediated ovarian follicular cell differentiation and female fertility. *Endocrinology* **2016**, *157*, 2160–2172. [CrossRef]

69. Damdimopoulou, P.; Chiang, C.; Flaws, J.A. Retinoic acid signaling in ovarian folliculogenesis and steroidogenesis. *Reprod. Toxicol.* **2019**, *87*, 32–41. [CrossRef]

70. Tahaei, L.S.; Eimani, H.; Yazdi, P.E.; Ebrahimi, B.; Fathi, R. Effects of retinoic acid on maturation of immature mouse oocytes in the presence and absence of a granulosa cell co–culture system. *J. Assist. Reprod. Gen.* **2011**, *28*, 553–558. [CrossRef]

71. Levi, L.; Ziv, T.; Admon, A.; Levavi-Sivan, B.; Lubzens, E. Insight into molecular pathways of retinal metabolism, associated with vitellogenesis in zebrafish. *Am. J. Physiol.-Endoc. Metab.* **2012**, *302*, 626–644. [CrossRef] [PubMed]

72. Kang, P.F.; Mao, B.; Fan, C.; Wang, Y.F. Transcriptomic information from the ovaries of red swamp crayfish (*Procambarus clarkii*) provides new insights into development of ovaries and embryos. *Aquaculture* **2019**, *505*, 333–343. [CrossRef]

73. Wang, J.J.; Li, J.T.; Ge, Q.Q.; Li, W.Y.; Li, J. Full–Length transcriptome sequencing and comparative transcriptomic analysis provide insights into the ovarian maturation of *Exopalaemon carinicauda*. *Front. Mar. Sci.* **2022**, *9*, 906730. [CrossRef]

Article

The Impact of Floating Raft Aquaculture on the Hydrodynamic Environment of an Open Sea Area in Liaoning Province, China

Kun Wang [1,†], Nan Li [1], Zhaohui Wang [1], Guangjun Song [1], Jing Du [1], Lun Song [1,†], Hengzhi Jiang [2,*] and Jinhao Wu [1,*]

[1] Liaoning Ocean and Fisheries Science Research Institute, Dalian 116023, China
[2] National Marine Environmental Monitoring Center, Dalian 116023, China
* Correspondence: jianghengzhi99@163.com (H.J.); jinhaow@126.com (J.W.)
† These authors contributed equally to this work.

Abstract: The sea area of Changhai County in Dalian City is a typical floating raft aquaculture area, located in Liaoning Province, China, where a key issue in determining the scale and spatial layout of the floating raft aquaculture is the assessment of the impact of aquaculture activities on the hydrodynamic environment. To address this issue, we established depth-averaged two-dimensional shallow water equations and three-dimensional incompressible Reynolds-averaged Navier–Stokes equations for the open sea area described in this paper. The impact of floating rafts for aquaculture on hydrodynamic force was reflected in the numerical model by changing the Manning number, where scenarios with different aquaculture densities were taken into account. Finally, the water exchange rate of the floating raft aquaculture area in the study area was calculated. It was found, through a comparison between the simulated value and the measured value obtained via layered observation, that the two values were in good agreement with each other, indicating that the model exhibits great accuracy. In addition, the calculation results for scenarios before and after aquaculture were compared and analyzed, showing that from low-density to high-density aquaculture zones, the variation in flow rate was greater than 80% at the peak of a flood tide. The water exchange rates of the water body after 1 day, 4 days, and 8 days of water exchange were also calculated, and the results show that they had been reduced by 17.92%, 13.59%, and 1.63%, respectively, indicating that the existence of floating rafts for aquaculture indeed reduced the water exchange capacity of the water body. The model described in this paper can serve as a foundation for other studies on aquaculture in open sea areas, and it provides a theoretical basis for the scientific formulation of marine aquaculture plans and the rational optimization of the spatial layout.

Keywords: floating raft aquaculture area; open sea area; hydrodynamic environment; water exchange capacity; numerical simulation

Citation: Wang, K.; Li, N.; Wang, Z.; Song, G.; Du, J.; Song, L.; Jiang, H.; Wu, J. The Impact of Floating Raft Aquaculture on the Hydrodynamic Environment of an Open Sea Area in Liaoning Province, China. *Water* **2022**, *14*, 3125. https://doi.org/10.3390/w14193125

Academic Editors: Xiangli Tian and Li Li

Received: 21 July 2022
Accepted: 26 September 2022
Published: 4 October 2022

1. Introduction

Changhai County, Dalian, is located on the eastern side of the Liaodong Peninsula in the northern waters of the Yellow Sea in Liaoning Province, China(as shown in Figure 1), and its geographic coordinates are 122°17′E–123°13′E, 38°55′N–39°35′N. As the only county completely located on islands in Northeast China, Changhai County has an area of water covering 10,324 square kilometers, which is an ideal habitat for temperate marine organisms, such as fish, shrimp, shellfish, and algae. In recent years, with the rapid development of the marine aquaculture industry, floating raft and cage aquaculture industries have emerged in this open sea area, which has brought not only a great deal of economic benefits to residents, but also huge challenges to marine hydrodynamics and ecological environment protection. Some aquaculture farmers excessively pursue high yields with a lack of scientific and reasonable justification, so they tend to increase the scale and density of aquaculture in a disorderly manner. Due to the over-crowded raft areas, the rafts and facilities have had a hindering effect on the hydrodynamic environment, inhibiting water exchange and weakening the substance transport and diffusion capacity of the water body. As a result, it is impossible for algae and bait to be evenly distributed with the hydrodynamic force, which would support the growth of marine organisms. This results in a phenomenon where the cultured organisms that were longitudinally arranged in a raft area appear to grow well, while those in the middle grow slowly or even die due to a lack of bait. Some aquaculture operators who have been cultivating scallops, oysters, or sea cucumbers in floating rafts and cages have gradually realized the severity of the problem. Thanks to such changes in their awareness, they are looking for a scientific and reasonable solution to the problem, with the ultimate goal of determining the degree of impact of the overall structure for aquaculture, including rafts, floaters, ropes and cages, and even the cultured organisms, on the hydrodynamic environment of an open sea area. The solution of this problem could provide technical support for the scientific formulation of a sowing density plan for cultured organisms, the rational selection of the location of an aquaculture area, and the precise placement of bait casting devices in bait-deficient zones.

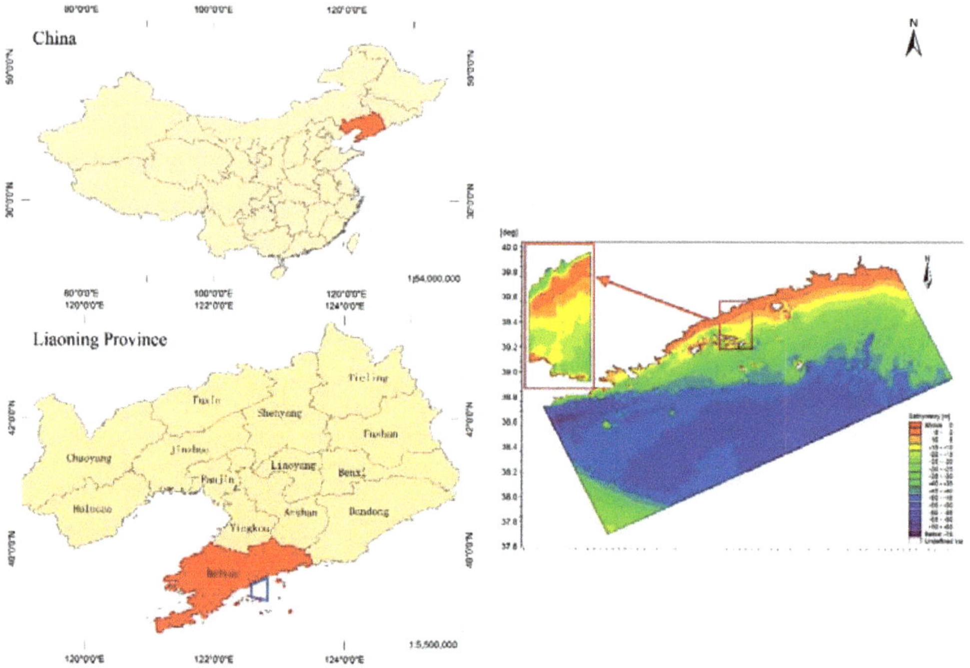

Figure 1. Location and scope of the aquaculture area.

In recent years, some researchers have carried out relevant studies on the mechanisms of interactions between raft placement and hydrodynamic environments in raft aquaculture areas; however, most of the studies have focused on the changes in water quality and the sediment environment or used field observations and model tests. For example, Zhao et al. [1] conducted a simulation-based assessment of the impact of the deep-sea cage aquaculture of *Lateolabraxjaponicus* on water quality and the sediment environment in the Yellow Sea of China based on a three-dimensional Lagrangian particle tracking model. Water quality simulations indicated that deep sea cages account for 26% of the total dissolved inorganic nitrogen and 19% of the active phosphorus content. The model results indicated that the installation of all deep-sea cages will lead to acceptable levels of water quality, but that sediments may become polluted. The coupled model can be used to predict the environmental impacts of deep-sea cage farming and provide a useful tool for designing the layout of the integrated multi-trophic aquaculture of organic extractive or inorganic extractive species. Klebertet et al. [2] carried out field monitoring and modeling for the three-dimensional deformation of a large circular flexible sea cage in high currents using an acoustic Doppler current profiler (ADCP) and an acoustic Doppler velocimeter (ADV). The results showed a reduction of 30% in the cage volume for a current velocity above 0.6 m/s. The measured current reduction in the cage was 21.5%. Moreover, a simulation model based on super elements describing the cage shape was applied, and the results showed good agreement with the cage deformations. Dong et al. [3] conducted an experimental study involving an internationally advanced experimental model of fluid–structure interactions, which described the fluid–structure interactions of flexible structures, in a study on the cage aquaculture of Thunnusorientalis. They measured the drag force, cage deformation, and flow field inside and around a scaled net cage model composed of different bottom weights under various incoming current speeds in a flume tank. Results indicated that the drag force and cage volume increased and decreased, respectively, with the bottom weight. Owing to the significant deformation of the flexible net cage, a complex fluid–structure interaction occurred and a strong negative correlation between the drag force and cage volume was obtained. Furthermore, an area where the current speed was often reduced was identified. The intensity of this reduction depended on the incoming current speed. The results of this study can be used to understand and design optimal flexible sea cage structures that can be used in modern aquaculture. In addition, a team led by Dong used model-scale test and full-scale sea test techniques [4] to determine the hydrodynamic characteristics of a sea area near a cage aquaculture area for silver salmon. In that study, the results of model-scale and full-scale tests were compared, showing that under the impact of lower currents, only bottom mesh deformation was found. As for the observed trends, the resistance, cage deformation, and cross-sectional area estimated based on the depth data from the full-scale test were generally consistent with the results converted from the model-scale test using the law of similarity. However, the resistance value of a full-sized cage converted from the model-scale test was larger than the depth estimated based on the depth data from the full-scale test. Conversely, the result from the model-scale test was smaller than the estimate from the full-scale test. In the future, cage deformation should be investigated at higher flow rates, and resistance should be measured at full scale to verify the results of model-scale tests and hydrodynamic model tests. Sintef et al. [5] also observed and investigated the turbulence and flow field changes in sea cages for commercial salmon aquaculture and their wakes in their study, where an acoustic Doppler current profiler (ADCP) installed on the seabed was used to measure the flow rate and turbulence on a layered basis, and an acoustic Doppler velocimeter(ADV) was used to measure the velocity inside the sea cages; dissolved oxygen sensors and echo sounders were also arranged in the sea cages to measure fish distribution, in order to facilitate the acquisition of data. The final results showed that a reduction in strong currents in the wakes near the cages and the existence of high-turbulence columns in the upper part of the water were both caused by the cages. Measurements performed in the cages indicated that although fish aggregation reduced water flow, there was no evidence that fish generated secondary radial and vertical flows within the cages.Ji et al. [6] observed, in a study on a gulf ecosystem for shellfish aquaculture, that in a crowded area with suspended shellfish, the sedimentation effect of organisms was very obvious, and the hydrodynamic effect was obviously insufficient. Hatcher et al. [7] conducted a measurement in the mussel aquaculture area located in the Upper South Cove, Canada, and found that the settlement of the raft aquaculture area was more than twice that of the control area without aquaculture. Bouchet and Sauriau [8] found, in an ecological quality assessment on a shellfish aquaculture area in the Pacific Ocean, that the suspended aquaculture system resulted in higher organic matter enrichment compared with a bottom sowing culture.

With the rapid development of computer technology, mathematical models have been widely adopted in numerical-simulation-based studies on marine aquaculture. Panchanget et al. [9] stated that the mathematical modeling of hydrodynamic force and particle tracking can be an effective method with which to study the laws of diffusion and transport of pollutants in aquaculture areas, and the fate and traceability of materials. Xing et al. [10] studied the impact of an aquaculture area on the distribution of the vertical structure of the water flow with a hydrodynamic model and found that the distribution of the vertical structure was mainly controlled by the bottom friction of the aquaculture area. Durateet et al. [11] calculated the hydrodynamic characteristics of the estuary in Galicia based on a three-dimensional numerical model. It was found, through the analysis of the residual current field, that raft aquaculture can reduce the flow rate of the residual current by at least 40%, which facilitates the development of harmful algal blooms, posing a serious threat to cultured organisms and the aquatic environment. Shiand Wei [12] simulated an aquaculture area in Sanggou Bay with an optimized POM and found that the high-density aquaculture and related facilities in Sanggou Bay reduced the flow rate by nearly 40% on average and increased the average half-exchange time by 71%. In summary, the valuable technical studies conducted by these researchers will greatly inspire our later studies.

The original intention of this work was to solve some problems with floating raft aquaculture areas. In this study, a typical floating raft aquaculture area located in Changhai County, Liaoning Province, was chosen as the research area on the basis of the successful establishment of the hydrodynamic model and tracer model in these area of Liaodong Bay, in order to quantitatively explain the impact of floating raft aquaculture on the hydrodynamic environment of an open sea area. Compared with the sea area of Liaodong Bay, the study area features a higher degree of openness. Aiming to comprehensively understand the temporal and spatial distribution and variation characteristics of hydrodynamic force in the waters near the floating raft aquaculture area located in Changhai County, Dalian, the project team simulated and analyzed the hydrodynamic field and water exchange rate in the sea area near the floating raft aquaculture area. In this study, depth-averaged two-dimensional shallow-water equations and three-dimensional incompressible Reynolds-averaged Navier–Stokes equations were established for the open sea area. We described the impact of rafts (floaters, ropes, cages, cultured organisms, etc.) on hydrodynamic force in the aquaculture area by changing the Manning number of the seabed. Finally, the model was verified with the observed hydrodynamic data, and the results show that the model has great accuracy, stability, and universality, and it can provide an accurate prediction of the hydrodynamic environment of aquaculture in the raft area.

2. Materials and Methods

2.1. Observational Data

The project team set up a temporary tide-level observation station, T1, in the coastal waters of Dalian, and conducted tide-level observations for three months, from 00:00, 1 August 2021 to 23:00, 31 October 2021. Two continuous observation stations, P1 and P2, were set up for ocean current observation, where a total of 25 h of layered and synchronous continuous ocean current observations were carried out, from 11:00, 13 September 2021 to 12:00, 14 September 2021. The specific coordinates of the stations are shown in Table 1, and their locations are shown in Figure 2. Refer to Section 3.1 for the specific observation values below.

Table 1. Coordinates of the hydrometric stations for hydrological tide tests.

Station	Longitude	Latitude
T1	121°41.12′ E	38°52.08′ N
P1	123°6.098′ E	39°7.111′ N
P2	123°12.539′ E	39°10.895′ N

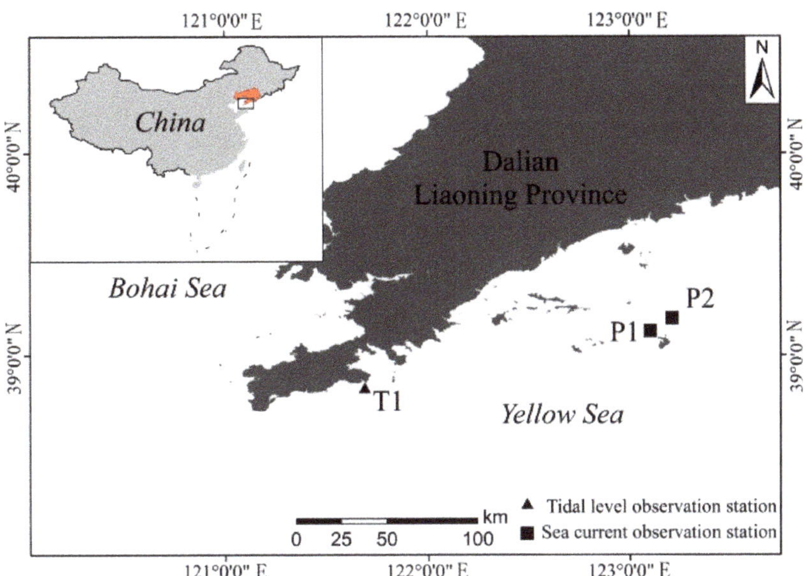

Figure 2. Locations of the observation stations.

2.2. Model and Methods

The model is based on the solution of the three-dimensional incompressible Reynolds-averaged Navier–Stokes equations. First, the integration of the horizontal momentum equations and the continuity equation over depth for the following two-dimensional shallow water equations was carried out [13–16]. Based on the aforesaid principle, the commercial model encapsulation platforms used in this study mainly included Hydro info, a water conservancy information system developed by Dalian University of Technology, China, and Mike, a commercial water simulation computing system developed by the Danish Hydraulic Institute (DHI).

Based on the above model and methods, the specific implementation process was completed, as follows. First, in order to accurately analyze the hydrodynamic conditions of the water area near Changhai County, two-dimensional models of the Yellow Sea and the Bohai Sea and the waters near Changhai County were created, where the open boundary of an open water area was driven by the time series file of the tidal level. Then, in order to reflect the hydrodynamic conditions of the sea area near the aquaculture area located in Changhai County in more detail, a three-dimensional model of a small area of interest in Changhai County was created with a nesting method [17,18] based on the tidal-level drive after the calibration of the two-dimensional model of Changhai County, where the calculation range mainly covered the area contained by the four control points C, D, E, and F shown in Table 2, and the locations of the control points are shown in Figure 3a.

Table 2. Coordinates of points in the range of the calculation domain.

Point	Longitude	Latitude
C	122.1380° E	39.1450° N
D	122.4400° E	38.4820° N
E	123.3560° E	38.8970° N
F	122.9990° E	39.6470° N

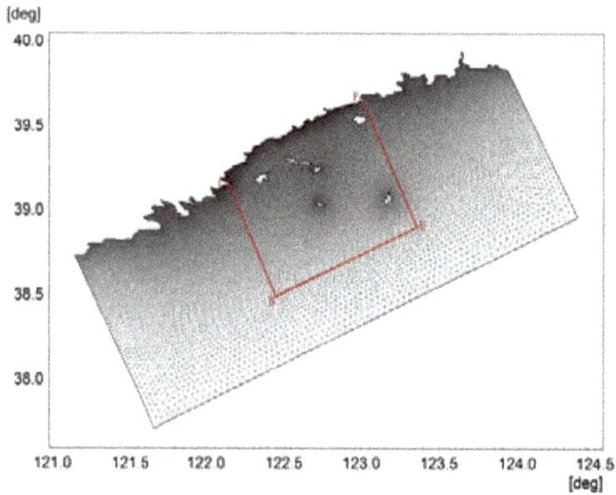

(**a**) Two-dimensional calculation range

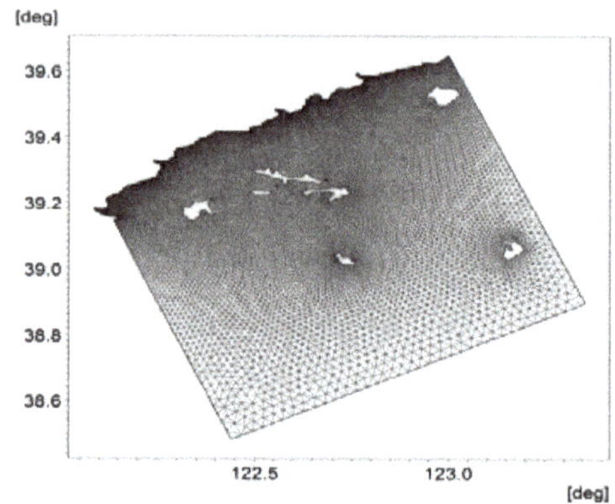

(**b**) Three-dimensional calculation range

Figure 3. Calculation ranges and grid distribution maps of the models.

2.2.1. Hydrodynamic Model

The study area is located along the northern coast of the Yellow Sea (see Figure 1), where tidal currents play a dominant role in various flow components. The three-dimensional Navier–Stokes equations for the free-surface flow of incompressible fluid in the Cartesian coordinate system were used for description; on this basis, the horizontal momentum equations and the continuity equation for the three-dimensional shallow water form were integrated in the range $H = \eta + h$ to obtain the following depth-averaged two-dimensional shallow water continuity equation:

$$\frac{\partial \eta}{\partial t} + \frac{\partial}{\partial x}(hu) + \frac{\partial}{\partial y}(hv) = 0 \tag{1}$$

Momentum equations

$$\frac{du}{dt} - \frac{\partial}{\partial x}\left(\gamma^h \frac{\partial u}{\partial x}\right) - \frac{\partial}{\partial y}\left(\gamma^h \frac{\partial u}{\partial y}\right) - fv + \frac{gu\sqrt{u^2 + v^2}}{C_z^2 H} = -g\frac{\partial \eta}{\partial x} \tag{2}$$

$$\frac{dv}{dt} - \frac{\partial}{\partial x}\left(\gamma^h \frac{\partial v}{\partial x}\right) - \frac{\partial}{\partial y}\left(\gamma^h \frac{\partial v}{\partial y}\right) + fu + \frac{gv\sqrt{u^2 + v^2}}{C_z^2 H} = -g\frac{\partial \eta}{\partial y} \tag{3}$$

where η represents the sea surface fluctuation (tidal level) relative to the still sea surface; h represents the still water depth (the distance from the seabed to the still sea surface); $H = \eta + h$ represents the total water depth; $C_z = n.\ H^{(1/6)}$ is the Chezy coefficient; $n = 1/M$ is Manning's roughness coefficient, and M represents the Manning number.

Equations (1)–(3) are the basic governing equations for solving the hydrodynamic elements. In order to comply with the uniqueness of solutions, the definite conditions must be given.

(1) Initial Conditions

The cold-start mode was used, meaning that the initial conditions were considered irrelevant to the final result of the calculation. In this study, the initial flow rate and tidal level were both determined as 0.

(2) Boundary Conditions

For the numerical model used in this study, two boundary conditions need to be given, including open boundary and closed boundary conditions. For a tidal flat near the island coastline, the position of the land–water interface changed with the fluctuation of the tide level, and the dry–wet variation of the grid nodes in the moving boundary was taken into consideration in this work [19–21].

① Open boundary condition:

The open boundary condition is also known as the water boundary condition, where, on this boundary, either the flow rate is given, or the time series condition of the tidal level is given. For the open boundary in this work, calculation was performed in the following form of tidal harmonic analysis:

$$\eta_i = \sum_i^m f_i H_i \cos[\omega_i t + (V_0 + u)_i - g_i] \tag{4}$$

where ω_i represents the angular velocity of the ith tidal constituent; f_i and u represent the intersection factor and epoch correction of the ith tidal constituent, respectively; H_i and g_i represent harmonic constants, which are the amplitude and epoch of each tidal constituent, respectively; V_0 represents the time angle of a tidal constituent. The time series data of the tide level at the open boundary of the model were also verified according to the global tide module.

② Closed boundary condition:

The normal flow rate at the shoreline of the given water body should be 0.

2.2.2. The Euler Model for Residual Current Calculation

As the most important environmental dynamic factor in coastal waters, the residual current plays a crucial role in the transport and diffusion of substances in seawater. In studies on ocean dynamics, Eulerian velocity is usually used to calculate residual currents. The Euler residual current in the ocean can be simply defined as the mean Eulerian velocity, which can be calculated with the following equation:

$$U_E = \frac{1}{nT} \int_{t_0}^{t_0+nt} u(x_0, t)dt, \ V_E = \frac{1}{nT} \int_{t_0}^{t_0+nt} v(x_0, t)dt \tag{5}$$

where U_E and V_E represent the mean Eulerian velocities in directions x and y, respectively; n represents the number of cycles used in the calculation; t_0 represents the start time of calculation; T represents the current cycle; $u(x_0,t)$ and $v(x_0, t)$ represent the component velocities in directions x and y. The numerically discrete form of Equation (6) is described below:

$$U_E = \frac{1}{N} \sum_{i=1}^{N} u_i, \ V_E = \frac{1}{N} \sum_{i=1}^{N} v_i \tag{6}$$

where $N = nT/\Delta t$ and Δt represent the time step of numerical simulation.

2.2.3. Mathematical Model of Water Exchange

The tracer method was used to simulate the degree of water exchange [22–24], where a dissolved non-degradable and conservative substance was set in the sea area, and its concentration diffusion under the action of hydrodynamic force was investigated. For the transport of the tracer, the convection–diffusion equation based on Eulerian substance transport was used, as shown below:

$$\frac{\partial hC}{\partial t} + \frac{\partial huC}{\partial x} + \frac{\partial hvC}{\partial y} = \frac{\partial}{\partial x}\left(hD_x \frac{\partial C}{\partial x}\right) + \frac{\partial}{\partial y}\left(hD_y \frac{\partial C}{\partial y}\right) - FhC + S \tag{7}$$

where C represents the substance concentration; D_x and D_y represent the substance diffusion coefficients in directions x and y, respectively; F represents the substance attenuation coefficient, which is zero ($F = 0$) for the conservative substance; S represents the point source concentration.

The substance diffusion coefficient was calculated with the following equation:

$$D_x = \frac{E_x}{\sigma_T}; \ D_y = \frac{E_y}{\sigma_T} \tag{8}$$

where $E_x = E_y$ represents the horizontal turbulent viscosity coefficient; σ_T represents the Prandtl number, which was determined as 1.0 in this study.

After a certain period of time, the percentage of the total amount of substance diffused from the system to open water divided by the total amount of initial substances in the system should be the water exchange rate of the overall system. The statistical calculation expression is provided below.

$$EX(t_j) = \left(1 - \frac{\sum\limits_{i=1}^{N} C_i(t_j) H_i(t_j)}{\sum\limits_{i=1}^{N} C_i(t_0) H_i(t_0)}\right) \times 100\% \tag{9}$$

where EX represents the water exchange rate; C represents the substance concentration; H represents the total water depth; i represents the node number in the statistical domain; n represents the total number of nodes in the statistical domain; j represents the time number.

2.3. Grid Creation and Parameter Setting

2.3.1. Grid Creation

The calculation grid was generated with the Surface Water Model System(SMS 10.1). This grid generation program can realize a flexible and variable resolution in the horizontal direction of the grid and a large gradient, and it can create a highly smooth grid at a location where a flow tends to be generated around an island. In addition, it can partially increase the density in areas with complex terrain, such as coastal areas, estuaries, and wetlands.

The entire two-dimensional simulated domain of Changhai County consists of 32,560 nodes and 63,579 triangular elements. Figure 3a shows the calculation domain and grid distribution of the established two-dimensional model of the sea area near Changhai County. The entire simulated domain of the three-dimensional model [25] consists of 12,539 nodes and 24,281 triangular elements. Figure 3b shows the calculation domain and grid distribution of the three-dimensional model of a small area of Changhai County.

2.3.2. Model Calculation Settings

(1) The calculation of hydrodynamic force

The calculation time step of the model was adjusted according to the CFL conditions to ensure that when the model calculation was converged, the minimum time step was 5.0 s. The seabed friction was controlled by the Manning number, with a specific value of 32–42 $m^{1/3}/s$. In many applications, a constant eddy viscosity can be used for the horizontal stress terms. Alternatively, Smagorinsky proposed to express sub-grid-scale transport by effective eddy viscosity related to a characteristic length scale. The sub-grid-scale eddy viscosity is given in [26], and the specific expression is shown below:

$$A = c_s^2 l^2 \sqrt{2 S_{ij} S_{ij}} \tag{10}$$

where c_s is a constant; l is a characteristic length; and the deformation rate is given by

$$S_{ij} = \frac{1}{2} \left(\frac{\partial u_i}{\partial x_j} + \frac{\partial u_j}{\partial x_i} \right) \tag{11}$$

The minimum calculation time step of the three-dimensional small-scale local model was 1.0 s. In the vertical grid, the sigma hierarchical function was adopted, and the impact of the rafts on the hydrodynamic force was described with a double-resistance model with the introduction of a secondary drag coefficient, where the frictional resistance of the seabed was controlled by secondary drag coefficient C_f, and the specific expression was determined by assuming a logarithmic profile between the seabed and a point at a distance of DZ_b above the seabed as follows:

$$C_f = [\frac{1}{\kappa} \ln\left(\frac{DZ_b}{Z_0}\right)]^{-2} \tag{12}$$

where $\kappa = 0.4$ is the von Kármán constant; Z_0 represents the length scale of the roughness of the riverbed; when the boundary surface is rough, Z_0 depends on the roughness height, where $Z_0 = mk_s$, the approximate value of m is 0.033, k_s is the roughness height, ranging between 0.01 m and 0.30 m, and the value was determined as 0.05 m. In summary, the average value of the secondary drag coefficient was 0.01.

(2) The calculation for the aquaculture area

The location of the selected aquaculture area, the range of the calculation domain, and the distribution of the seabed topography are shown in Figure 1. The aquaculture area is located in a sea area near the Changshan Archipelago in the southeast of Changhai County, and its boundaries are shown in Table 3 below.

Table 3. Scope of the aquaculture area.

Longitude	Latitude
122.7492° E	39.2382° N
122.7642° E	39.2585° N
122.7527° E	39.2697° N
122.7363° E	39.2480° N

In the post-aquaculture model, unstructured grids were also used to divide the horizontal calculation domain and locally densify the sea area where the aquaculture area was located, with a grid scale of 30 m. In other areas along the shoreline, the grid scale ranged between 50 m and 100 m; in sea areas far away from the aquaculture area, the maximum grid scale was 400 m, and the calculation domain contained 17,642 triangular grids and 9011 nodes. The Manning field considering aquaculture areas of different densities is shown in Figure 4 below.

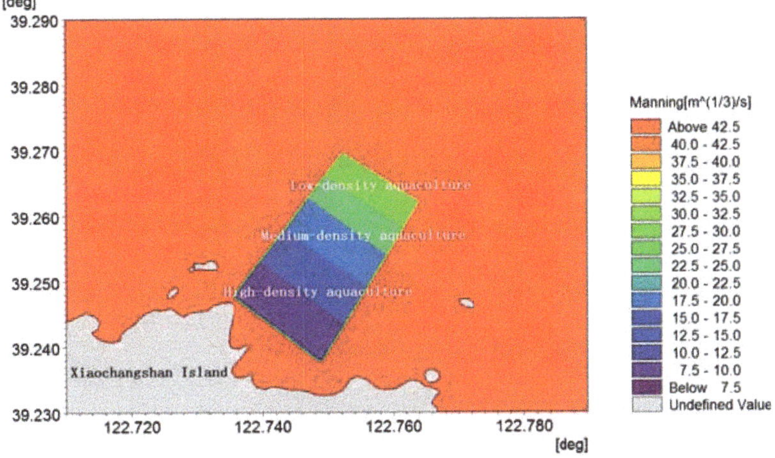

Figure 4. Distribution of Manning field after the implementation of aquaculture activities.

(3)   Assessment of water exchange capacity

In this study, the water exchange rate was used as an index to describe the water exchange capacity of the aquaculture area. A dissolved conservative substance was placed in the sea area where the aquaculture area was located, which would be carried by the water body and could not be degraded. The convection and diffusion of the conservative matter directly reflect the form of movement of the water body. Based on the above considerations, in this study, a conservative substance with a concentration of 1.0 was placed in the aquaculture area; the concentration of substances in open water was set at 0.0; the attenuation coefficient was set as $F = 0$, and the point source concentration was set as $S = 0$. The substance diffusion coefficient was equal to the turbulent viscosity coefficient of water flow ($\sigma_T = 1.0$).

3. Results

3.1. Model Verification Results

The actual calculation and simulation period of the model was from 0:00, 1 August 2021 to 23:00, 31 October 2021. Figure 5 shows a time-curve-based comparison between the simulated and measured values of the tidal level during the period from 0:00, 8 August 2021 to 23:00, 15 September 2021. Figure 6 shows the comparison between P1 and P2 in terms of flow rate, flow direction, and measured value during the period from 11:20, 13 September 2021 to 12:30, 14 September 2021. To fully represent the calculation results of the numerical model, Figure 7 shows the flow rates and flow direction fields of the top, middle, and bottom layers at the same moment during a spring tide and a neap tide in the three-dimensional calculation model for Changhai County.

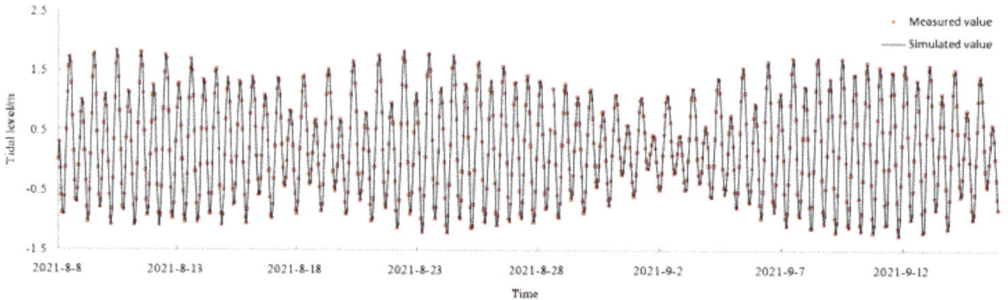

Figure 5. Tidal-level verification map of T1.

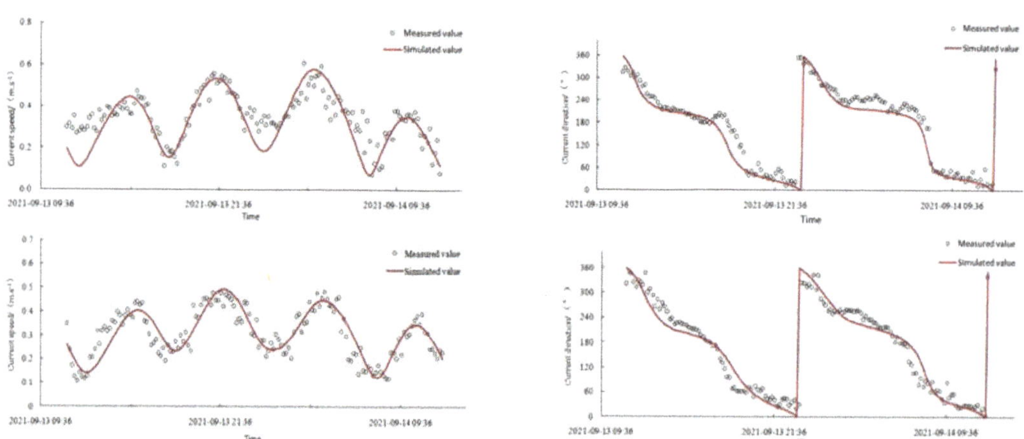

Figure 6. Verification of flow rates and directions at P1 and P2.

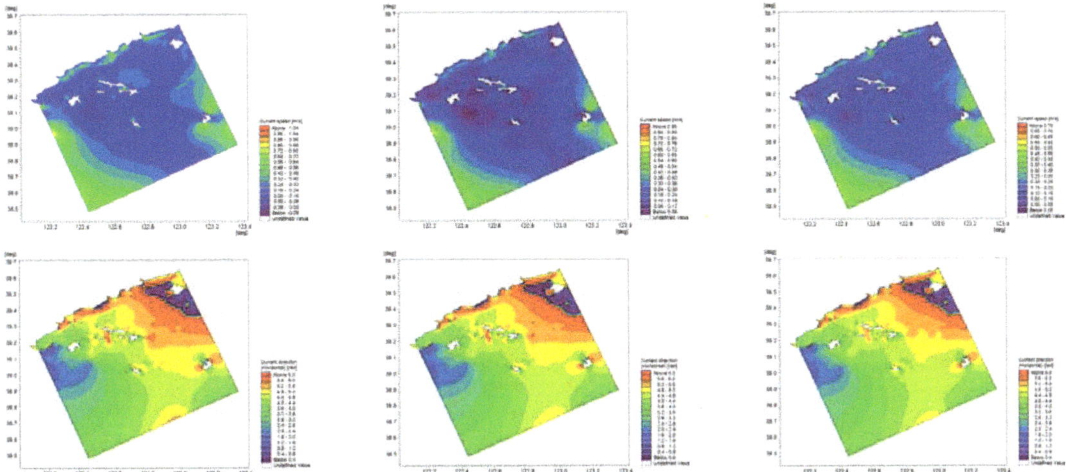

Figure 7. Flow field distribution maps of the top, middle, and bottom layers during a spring tide in the three-dimensional calculation domain of a small area of Changhai County.

A comparison between Figures 5 and 6 shows that the results calculated by the model are in good agreement with the measured values, where the error is within an acceptable range. According to the verification results of tidal currents and the flow field distribution maps at different moments, as shown in Figure 5, the mathematical model can reflect the flow field in the sea area near Changhai County in a more realistic manner, indicating that the model has reasonable boundaries and parameters and can be used for the calculation of subsequent working conditions. Figure 6 indicates that the trends of variation in the simulated flow rate and flow direction are generally the same as those of the measured values, where the measured maximum flow rates of P1 and P2 during 13 September and 14 September are 0.606 m/s and 0.491 m/s, respectively, and the simulated maximum flow rates are 0.577 m/s and 0.493 m/s, respectively. In general, the simulated calculated values are in good agreement with the measured values. Figure 7 shows that although grids of different scales were used in the calculation of the two-dimensional and three-dimensional models, the flow rate and flow direction in the entire spatial calculation domain reasonably changed, featuring a strong gradient and no sudden change, which indicates that the model is stable and can be used as the basis for the calculation of subsequent working conditions.

In order to verify the accuracy, the root-mean-square error (R) was used to quantitatively analyze the error between the calculation results of the model and the measured values, where R represents the mean deviation between the results of the model and the observed data. R is calculated as follows:

$$R = \sqrt{\frac{\sum_{i=1}^{n}(M_i - O_i)^2}{n}} \tag{13}$$

where M_i represents the calculated value of the model; O_i represents the observed value; and n represents the number of observed values. After calculation, the root-mean-square error in the tidal level of T1 is less than 0.18 m, generally indicating that the simulated value is in good agreement with the measured value, and the model has great accuracy and reliability.

Tracer model verification is indeed a relatively important part of the assessment of water exchange capacity, which has been verified in other studies [27,28], as described below. In these studies, one of the major tasks was the numerical simulation of the transportation of water pollutants in Liaodong Bay, the northernmost bay in the Bohai Sea in China. On the basis of a comprehensive understanding of the natural conditions of the sea area of Liaodong Bay, the impact of point source input was introduced to establish a convection-diffusion model for pollutant transport in the sea area of Liaodong Bay, with which the distribution of different nutrient elements in the sea area was simulated, and where major water quality indicators included NH_3-N and COD. The accuracy and stability of the model were verified through a comparison between the simulated values and the concentration levels of the elements obtained by field sampling and analysis in the sea area, indicating that the model features a great ability to reproduce and predict the concentration field in the sea area of Liaodong Bay. In respect of the following few years, the distribution of PO4-P, a water quality indicator, in Liaodong Bay, was reproduced, and model verification was performed, providing the simulation results of the hydrodynamic field, half exchange time, and concentration field in Liaodong Bay at different typical moments. Finally, this model was adopted in the simulation and assessment for the identification of marine pollution accidents, and it delivered satisfactory results.

3.2. Simulation Results of Tidal Current and Residual Current in the Aquaculture Area

The results regarding the distribution of the flow field in the sea area near the aquaculture area are shown in Figures 8–11, indicating that the flow field in the aquaculture area exhibits the characteristics of reciprocating motion, where the main flow direction is NW–SE, and the flow rate magnitude during a flood tide and an ebb tide is 1.0 m/s. In accordance with the simulation result of the hydrodynamic field in the sea area of Changhai County, the Euler residual current field in the sea area near the aquaculture area was obtained. Figure 12 shows the Euler residual current fields in the cycle of a spring tide before and after the implementation of aquaculture activities, indicating that the mean residual current intensity in the aquaculture area was approximately 0.018 m/s; the direction was NE; and there was generally no significant variation before and after the implementation of aquaculture activities.

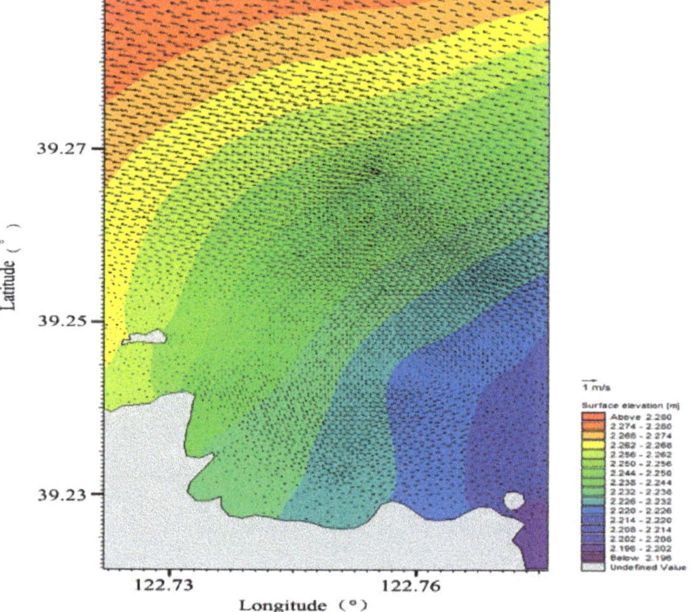

Figure 8. Flow field distribution at high tide.

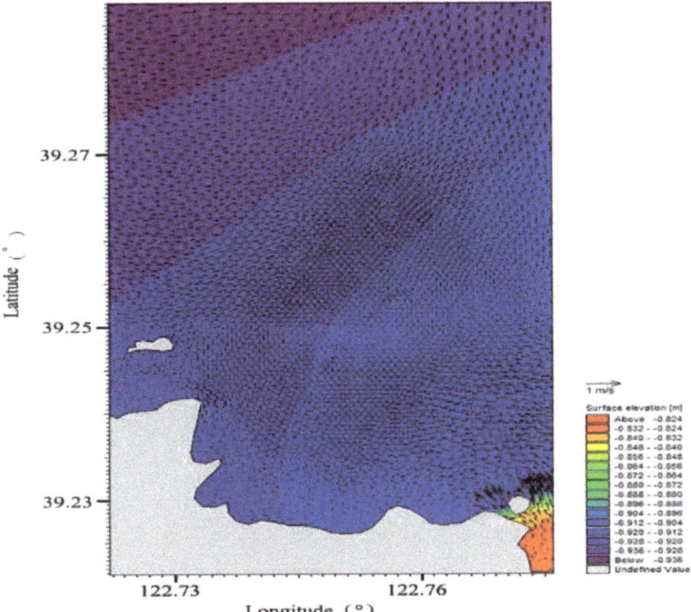

Figure 9. Flow field distribution at low tide.

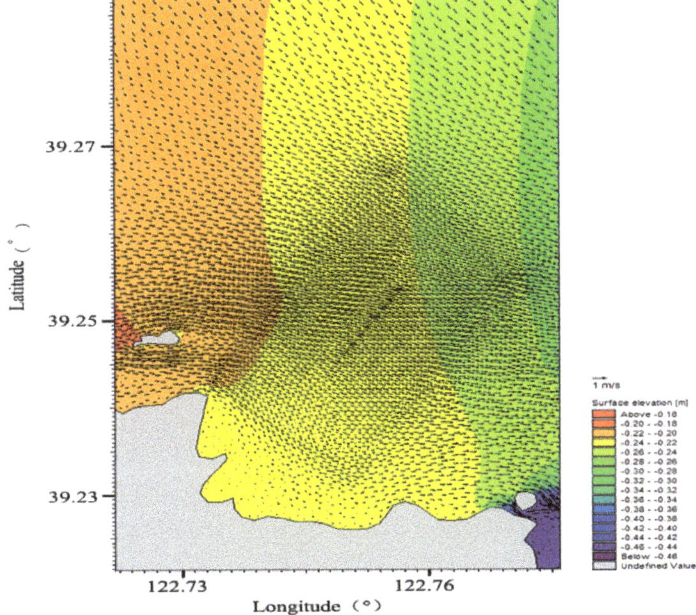

Figure 10. Flow field distribution at the peak of an ebb tide.

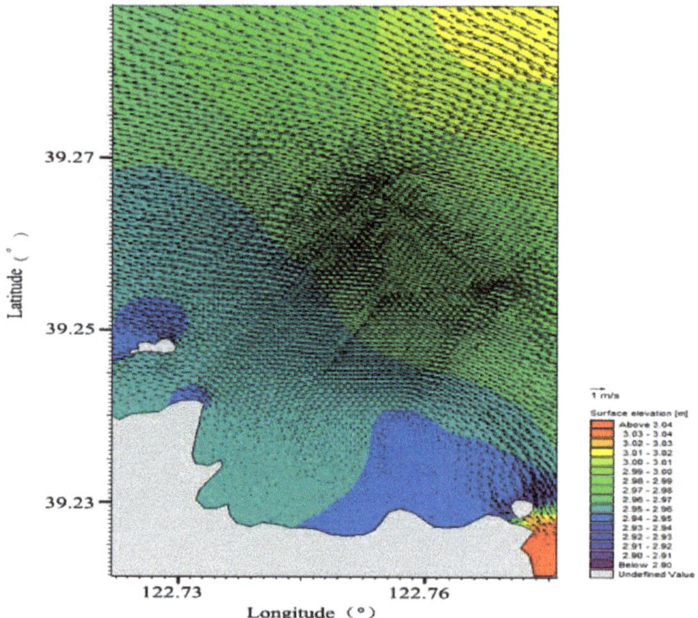

Figure 11. Flow field distribution at the peak of a flood tide.

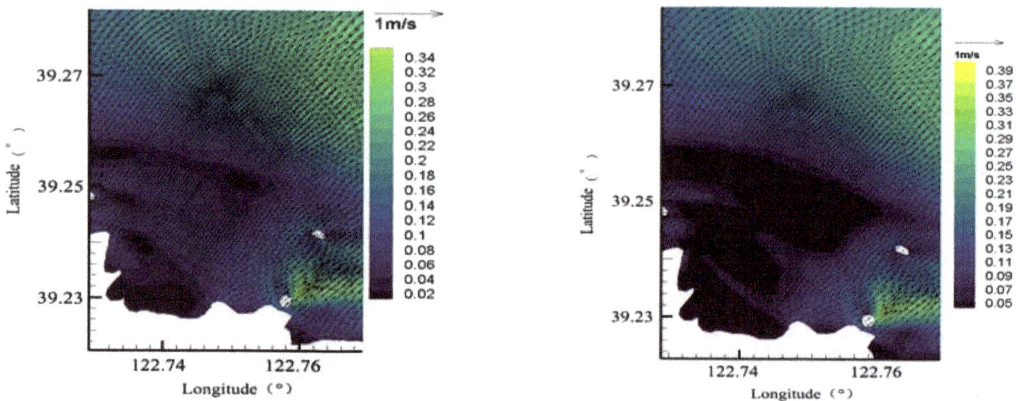

Figure 12. Residual current distribution in the sea area near the aquaculture area ((**left**): before aquaculture; (**right**): after aquaculture).

In order to quantitatively and clearly reflect the impact of surface roughness (Manning) on the calculation results, the sensitivity of the Manning number was analyzed. For the sea area near the floating raft aquaculture area, which is 180–3650 m from the shoreline, the simulated values of flow rate and flow direction at the peak of a flood tide and at the peak of an ebb tide during a spring tide in one tidal cycle were compared before and after the implementation of aquaculture, where the Manning settings under the two working conditions were as shown in Part 2. The results show that the changes in flow rate and flow direction were generally significant in the study area, especially during the flood tide, wherein the flow rate changed by more than 80% within 750 m in the aquaculture area; the mean change in flow rate was approximately 10%, and the number of points where the flow direction changed by more than 45° accounted for around 20% of the total number. This indicates that, after the establishment of aquaculture activities, as the Manning number of the aquaculture area decreased and the roughness increased, which, together with the effect of flow resistance arising from the raft net, indeed significantly affected the flow rate and flow direction of the sea area near the aquaculture area.

3.3. Simulation Results of Water Exchange Rate

Based on the aforesaid hydrodynamic simulation results, a mathematical model of water exchange, described in Equations (7)–(9), was used to study the water exchange capacity of the aquaculture area. In addition, in order to enable the model to reflect the variation in water exchange capacity before and after the implementation of aquaculture activities in a clearer and more sensitive manner, the scope of the aquaculture area was appropriately magnified according to the actual sea area, and the calculation domain and grid were rearranged and refined, where the average grid scale was 50 m, and the minimum grid scale for the key areas of interest in the aquaculture area was 30 m.

Figures 13 and 14 show the initial field distribution of the tracer concentration distribution, and the overall water exchange rate–time curves before and after the implementation of aquaculture activities, respectively. Table 4 shows the statistics for the overall water exchange rate–time curves before and after the implementation of aquaculture activities, indicating that the water exchange rate after the implementation of aquaculture decreased compared with that before implementation. Before the implementation of aquaculture, the water exchange rates after 1, 4, and 8 days of water exchange were 27.90%, 61.83%, and 76.48%, respectively; after the implementation of aquaculture, the water exchange rates after 1, 4, and 8 days of water exchange were 22.90%, 53.43%, and 75.23%, respectively.

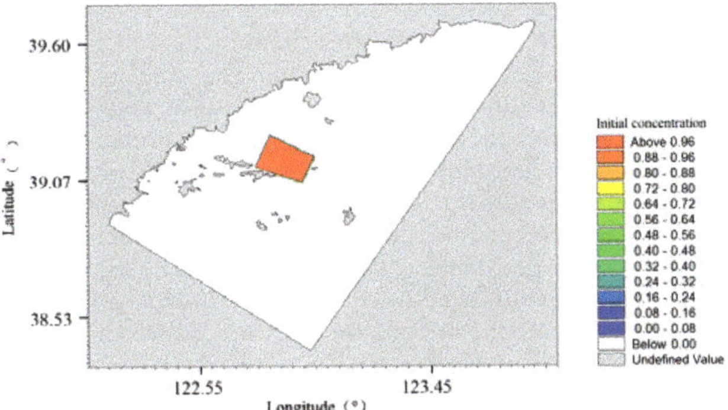

Figure 13. Tracer concentration distribution at the initial time point (the red area is the statistical area for water exchange rate).

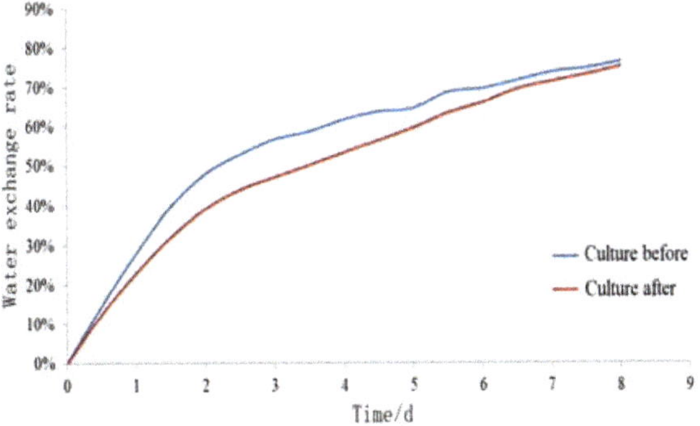

Figure 14. Water exchange rate–time curves before and after the implementation of aquaculture.

Table 4. Statistics for the water exchange rate–time curves before and after the implementation of aquaculture.

Working Condition/Time	0 d	1 d	2 d	4 d	6 d	8 d
Before the implementation of aquaculture activities	0	27.90%	47.95%	61.83%	69.71%	76.48%
After the implementation of aquaculture activities	0	22.90%	39.05%	53.43%	66.04%	75.23%

4. Discussion

4.1. Tidal Current Conditions

The hydrodynamic force calculation results indicate that the hydrodynamic force in the waters near Changhai County is mainly in the NW–SE direction during a spring tide, during which the tidal currents of flood and ebb tides rotate counterclockwise. During the spring tide, the tidal field shows that the tidal direction in the open waters of Changhai County was NW, the tidal field was stable, and the flow rate generally ranged between 0.50 m/s and 0.85 m/s. The nearshore current is a coastal current in essentially the same direction as the shoreline and has a lower flow rate, ranging between 0.2 m/s and 0.4 m/s. This is mainly because the flow rate is significantly reduced by the bottom friction due to the shallow water in the near-shore area. The local coastal waters are affected by the coastline, with a maximum flow rate of 1.2 m/s during a flood tide; during an ebb tide, the flow direction in the open waters is SE, and the flow rate ranges between 0.45 m/s and 0.90 m/s. Affected by the topography, coastal waters generally feature lower flow rates, with a maximum flow rate of approximately 1.1 m/s.

4.2. Conditions of the Aquaculture Area

The calculation of the characteristics of the flow rate in the aquaculture area shows that flow rates in the aquaculture area usually range between 0.2 m/s and 0.4 m/s. There is a small island in the SE direction in the aquaculture area, so the maximum flow rate in the aquaculture area is up to 0.7 m/s; the flow rate gradually increases from the shoreline to the sea, and it reaches 1.0 cm/s in the part of the aquaculture area that is closer to the shoreline. A comparison with the flow rate before the implementation of aquaculture activities shows that, with regard to the degree of flow resistance imposed by the floating rafts, the variation in flow rate ranges between 2.87% and 84.58% at the peak of a flood tide, and between 2.65% and 20.89% at the peak of an ebb tide from a low-density zone to a high-density zone of the aquaculture area. This indicates that the variation in flow rate caused by the floating rafts in the sea area near the aquaculture area of Changhai County is significantly greater during a flood tide than that during an ebb tide, and the flow resistance rate at the peak of a flood tide is greater than 80%. Therefore, aquaculture operators and marine environmental protection workers should pay attention to the impact of floating rafts for aquaculture. Even in open sea areas, during the setting of the orientation and density of a floating raft aquaculture area, it is crucial to first investigate the hydrodynamic conditions and the impact of aquaculture activities on the hydrodynamic conditions in the sea area, in order to scientifically implement aquaculture activities and rationally determine the layout while protecting the marine environment.

4.3. Residual Current Conditions

Residual current distribution plays a decisive role in the transport and diffusion of bait, nutritive salts, and other related substances in an aquaculture area. According to a comparison with the residual current before the implementation of aquaculture activities, the extent of variation ranged between 3.01% and 84.74% during a spring tide, and it ranged between 9.46% and 78.50% during a neap tide, indicating that the extents of variation in residual current during tides are essentially the same; they should not be underestimated. Therefore, to accurately understand the distribution of algae and bait in the floating raft aquaculture area, we must calculate and analyze the residual current in the sea area based on accurate hydrodynamic analysis. In this way, we can understand the characteristics of the transport and diffusion of floating and suspended substances in the sea area in real time, thereby providing guidance for the formulation of aquaculture plans and production activities.

4.4. Water Exchange Conditions

The quantitative calculation shows that, due to the aquaculture activities, the water exchange rates of the open sea area decreased by 17.92%, 13.59%, and 1.63% compared with those before implementation; moreover, the half-exchange cycle of the water body appeared in 2.3 d and 3.9 d, respectively, before and after the implementation of aquaculture. This indicates that even floating rafts for aquaculture located in an open sea area have a certain impact on the water exchange capacity, and the specific extent of such impact is closely related to various factors, such as the density, size, scope, and location of rafts in the aquaculture area.

5. Conclusions

In this study, a numerical simulation method was applied to a floating raft aquaculture sea area to quantitatively calculate and assess the changes in the hydrodynamic environment of the open sea area. The model is based on the solution of the three-dimensional incompressible Reynolds-averaged Navier–Stokes equations. Then, the integration of the horizontal momentum equations and the continuity equation over depth for the two-dimensional shallow water equations was carried out. In the hydrodynamic model, in order to generalize the impact of rafts on the hydrodynamic force in the aquaculture area, the Manning number of the seabed—namely the seabed roughness—in the two-dimensional mode was changed; in the three-dimensional mode, a double-resistance model of the top and bottom layers was used, with the introduction of a secondary drag coefficient. The final verification and results show that the numerical model proposed in this paper can satisfactorily simulate and predict the hydrodynamic conditions of the sea area near the aquaculture area in Changhai County; the three-dimensional flow field can reflect the variation in the spatially stratified hydrodynamic indexes of the dynamic environment in a more realistic way, which can also reflect the hindering effect of rafts on hydrodynamic force in a more accurate way, and the model features great accuracy and stability. According to the working conditions before and after the implementation of aquaculture activities, the impact of the floating rafts on the hydrodynamic environment and water exchange capacity was compared and analyzed. The results indicate that the flow resistance rate was greater than 80%; the maximum decrease in the water exchange rate was close to 20%. The quantitative results sufficiently show that, even if floating rafts are arranged with a certain density in a completely open sea area, they have a great impact on the hydrodynamic conditions of the sea area. Therefore, aquaculture operators and marine environmental protection workers must pay sufficient attention to the impact of floating rafts for aquaculture on the hydrodynamic conditions of sea area. The establishment of the method in this paper provides a basic model for the rational arrangement of a fully open raft aquaculture area and the scientific determination of breeding density, and it offers a quantitative numerical calculation method for the assessment of the water exchange capacity in aquaculture areas containing flexible objects [29] (such as rafts, vegetation, etc.).It also provides aquaculture operators with technical support in making scientific and effective decisions regarding aquaculture.

In future studies, spatial modeling for floating rafts, mainly including floaters, external aquaculture nets for hanging cages and organisms, will be added, and a fluid–structure interaction-based multiphase flow (volume of fluid, VOF) model will be used to simulate the impact of floating rafts for aquaculture on the dynamic environment and water exchange, in order to provide more accurate and comprehensive technical support for the rational arrangement of aquaculture orientation and the scientific setting of aquaculture density. Furthermore, after the accurate determination of the impact of a raft aquaculture area on the hydrodynamic conditions, it is also possible for aquaculture operators to reasonably select a site for the installation of bait casting and distribution devices, which can thus help to considerably increase the production efficiency of raft aquaculture, guarantee a stable income for aquaculture operators, and improve the social and economic benefits of raft aquaculture in sea areas in Changhai County and even in other open sea areas with floating raft aquaculture.

Author Contributions: Conceptualization, K.W. and L.S.; methodology, K.W. and H.J.; software, K.W. and H.J.; validation, K.W., H.J. and J.W.; formal analysis, N.L.; investigation, Z.W. and G.S.; resources, K.W.; data curation, K.W.; writing—original draft preparation, K.W.; writing—review and editing, K.W., H.J., N.L. and J.D.; visualization, K.W. and Z.W.; supervision, K.W. and L.S.; project administration, L.S. and J.W.; funding acquisition, K.W. and L.S. All authors have read and agreed to the published version of the manuscript.

Funding: This research was funded by the Modern Agro-Industry Technology Research System (grant number CARS-49), National Key R&D Program of China (grant number 2018YFD0901604), 2021 Special Program for Marine Economic Development of Liaoning Province, Science and Technology Innovation Fund Program of Dalian (grant number 2021JJ13SN74) and Outstanding Young Sci-Tech Talent Program of Dalian (grant number 2019RJ09).

Data Availability Statement: This research did not report any data that are linked to publicly archived datasets analyzed or generated during the study.

Acknowledgments: Thanks to the hard work of the field survey researchers, we obtained detailed model verification data. Thanks to the great support and cooperation of the writing and checking researchers, the numerical simulation results in the article could be displayed. The authors would also like to thank all the editors and anonymous reviewers for their helpful comments that greatly improved the quality of the manuscript.

Conflicts of Interest: No potential conflicts of interest were reported by the authors. The funders had no role in the design of the study; in the collection, analyses, or interpretation of data; in the writing of the manuscript, or in the decision to publish the results.

References

1. Zhao, Y.; Zhang, J.; Liu, Y.; Sun, K.; Zhang, C.; Wu, W.; Teng, F. Numerical assessment of the environmental impacts of deep sea cage culture in the Yellow Sea, China. *Sci. Total Environ.* **2019**, *706*, 135752.1–135752.10. [CrossRef] [PubMed]
2. Klebert, P.; Patursson, O.; Endresen, P.C.; Rundtop, P.; Birkevold, J.; Rasmussen, H.W. Three-dimensional deformation of a large circular flexible sea cage in high currents: Field experiment and modeling. *Ocean Eng.* **2015**, *104*, 511–520. [CrossRef]
3. Dong, S.; You, X.; Hu, F. Experimental investigation on the fluid–structure interaction of a flexible net cage used to farm Pacific bluefin tuna (*Thunnus orientalis*). *Ocean Eng.* **2021**, *226*, 1–12. [CrossRef]
4. Dong, S.; Park, S.; Kitazawa, D.; Zhou, J.; Yoshida, T.; Li, Q. Model tests and full-scale sea trials for drag force and deformation of a marine aquaculture net cage. *Ocean Eng.* **2021**, *240*, 1–17. [CrossRef]
5. Klebert, P.; Su, B. Turbulence and flow field alterations inside a fish sea cage and its wake. *Appl. Ocean. Res.* **2020**, *98*, 1–15. [CrossRef]
6. Ji, R.B.; Mao, X.H.; Zhu, M.Y. The impact of shellfish aquaculture on the gulf ecosystem. *J. Oceanogr. Huanghai Bohai Seas* **1998**, *1*, 21–27.
7. Hatcher, A.; Grant, J.; Schofield, B. Effects of suspended mussel culture (*Mytilus* spp.) on sedimentation, benthic respiration and sediment nutrient dynamics in a coastal bay. *Mar. Ecol. Prog. Ser.* **1994**, *115*, 219–235. [CrossRef]
8. Bouchet, V.M.P.; Sauriau, P. Influence of oyster culture practices and environmental conditions on the ecological status of intertidal mudflats in the Pertuis Charentais (SW France): A multi-index approach. *Mar. Pollut. Bull.* **2008**, *56*, 1898–1912. [CrossRef]
9. Panchang, V.; Cheng, G.; Newell, C. Modeling Hydrodynamics and Aquaculture Waste Transport in Coastal Maine. *Estuaries* **1997**, *20*, 14–41. [CrossRef]
10. Fan, X.; Wei, H.; Yuan, Y.; Zhao, L. Vertical structure of tidal current in a typically coastal raft-culture area. *Cont. Shelf Res.* **2009**, *29*, 2345–2357. [CrossRef]
11. Duarte, P.; Alarez-Salgado X., A.; Fernandez-Reiriz, M.J.; Piedracoba, S.; Labarta, U. A modeling study on the hydrodynamics of a coastal embayment occupied by mussel farms (Ria de Ares-Betanzos, NW Iberian Peninsula). *Estuar. Coast. Shelf Sci.* **2014**, *147*, 42–55. [CrossRef]
12. Jie, S.; Hao, W. Numerical simulation of the hydrodynamic field in a semi-enclosed high-density raft aquaculture sea area. *Period. Ocean Univ. China* **2009**, *39*, 1181.

13. Shi, F.; Dong, X. Three-dimension numerical simulation for vulcanization process based on unstructured tetrahedron mesh. *J. Manuf. Process.* **2016**, *22*, 1–6. [CrossRef]

14. Zhang, P.; Zhang, R.J.; Huang, J.M.; Sheng, Z.; Wang, S. FVCOM model-based study on tidal prism and water exchange capacity of Haizhou Bay. *Water Resour. Hydropower Eng.* **2021**, *52*, 143–151.

15. Chen, C.S.; Liu, H.D.; Beardsley, R.C. An unstructured grid, finite-volume, three-dimensional, primitive equations ocean model: Application to coastal ocean and estuaries. *Atmos. Ocean. Technol.* **2003**, *20*, 159–186. [CrossRef]

16. Zhao, C.; Ren, L.; Yuan, F.; Zhang, L.; Jiang, S.; Shi, J.; Chen, T.; Liu, S.; Yang, X.; Liu, Y.; et al. Statistical and Hydrological Evaluations of Multiple Satellite Precipitation Products in the Yellow River Source Region of China. *Water* **2020**, *12*, 3082. [CrossRef]

17. Zhao, S.; Jin, S.; Ai, C.; Zhang, N. Visual analysis of three-dimensional flow field based on WebVR. *J. Hydroinformat.* **2019**, *21*, 671–686. [CrossRef]

18. Casulli, V.; Walters, R.A. An unstructured grid, three dimensional model based on the shallow water equations. *Int. J. Numer. Methods Fluids* **2000**, *32*, 331–348. [CrossRef]

19. Minhaz, F.A.; Mazlin, B.M.; Chen, K.L.; Nuriah, A.M. Identification of Water Pollution Sources for Better Langat River Basin Management in Malaysia. *Water* **2022**, *14*, 1904.

20. Liu, X.; Ma, D.G.; Zhang, Q.H. A higher-efficient non-hydrostatic finite volume model for strong three-dimensional free surface flows and sediment transport. *China Ocean. Eng.* **2017**, *31*, 736–746. [CrossRef]

21. Stanovoy, V.V.; Eremina, T.R.; Isaev, A.V.; Neelov, I.A.; Vankevich, R.E.; Ryabchenko, V.A. Modeling of oil spills in ice conditions in the Gulf of Finland on the basis of an operative forecasting system. *Marine Phys.* **2012**, *52*, 754–759. [CrossRef]

22. Wang, S.; Zhang, P.; Xi, Y.B.; Ding, J.; Han, Y.; Zhang, R. Study on the effect of culturing floating raft on water exchange capacity in Haizhou Bay. *Water Resour. Hydropower Eng.* **2021**, *52*, 109–120.

23. Shi, J.; Wei, H.; Zhao, L.; Yuan, Y.; Fang, J.; Zhang, J. A physical–biological coupled aquaculture model for a suspended aquaculture area of China. *Aquaculture* **2011**, *318*, 412–424. [CrossRef]

24. Bilgili, A.; Proehl, J.A.; Lynch, D.R.; Smith, K.W.; Swift, M.R. Estuary/ocean exchange and tidal mixing in a Gulf of Maine Esturay: A Lagrangian modeling study. *Estuar. Coast. Shelf Sci.* **2005**, *65*, 607–624. [CrossRef]

25. Park, J.C.; Kim, M.H.; Miyata, H. Filly nor linear free-surface simulations by a 3D viscous numerical wave tank. *Int. J. Numer. Methods Fluids* **1999**, *29*, 685–703. [CrossRef]

26. Smagorinsky, J. General circulation experiments with the primitive equations: I. The basic experiment. *Mon. Weather Rev.* **1963**, *91*, 99–164. [CrossRef]

27. Wang, K.; Wang, N.B. Numerical Simulation of Water Pollutant Transport in Liaodong Bay. *J. Hydrodyn.* **2010**, *25*, 493–498.

28. Wang, K.; Guo, N.; Wang, N.; Leng, C.; Wang, Z.; Li, A.; Xu, X. Simulation of water exchange capacity in Liaodong Bay and application of conservative water quality model. *Fish. Sci.* **2013**, *32*, 475–481.

29. Wang, C.; Zhu, P.; Wang, P.F. Effects of aquatic vegetation on flow in the Nansi Lake and its flow velocitymodeling. *J. Hydrodyn.* **2006**, *18*, 640–648. [CrossRef]

Article

Effects of Temperature on Growth, Molting, Feed Intake, and Energy Metabolism of Individually Cultured Juvenile Mud Crab *Scylla paramamosain* in the Recirculating Aquaculture System

Jiahao Liu [1,2], Ce Shi [1,2,3,*], Yangfang Ye [1,2], Zhen Ma [4,5], Changkao Mu [1,2], Zhiming Ren [1,2], Qingyang Wu [1,2] and Chunlin Wang [1,2]

[1] Key Laboratory of Aquacultural Biotechnology, Ningbo University, Chinese Ministry of Education, Ningbo 315211, China
[2] Collaborative Innovation Center for Zhejiang Marine High-Efficiency and Healthy Aquaculture, Ningbo 315211, China
[3] Marine Economic Research Center, Donghai Academy, Ningbo University, Ningbo 315211, China
[4] Key Laboratory of Environment Controlled Aquaculture (KLECA), Ministry of Education, 52 Heishijiao Street, Dalian 116023, China
[5] College of Marine Technology and Environment, Dalian Ocean University, Dalian 116023, China
* Correspondence: shice3210@126.com

Citation: Liu, J.; Shi, C.; Ye, Y.; Ma, Z.; Mu, C.; Ren, Z.; Wu, Q.; Wang, C. Effects of Temperature on Growth, Molting, Feed Intake, and Energy Metabolism of Individually Cultured Juvenile Mud Crab *Scylla paramamosain* in the Recirculating Aquaculture System. *Water* **2022**, *14*, 2988. https://doi.org/10.3390/w14192988

Academic Editors: Xiangli Tian and Li Li

Received: 20 August 2022
Accepted: 12 September 2022
Published: 23 September 2022

Publisher's Note: MDPI stays neutral with regard to jurisdictional claims in published maps and institutional affiliations.

Abstract: An eight-week experiment was conducted to investigate the effects of temperature (20, 25, 30, and 35 °C) on growth performance, feed intake, energy metabolism, antioxidant capacity, and the stress response of juvenile *Scylla paramamosain* in a recirculating aquaculture system. The results showed that the survival rate of the 35 °C group was 80.36 ± 5.92%, significantly lower than that of the other three groups (100%). The high molt frequency of mud crabs was observed in high-temperature groups, accompanied by a higher ecdysone level and ecdysone receptor gene expression but lower molt inhibitory hormone gene expression. However, the molt increment (73.58 ± 2.18%), food intake, and feed conversion efficiency showed a parabolic trend, with the lowest value found in the 35 °C group. Oxygen consumption rate and ammonia excretion rate increased with the increasing temperature, and oxygen-nitrogen ratio, lactic acid, triglyceride, total cholesterol, glucose, and cortisol peaked at 35 °C. Temperature also significantly affected the antioxidant system of *S. paramamosain*. Crabs in the 25 °C and 30 °C had a significantly higher total antioxidant capacity and lower malondialdehyde compared with the 35 °C group ($p < 0.05$). Although the high temperature promoted molting, it decreased the feeding rate and growth performance, leading to oxidative stress and functional hypoxia. The quadratic function model demonstrated the optimum temperature for the specific growth rate of juvenile *S. paramamosain* was 28.5–29.7 °C.

Keywords: mud crab; temperature; molting; energy metabolism

1. Introduction

Throughout tropical and temperate waters, the mud crab (*Scylla paramamosain*) is a common and economically important marine crab [1,2]. There is a long history of mud crab farming in China, Japan, and the Philippines [3]. China's 2020 mud crab production is 159,433 tonnes [4]. Mud crab is mainly cultured in ponds, but the unit output is low due to serious cannibalism. In recent years, aquaculturists have tried to culture the mud crab in recirculating aquaculture system (RAS). Unlike traditional pond culture, RAS can prevent cannibalism from the early developing stages and effectively improve the survival rate of mud crabs, even in the nursery. Although researchers have studied the factors such as tank bottom area [5] and tank color [6], they are still insufficient compared to the rapid development of the industry.

Temperature is a ubiquitous factor in the life history of aquatic animals, and the high specific heat and heat conduction of water create a challenging thermal environment for aquatic animals [7]. The temperature of the water plays a major part in the survival and growth of crustaceans, according to a number of studies. For example, the mortality rate of ornamental red cherry shrimp (*Neocaridina heteropoda heteropoda*) at 32 °C was significantly higher than at 24 and 28 °C [8]. The survival rate of *Macrobrachium amazonicum* was also affected by temperature, and the survival rate at 28 °C was higher than that at 30 and 32 °C [9]. On growth, the red king crab (*Paralithodes camtschaticus*) grows exponentially with temperature [10]. Crustaceans achieve faster growth by increasing the molt increment (MI) or molt frequency (MF). Synthesized in the Y organ, ecdysone diffuses across the cell membrane and releases into the hemolymph, where it is converted to 20-HE and binds to the EcR-RXR-ecdysone complex to regulate molting [11]. However, the *mih* gene inhibits molting by inhibiting ecdysone in the hemolymph [12].

According to the principle of thermodynamics, the increase in temperature stimulates the metabolic process of the organism [13]. Ambient temperature can significantly affect aquatic animals' metabolic levels and physiological regulation mechanisms. The optimal temperature strongly supports the physiological and biochemical processes of the organisms. At the same time, it can provide maximum energy efficiency [14,15]. The Oxygen consumption rate (OCR) of most crustaceans in the thermophilic range increases gradually with increasing temperature [16]. When the temperature is too high, the metabolic level of crustaceans decreases [17]. For example, in the southern rock lobster (*Jasus edwardsii*), the OCR positively correlates to the temperature at 18–22 °C. However, when the temperature increased further, the metabolic level decreased instead [17]. Furthermore, since ammonia production is the result of amino acid deamination, ammonia excretion rate (AER) can be used to estimate protein utilization by aquatic organisms [18]. Compared with fishes, crustaceans had an open-vessel circulatory system and transported nutrients through the hemolymph [19]. Hemolymph metabolites could reflect the morphological and physiological adaptation of crustaceans to the environment [20]. Total cholesterol (T-CHO), triglyceride (TG), and glucose (GLU) in hemolymph can evaluate the energy metabolism of crustaceans [21].

An increase in temperature is associated with a higher metabolic rate (Q10 effect), resulting in increased oxygen consumption, increased flux at the electron transport chain level, and more reactive oxygen species (ROS) [22]. ROS can oxidize surrounding molecules, impair cellular functions, and lead to oxidative stress [23,24]. Oxidative enzymes and non-enzymatic antioxidants are used by aquatic organisms to remove excess ROS. The antioxidant enzymes superoxide dismutase (SOD), catalase (CAT), and glutathione peroxidase (GPX) are among them. Non-enzymatic antioxidants include fat-soluble vitamins (e.g., alpha-tocopherol) and small water-soluble molecules such as glutathione (GSH) [25]. Total antioxidant capacity (T-AOC) reflects the metabolic ability of antioxidant enzymatic and non-enzymatic systems under external stress [26]. Malondialdehyde (MDA), as the end product of lipid peroxidation, reflects the degree of cellular oxidative damage [27]. Studies have shown that long-term stress in crustaceans inhibits antioxidant enzyme activity and produces more MDA [5,28,29], accompanied by reduced food intake (FI) [30].

Therefore, this study aimed to evaluate the optimal temperature range for juvenile mud crabs in RAS in terms of growth, molting, energy metabolism, antioxidant capacity, and stress response. This study could help define the management protocol of mud crab and support the design of crab RAS.

2. Material and Methods

2.1. Experimental Design

Four temperatures, i.e., 20, 25, 30, and 35 °C, were set with four independent RAS. Each RAS consisted of 7 square tanks (0.5 × 0.3 × 0.2 m^3) with six compartments (0.1 × 0.1 × 0.13 m^3) in each tank. Thus, each RAS had 42 compartments with one crab per compartment. The crabs were distributed to three replicates for each treatment, with 14 crabs per replicate. The experiment lasted eight weeks at Ningbo University.

2.2. Experimental Animal and Rearing Conditions

The experiment was conducted in July–August 2021 in the Intelligent Aquaculture Laboratory (Ningbo City, Zhejiang province, China). Selection of healthy and uniformly sized juvenile mud crabs bought from mud crab nursery farm (Ningbo City, Zhejiang province, China). Before the experiment, the crabs were domesticated in the laboratory for one week. During this period, the commercial feed was overfed daily at 20:00 (Table 1). The excessive commercial feed was siphoned, and 1/3 of isothermal seawater was exchanged at 18:00 every day.

Table 1. Specific primers were used for real-time PCR in this study.

Gene	Sequence (5′ to 3′)	Size (bp)	Access Number
ecr	F: TAAGTGATGACGACTCGGATGC R: ACGAGCAAGCCTTTAGCAGTG	150	KC354381
mih	F: TATCAAGTGCAGGAACTCAG R: GGAACATACAAGCCTAAACA	110	EU869539
β-actin	F: CGAAACCTTCAACACTCCCG R: GATAGCGTGAGGAAGGGCATA	154	FJ641977

ecr: Ecdysone receptor, *mih*: Molt-inhibiting hormone.

When the experiment started, the 168 juvenile *S. paramamosain* with complete appendages and good vitality were weighed and then distributed to 4 RAS (weight: 0.36 ± 0.09 g). Starting from the room temperature of 28 °C, the RAS was adjusted to 20 °C (20.11 ± 0.43 °C), 25 °C (24.98 ± 0.23 °C), 30 °C (30.08 ± 0.11 °C), and 35 °C (34.88 ± 0.39 °C) at a rate of $1\ °C\ d^{-1}$.

During the experiment, 20 and 25 °C were achieved with a refrigerating machine (AO LING HENG YE, LA-160, China). 30 and 35 °C groups were achieved by the heater (SUN SUN, AR-450, China). The same feeding protocol was used as described above, and the residual feed was counted daily. The seawater salinity was 25 ppt, and the photoperiod was 14L:10D.

2.3. Sampling

At the end of the 8-week breeding experiment, the crabs were starved for 24 h and then sampled. The crabs were anesthetized on ice, weighed, and then dissected. The eyestalk, hepatopancreas, and muscle were isolated with tweezers quickly. The hemolymph was collected with a disposable syringe from the pericardial sinus of the crab and kept at 4 °C overnight, then centrifuged at 3500 RCF for 15 min. The supernatant and all the other tissues were stored at −80 °C.

2.4. Ecdysone Content and Molting-Related Genes

Six crabs in each group were randomly selected to extract hemolymph to determine ecdysone content using a crab ecdysone-specific enzyme-linked immunosorbent assay kit (Enzyme Link Biotechnology, Shanghai, China). Every two crabs as a repeat. Most of the mud crab's hormones were secreted and synthesized by the eyestalk. Therefore, the eyestalks of 2 crabs were selected in each replicate of each treatment to examine the relative expression of molting-related genes (total of 6/treatment). After adding liquid nitrogen to the mortar, the eyestalk was ground into powder, and the powder was added to 1 ml of Trizol reagent (Invitrogen, Waltham, MA, USA). After rapid shaking and mixing, put into liquid nitrogen flash freezing, and the total RNA was extracted after. The total RNA product was aspirated and subjected to 1% agarose denaturing gel electrophoresis to detect RNA integrity. Synthesis of cDNA using HiFiScript cDNA synthesis kit (CW Biotech. Co. Lid., Shanghai, China) with total RNA as a template by reverse transcription.

The expression of *mih* and *ecr* were detected by real-time PCR (LightCycler480 II, ROCHE, Basel, Switzerland) with *β-actin* as an internal reference gene. The relative expression levels of *ecr* and *mih* gene were calculated by $2^{-\Delta\Delta CT}$ method [31].

2.5. OCR and AER

The OCR and AER were measured on the sampling day. The water used in the experiments was fully aerated to saturation and was recorded as initial dissolved oxygen by YSI Pross handheld multiparameter water quality analyzer (USA). Six juvenile crabs of similar body weight and intact appendages from each treatment were carefully transferred to conical flasks with 0.1 L of aerated water. To prevent gas exchange, the mouth of the conical flasks was utterly sealed with plastic film. The conical flasks were sufficiently immersed in each RAS to keep a constant temperature. To exclude the interference of water respiration, a control group without crab was set for each treatment group. The experiment lasted for 60 min, and the final dissolved oxygen was measured [32]. The experimental method of AER is the same as OCR. The experiment lasted for 6 h. The AER was calculated according to the ammonia nitrogen concentration change before and after the experiment (HACH, 2604545) [33].

2.6. Antioxidant Capacity

Six crabs were randomly selected from each group, and every two crabs were used as one replicate for hepatopancreas antioxidant capacity measurement. The samples were centrifuged at 3500 rcf at 4 °C for 15 min after averaging in ice-cold physiological saline. T-AOC (A015-2-1), MDA (A003-1-2), SOD (A007-1-1), CAT (007-1-1), and GSH (A006-2-1) were analyzed using commercial kits (Nanjing Jiancheng Bioengineering Institute, Nanjing, China) according to the manufacturer's instructions [28].

2.7. Hemolymph Indexes

Six crabs were randomly selected from each group, the hemolymph samples were determined, and every two crabs were used as one replicate. The determination of cortisol in crab hemolymph with a crab's specific cortisol ELISA kit (Enzyme-linked Biotechnology, Shanghai, China). T-CHO (A111-1-1), UA (C012-2-1), GLU (A154-1-1), TG (A110-1-1), and LD (A109-2-1) content were determined by commercial kits (Nanjing Jiancheng Bioengineering Institute, Nanjing, China) [34].

2.8. Data Collection and Calculation

Survival rate (%) = 100 × (final number of crabs)/(initial number of crabs)
Specific growth rate (SGR, % day^{-1}) = 100 × (Ln Wf − Ln Wi)/t
Feed conversion efficiency (FCE, %) = $(W_f - W_i)$ × 100/FC
Molt frequency (MF) = $\Sigma((C_n - 1) \times N_n)/N_t$
OCR = $[(O_1 - O_2) \times V]/(W_f \times T)$
AER = $[(N_1 - N_2) \times V]/(W_f \times T)$
O: N ratio = (OCR/16)/(AER/14)
FI (% body weight d^{-1}) = FC/[T × $(W_f + W_i)/2$] × 100

W_f, final weight (g); W_i, initial weight (g); t, duration of the experiment (d); C_n, the developmental stage of crab; N_n, the number of molting stages; N_t, total number of survival crabs; O_1, dissolved oxygen in the blank group (mg L^{-1}); O_2, test group dissolved oxygen (mg L^{-1}); V, the volume of water in a beaker (L); T, metabolism time (h); N_1, ammonia nitrogen in the control group (mg L^{-1}); N_2, ammonia nitrogen in the experimental group (mg L^{-1}); FC, the weight of food ingested during the experiment (dry weight, g).

2.9. Statistics

Using SPSS 25.0 statistical software, all results were expressed as mean ± standard deviation ($n = 3$). Before analysis, the Kolmogorov - Smirnov and Levene tests were used to test the distribution normality and homogeneity of variance of the original data. One-way

ANOVA and Tukey's multiple comparison post hoc test were used to determine whether significant variation existed between the treatments. The *ecr* gene expression and CAT could not achieve normality and homogeneity and were analyzed with the nonparametric Kruskal-Wallis test. A quadratic model was used to fit the SGR to obtain the optimal growth temperature of the juvenile mud crabs. To determine the relationship between temperature, growth, ingestion, OCR, AER, molting, and the expression of *mih* and *ecr*, Pearson correlation analysis and t-test were used. A statistically significant level of $p < 0.05$ was applied in the present study.

3. Results

3.1. Survival, Growth Performance, Molting, and Feeding

The survival rate and hepatopancreatic index of the 35 °C group were significantly lower than the other groups ($p < 0.05$) (Figure 1A,B). The temperature significantly affected the growth, and SGR showed a parabolic trend (Figures 1C and 2A). Juvenile mud crabs had the best growth performance between 28.5 °C and 29.7 °C in accordance with the quadratic regression model analysis (Figure 2).

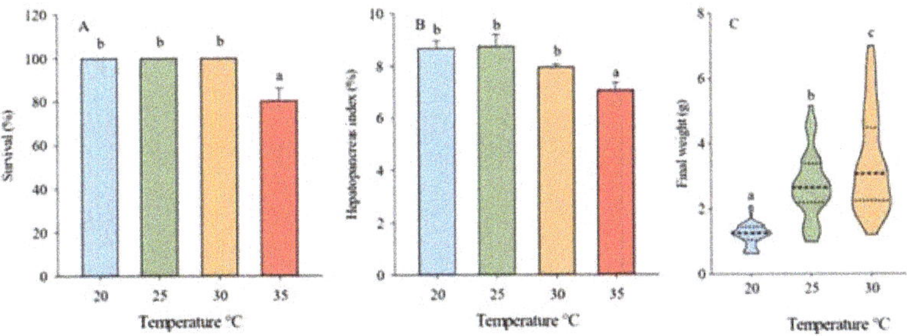

Figure 1. Effects of different temperatures (20–35 °C) on survival rate (**A**), hepatopancreas index (**B**), and final weight (**C**) of juvenile mud crabs. Values are expressed as mean $\pm$ SD ($n = 3$). Different superscripts indicate significant differences between treatments ($p < 0.05$).

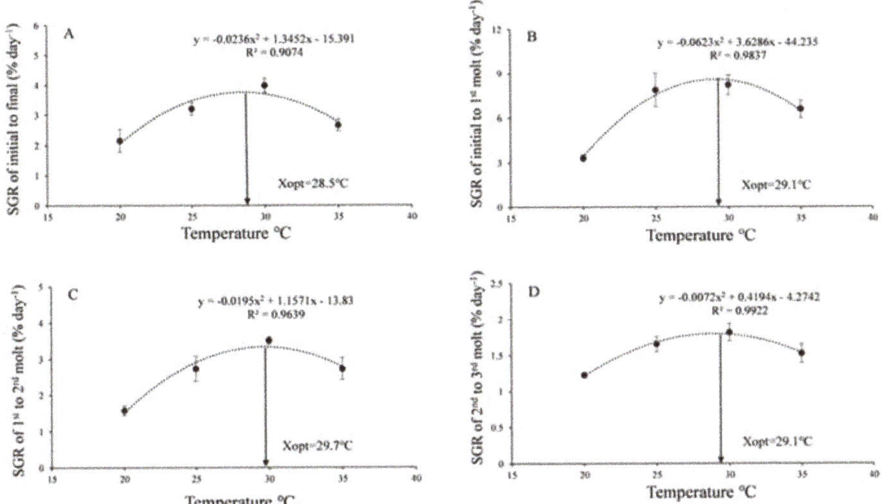

Figure 2. Relationship between the temperature and the SGR of initial to 1st molt (**A**), 1st to 2nd molt (**B**), 2nd to 3rd molt (**C**), and initial to final (**D**), respectively. Where Xopt means the optimal temperature for the maximum SGR.

The MF of the 30 °C (19.80 ± 1.33) group was significantly higher than the 20 °C (8.40 ± 1.05), 25 °C (13.21 ± 0.51), and 35 °C (14.00 ± 1.54) ($p < 0.05$). However, the MI of the 35 °C groups (73.58 ± 2.18%) was significantly lower than that of the 20 °C (87.13 ± 9.10%), 25 °C (93.44 ± 5.16%), and 30 °C (86.58 ± 4.36%) groups ($p < 0.05$) (Figure 3).

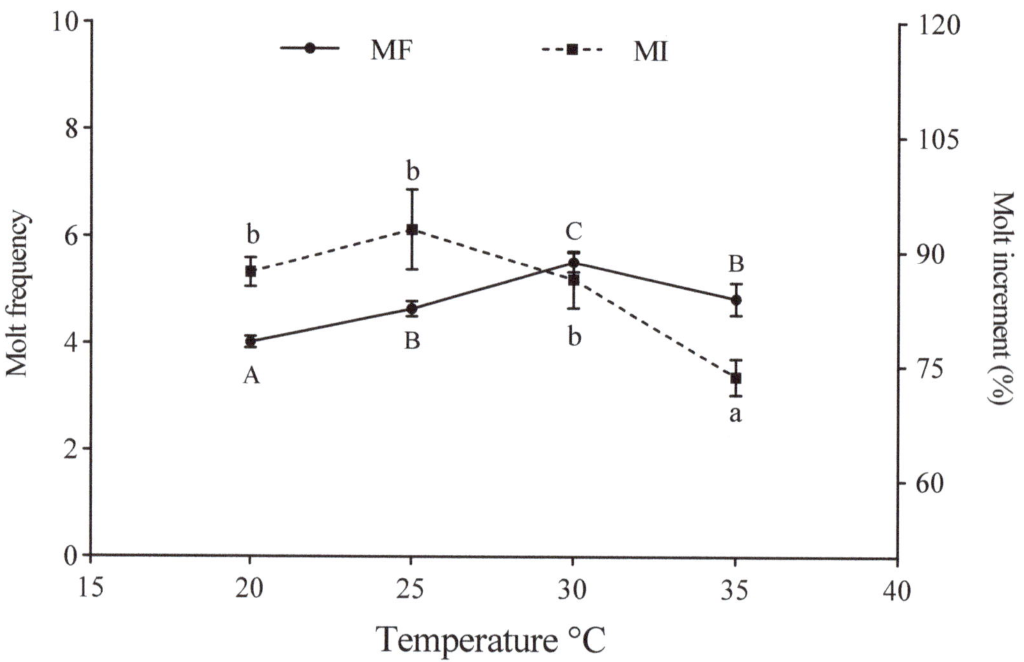

Figure 3. Effects of different temperatures (20–35 °C) on the molt frequency and molt increment of juvenile mud crabs. Values are expressed as mean ± SD ($n = 3$). Different superscripts indicate significant differences between treatments ($p < 0.05$).

With the increase in temperature, the average daily food intake showed a trend of first increase and then decrease, and the difference was significant ($p < 0.05$) (Figure 4A). The feed conversion efficiency at 25 °C and 30 °C was significantly higher than that in the 20 and 35 °C groups ($p < 0.05$), the 35 °C group was significantly lower than the other three groups ($p < 0.05$), and the FCE in the 30 °C group was the highest (45.48 ± 1.21%), the FCE of the 35 °C group was the lowest (21.00 ± 3.00%) (Figure 4B).

3.2. Ecdysone Content and Expression of Molting-Related Genes

The content of ecdysone in crab hemolymph increased significantly with the increasing temperature ($p < 0.05$) (Figure 5A). The expression level of the *ecr* gene in the 35 °C group was the highest and was significantly higher than that in the other three groups ($p < 0.05$) (Figure 5B). The expression of the *mih* gene decreased with the increasing temperature (Figure 5C).

3.3. OCR, AER, and O: N

Between 20–30 °C, the OCR increased significantly with the increasing temperature ($p < 0.05$), but no significant difference was found between the 30 °C and 35 °C groups ($p > 0.05$) (Figure 6A). The AER peaked at 35 °C and was significantly higher than the 20–30 °C group ($p < 0.05$) (Figure 6B). The O: N showed a parabolic trend and peaked at 30 °C (Figure 6C).

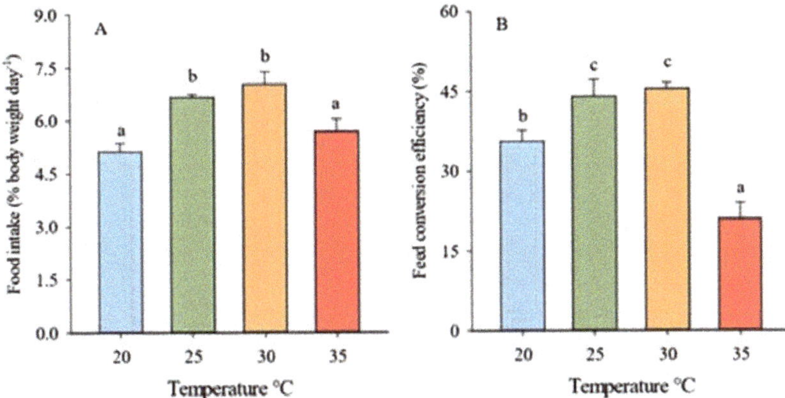

Figure 4. Effects of different temperatures (20–35 °C) on the average daily food intake (**A**) and feed conversion efficiency (**B**) of juvenile mud crabs. Values are expressed as mean ± SD (n = 3). Different superscripts indicate significant differences between treatments ($p < 0.05$).

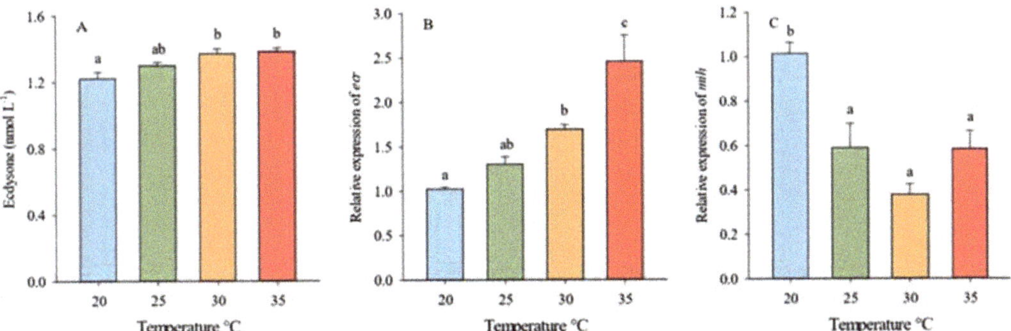

Figure 5. The content of ecdysone (**A**) in the hemolymph. Expression of the molting-related gene (*ecr*) (**B**) and molt-inhibiting hormone (*mih*) (**C**) in eyestalks. Values are presented as mean ± SD (n = 3). Different superscripts indicate significant differences between treatments ($p < 0.05$).

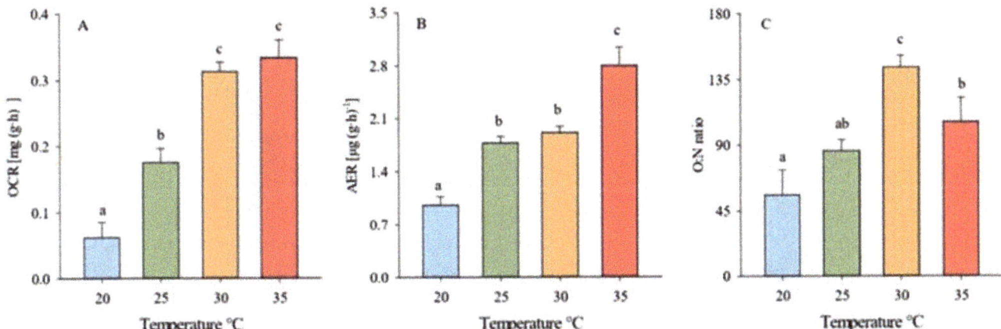

Figure 6. Oxygen consumption rate (OCR) (**A**), ammonia excretion rate (AER) (**B**), and oxygen-nitrogen ratio (O: N ratio) (**C**) of juvenile mud crabs at different temperatures (20–35 °C). Values are expressed as mean ± SD (n = 3). Different superscripts indicate significant differences between treatments ($p < 0.05$).

3.4. Antioxidant Capacity

T-AOC activity was the highest at 25 and 30 °C (Figure 7A). The activity of CAT showed a significant upward trend with the increasing temperature ($p < 0.05$) (Figure 7B). The GSH and SOD activities peaked at 30 °C (Figure 7C,D). Moreover, the MDA content was significantly lower in the 25 and 30 °C groups compared to the 35 °C group (Figure 7E).

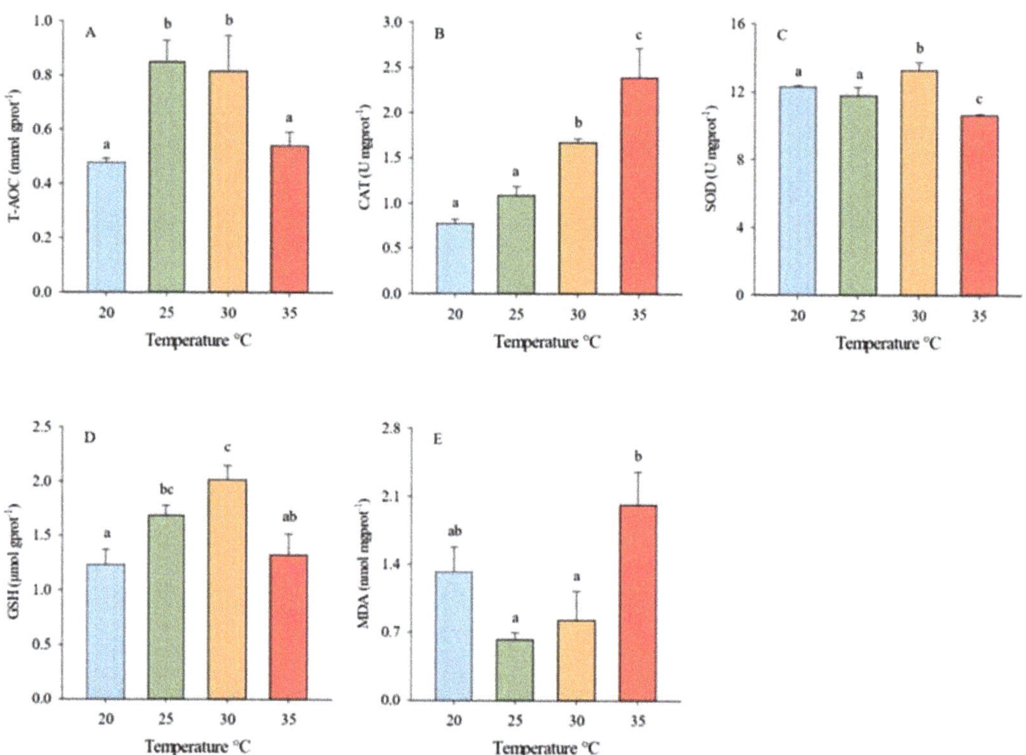

Figure 7. Effects of different temperatures on the activities total of antioxidant capacity (T-AOC) (**A**), catalase (CAT) (**B**), superoxide dismutase (SOD) (**C**), glutathione (GSH) (**D**), and malondialdehyde (MDA) (**E**) content of juvenile mud crabs. Values are expressed as mean ± SD (n = 3). Different superscripts indicate significant differences between treatments ($p < 0.05$).

3.5. Hemolymph Index

In comparison with other groups, the hemolymph of the 35 °C group contained significantly more cortisol and LD ($p < 0.05$), and there was no significant difference between the 20–30 °C groups ($p > 0.05$) (Figure 8A,B). The uric acid content increased with the increasing temperature (Figure 8F). Glucose and total cholesterol showed a parabolic trend (Figure 8C,E). Significantly higher TG content was found in the 35 °C group than in either the 30 °C or 20 °C groups ($p < 0.05$) (Figure 8D).

3.6. Correlation Analysis

Correlations between temperature and SGR, MF, MI, OCR, AER, FI, FCE, and relative expression of *mih* and *ecr* parameters were analyzed by Pearson correlation coefficient (Figure 9). Overall, *ecr* gene expression, AER, OCR, and MF were positively correlated with temperature, while *mih* gene expression and MI were negatively correlated. Interestingly, the *mih* gene expression was also negatively correlated with SGR, MF, OCR, AER, and FI.

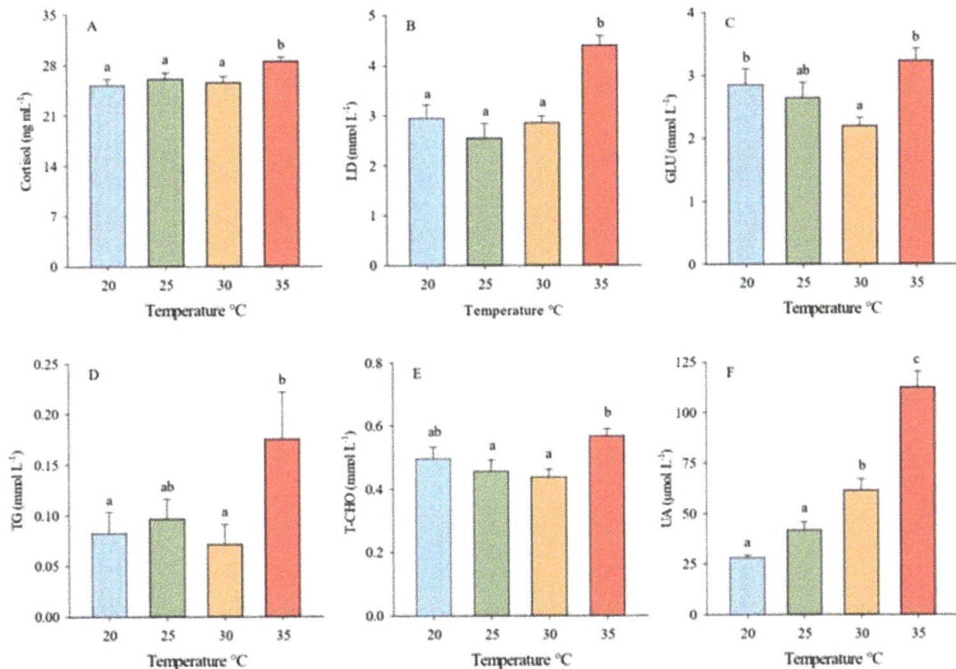

Figure 8. Effects of different temperatures on cortisol (**A**), lactic acid (LD) (**B**), glucose (GLU) (**C**), triglyceride (TG) (**D**), total cholesterol (T-CHO) (**E**), and uric acid (UA) (**F**) content in juvenile mud crabs. Values are presented as mean $\pm$ SD ($n = 3$). Different superscripts indicate significant differences between treatments ($p < 0.05$).

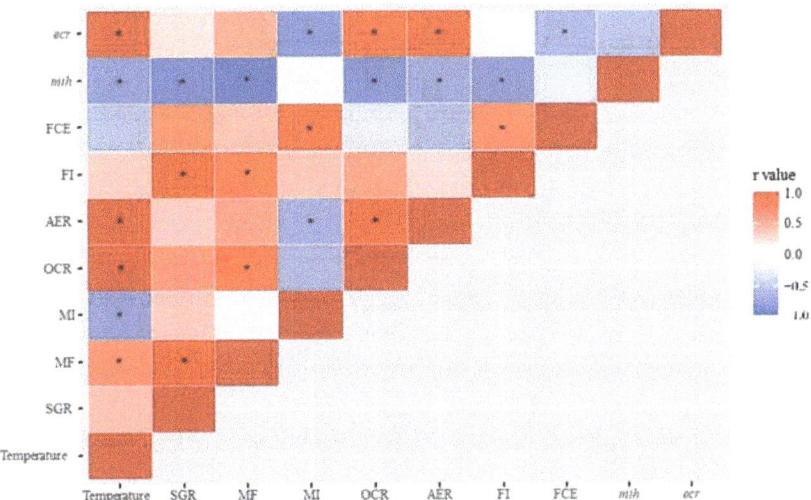

Figure 9. Correlation analyses among temperature SGR, MF, MI, OCR, AER, FI, FCE, and relative expression of *mih* and *ecr* gene ($n = 3$, 2 individuals per replicate for AER, OCR, *ecr*, and *mih*; $n = 3$ replicate, and 31, 42, 42, and 42 individuals per treatment of SGR, MF, FI, and FCE). "*" indicates $p < 0.05$.

4. Discussion

Temperature significantly affected the survival and growth of the mud crab. For crabs reared at 20–30 °C, the survival rate was 100%, while high water temperature (35 °C) caused mortality. The same trend was also observed in growth performance. According to the quadratic regression analysis, the optimal water temperature for the growth of mud crab was 28.5–29.7 °C. A similar study also suggested that the most suitable temperature in Crablet 1 to Crablet 2 phase was 28–32 °C [35]. However, another recent research observed that the SGR of the juvenile *S. paramamosain* in the 36–37 °C ($14.65 \pm 0.23\%$ day^{-1}) group was significantly higher than the 27–28 °C ($12.30 \pm 0.42\%$ day^{-1}), 30–31 °C ($11.58 \pm 0.14\%$ day^{-1}), and 33–34 °C ($10.11 \pm 0.06\%$ day^{-1}) groups [35]. The difference could be partly attributed to the fact that the mud crabs were cultured in groups of 30 individuals per tank [36]; thus, temperature-induced cannibalism could potentially contribute to the growth performance. In addition, the speculation was also supported by the markedly lower survival rate ($28.9 \pm 2.94\%$) and higher SGR ($14.65 \pm 0.23\%$ day^{-1}) at 36–37 °C [35] compared to the present study ($80.36 \pm 5.92\%$ and $3.00 \pm 0.26\%$ day^{-1} at 35 °C).

The growth of organisms was closely related to feeding [37]. Generally, FI increases with increasing temperature and decreases when the temperature exceeds the optimum temperature range [38]. In this study, both low and high temperatures inhibited FI. Low temperatures inhibit the metabolic capacity of crabs, thereby affecting their appetite and energy balance. On the contrary, long-term high-temperature stress caused heat stress and decreased FI. In the present study, the FCE and FI had a consistent trend, indicating that a suitable temperature not only stimulates FI but also effectively improves the assimilation of food. This result was also observed in turbot (*Scophthalmus maximus*) [39], *Penaeus japonicus* [40], and Atlantic salmon (*Salmo salar*) [41]. Therefore, the decline in FI and FCE could be one of the reasons for the suboptimum growth performance for the 20 and 35 °C group.

Molting is essential in crustacean growth [42]. High water temperature inhibited the expression of the *mih* gene but consequently promoted ecdysone synthesis and *ecr* gene expression. These factors could jointly promote the synthesis of the EcR-RXR-ecdysone complex, resulting in a significantly increasing MF. Crustacean growth and molting are highly correlated in most studies [5,43–45]. Interestingly, in this study, both MI and MF dropped simultaneously in the 35 °C group compared with the 30 °C group. A similar phenomenon was observed in *Penaeus japonicus* [40]. Nonetheless, the underlying physiological process is poorly understood. A possible explanation is crab's metabolic capacity is temperature-dependent.

The metabolic rate of animals, measured by OCR and AER, varies directly with temperature [46,47]. O: N could represent the ratio of protein, fat, and carbohydrate catabolism of aquatic organisms [48]. O: N < 10 suggests the respiratory substrate mainly consisted of protein. On the contrary, a high O: N means the substrate is mainly provided by fats and carbohydrates [49]. The OCR of crabs at 30–35 °C was similar, but AER raised dramatically at 35 °C, leading to a lower O: N (105.66 ± 16.90) compared with 30 °C (143.46 ± 7.97). Though the value is much higher than 10, the results suggest a higher proportion of proteins consumed by respiration than tissue growth at 35 °C. The conjecture could further be confirmed by the high hemolymph uric acid level.

The stress response of an organism can be divided into three levels [50]. First-order stress responses include elevated cortisol levels [51]. Secondary stress includes increased energy mobilization, manifested primarily by increases in blood glucose and circulating lipids, which provide the organism with the energy necessary to resist stress [52]. Primary and secondary stress responses trigger tertiary stress responses that ultimately affect animal growth and survival [53]. In this study, a significant increase in cortisol, GLU, TG, and T-CHO in the hemolymph at high temperatures was observed. The results indicate that elevated temperature caused cortisol secretion, increased protein catabolism, and plasma cholesterol, and promoted GLU by inhibiting GLU breakdown. Furthermore, mud crabs raised at 35 °C also had higher lactic acid and significantly lower T-AOC, SOD, and GSH, suggesting an inferior thermal resistance [54,55] and antioxidant capacity, leading to down-regulated survival rates.

5. Conclusions

The effects of rearing temperature on mud crab *S. paramamosain* were estimated by growth, molting, feeding intake, energy metabolism, and stress responses. The result showed that the crabs at high temperature (35 °C) had a significantly higher ecdysone level and expression of the *ecr* gene but lower MI, MF, and survival rate. The effect could be mediated through respiration, energy metabolism, and antioxidant pathways. It was found that 28.5–29.7 °C provided the best conditions for mud crab growth. These findings could help regulate the temperature in mud crab RAS and provide a thread for the thermal adaptation of crustaceans.

Author Contributions: Conceptualization, C.S. and J.L.; methodology, J.L.; formal analysis, J.L.; investigation, J.L.; Funding acquisition, C.S.; writing—original draft preparation, J.L.; writing—review and editing, C.S., Y.Y., Z.R., Q.W., Z.M., C.M.; project administration, C.W. All authors have read and agreed to the published version of the manuscript.

Funding: The research was sponsored by the National Natural Science Foundation of China (Grant Nos. 32172994 and 31972783), the National Key Research and Development Program of China (Project No. 2019YFD0901000), the Province Key Research and Development Program of Zhejiang (2021C02047), Key Scientific and Technological Grant of Zhejiang for Breeding New Agricultural Varieties (2021C02069-6), 2025 Technological Innovation for Ningbo (2019B10010), China Agriculture Research System of MOF and MARA, SanNongLiuFang Zhejiang Agricultural Science and Technology Cooperation Project (2021SNLF029), K. C. Wong Magna Fund in Ningbo University and the Scientific Research Foundation of Graduate School of Ningbo University (IF2020145).

Institutional Review Board Statement: Our study did not involve endangered or protected species. In China, breeding and catching mud crabs, *Scylla paramamosain*, does not require specific permits. All efforts were made to minimize animal suffering and discomfort. The animal study protocol was approved by the Animal Ethics Committee of Ningbo University.

Data Availability Statement: The data presented in this study are not publicly available but are available upon request from the corresponding author.

Acknowledgments: Thanks to Qingsong Zhao for providing the site for this experiment.

Conflicts of Interest: The authors declare no conflict of interest.

References

1. Keenan, C.P. Aquaculture of the mud crab, genus Scylla-past, present and future. In *Aciar Proceedings*; Australian Centre for International Agricultural: Washington, DC, USA, 1999; pp. 9–13.
2. Le Vay, L. Ecology and Management of *Mud Crab Scylla* spp. *Asian Fish. Sci.* **2001**, *14*, 101–112. [CrossRef]
3. Cowan, L. Crab farming in Japan, Taiwan and the Philippines. *Qld. Dep. Prim. Ind. Inf. Ser. Q* **1984**, *85*, 184009.
4. Liu, X. *Fisheries Bureau of Agriculture Ministry of China*; China Agriculture Press: Beijing, China, 2021.
5. Yu, K.; Shi, C.; Liu, X.; Ye, Y.; Wang, C.; Mu, C.; Song, W.; Ren, Z. Tank bottom area influences the growth, molting, stress response, and antioxidant capacity of juvenile mud crab *Scylla paramamosain*. *Aquaculture* **2022**, *548*, 737705. [CrossRef]
6. Ma, X.; Zhang, G.; Gao, G.; Wang, C.; Mu, C.; Ye, Y.; Shi, C. Effects of culture container color on growth, stress and carapace color of juvenile mud crabs *Scylla paramamosain*. *J. Ningbo Univ. Sci. Technol. Ed.* **2021**, *34*, 43–49.
7. Beitinger, T.L.; Fitzpatrick, L.C. Physiological and Ecological Correlates of Preferred Temperature in Fish. *Am. Zool.* **1979**, *19*, 319–329. [CrossRef]
8. Tropea, C.; Stumpf, L.; Greco, L.S.L. Effect of Temperature on Biochemical Composition, Growth and Reproduction of the Ornamental Red Cherry Shrimp *Neocaridina heteropoda heteropoda* (Decapoda, Caridea). *PLoS ONE* **2015**, *10*, e0119468. [CrossRef]
9. Bastos, A.M.; Lima, J.F.; Tavares-Dias, M. Effect of increase in temperature on the survival and growth of *Macrobrachium amazonicum* (Palaemonidae) in the Amazon. *Aquat. Living Resour.* **2018**, *31*, 21. [CrossRef]
10. Stoner, A.W.; Ottmar, M.L.; Copeman, L.A. Temperature effects on the molting, growth, and lipid composition of newly-settled red king crab. *J. Exp. Mar. Biol. Ecol.* **2010**, *393*, 138–147. [CrossRef]
11. Hyde, C.J.; Elizur, A.; Ventura, T. The crustacean ecdysone cassette: A gatekeeper for molt and metamorphosis. *J. Steroid Biochem. Mol. Biol.* **2019**, *185*, 172–183. [CrossRef] [PubMed]

12. Zhang, X.; Huang, D.; Jia, X.; Zou, Z.; Wang, Y.; Zhang, Z. Functional analysis of the promoter of the molt-inhibiting hormone (mih) gene in mud crab *Scylla paramamosain*. *Gen. Comp. Endocrinol.* **2018**, *259*, 131–140. [CrossRef]

13. Lushchak, V.I. Environmentally induced oxidative stress in aquatic animals. *Aquat. Toxicol.* **2011**, *101*, 13–30. [CrossRef]

14. Hernández-Sandoval, P.; Díaz-Herrera, F.; Díaz-Gaxiola, J.M.; Martínez-Valenzuela, C.; García-Guerrero, M. Effect of temperature on growth, survival, thermal behavior, and critical thermal maximum in the juveniles of *Macrobrachium occidentale* (Holthuis, 1950) (Decapoda: Caridea: Palaemonidae) from Mexico. *J. Crustacean Biol.* **2018**, *38*, 483–488. [CrossRef]

15. Re, A.D.; Diaz, F.; Sierra, E.; Rodríguez, J.; Perez, E. Effect of salinity and temperature on thermal tolerance of brown shrimp *Farfantepenaeus aztecus* (Ives) (Crustacea, Penaeidae). *J. Therm. Biol.* **2005**, *30*, 618–622. [CrossRef]

16. González, R.A.; Díaz, F.; Licea, A.; Re, A.D.; Sánchez, L.N.; García-Esquivel, Z. Thermal preference, tolerance and oxygen consumption of adult white shrimp *Litopenaeus vannamei* (Boone) exposed to different acclimation temperatures. *J. Therm. Biol.* **2010**, *35*, 218–224. [CrossRef]

17. Thomas, C.W.; Crear, B.J.; Hart, P.R. The effect of temperature on survival, growth, feeding and metabolic activity of the southern rock lobster, *Jasus edwardsii*. *Aquaculture* **2000**, *185*, 73–84. [CrossRef]

18. Frisk, M.; Steffensen, J.F.; Skov, P.V. The effects of temperature on specific dynamic action and ammonia excretion in pikeperch (*Sander lucioperca*). *Aquaculture* **2013**, *404*, 65–70. [CrossRef]

19. Hu, Y.; Tan, B.; Mai, K.; Ai, Q.; Zheng, S.; Cheng, K. Growth and body composition of juvenile white shrimp, *Litopenaeus vannamei*, fed different ratios of dietary protein to energy. *Aquac. Nutr.* **2008**, *14*, 499–506. [CrossRef]

20. Pascual, C.; Gaxiola, G.; Rosas, C. Blood metabolites and hemocyanin of the white shrimp, *Litopenaeus vannamei*: The effect of culture conditions and a comparison with other crustacean species. *Mar. Biol.* **2003**, *142*, 735–745. [CrossRef]

21. Ciaramella, M.A.; Battison, A.L.; Horney, B. Measurement of tissue lipid reserves in the American lobster (*Homarus americanus*): Hemolymph metabolites as potential biomarkers of nutritional status. *J. Crustac. Biol.* **2014**, *34*, 629–638. [CrossRef]

22. Halliwell, B.; Gutteridge, J.M. *Free Radicals in Biology and Medicine*; Oxford University Press: New York, NY, USA, 2015.

23. Madeira, D.; Narciso, L.; Cabral, H.N.; Vinagre, C.; Diniz, M.S. Influence of temperature in thermal and oxidative stress responses in estuarine fish. *Comp. Biochem. Physiol. Part A Mol. Integr. Physiol.* **2013**, *166*, 237–243. [CrossRef] [PubMed]

24. Paital, B.; Chainy, G.B.N. Effects of temperature on complexes I and II mediated respiration, ROS generation and oxidative stress status in isolated gill mitochondria of the mud crab Scylla serrata. *J. Therm. Biol.* **2014**, *41*, 104–111. [CrossRef] [PubMed]

25. Kong, X.; Wang, G.; Li, S. Effects of low temperature acclimation on antioxidant defenses and ATPase activities in the muscle of mud crab (*Scylla paramamosain*). *Aquaculture* **2012**, *370–371*, 144–149. [CrossRef]

26. Feng, G.; Shi, X.; Huang, X.; Zhuang, P. Oxidative Stress and Antioxidant Defenses after Long-term Fasting in Blood of Chinese Sturgeon (*Acipenser sinensis*). *Procedia Environ. Sci.* **2011**, *8*, 469–475. [CrossRef]

27. Viarengo, A.; Canesi, L.; Pertica, M.; Poli, G.; Moore, M.N.; Orunesu, M. Heavy metal effects on lipid peroxidation in the tissues of mytilus gallopro vincialis lam. *Comp. Biochem. Physiol. Part C Comp. Pharmacol.* **1990**, *97*, 37–42. [CrossRef]

28. Chen, S.; Migaud, H.; Shi, C.; Song, C.; Wang, C.; Ye, Y.; Mu, C. Light intensity impacts on growth, molting and oxidative stress of juvenile mud crab *Scylla paramamosain*. *Aquaculture* **2021**, *545*, 737159. [CrossRef]

29. Gao, Y.; He, Z.; Zhao, B.; Li, Z.; He, J.; Lee, J.Y.; Chu, Z. Effect of stocking density on growth, oxidative stress and HSP 70 of pacific white shrimp *Litopenaeus vannamei*. *Turk. J. Fish. Aquat. Sci.* **2017**, *17*, 877–884. [CrossRef]

30. Maniam, J.; Morris, M.J. The link between stress and feeding behaviour. *Neuropharmacology* **2012**, *63*, 97–110. [CrossRef] [PubMed]

31. Chen, S.; Shi, C.; Migaud, H.; Song, C.; Mu, C.; Ye, Y.; Wang, C.; Ren, Z. Light Spectrum Impacts on Growth, Molting, and Oxidative Stress Response of the Mud Crab *Scylla paramamosain*. *Front. Mar. Sci.* **2022**, *9*, 840353. [CrossRef]

32. Shi, C.; Wang, J.; Peng, K.; Mu, C.; Ye, Y.; Wang, C. The effect of tank colour on background preference, survival and development of larval swimming crab *Portunus trituberculatus*. *Aquaculture* **2019**, *504*, 454–461. [CrossRef]

33. Luo, H.; Wu, Q.; Lin, J.; He, L.; Chen, X.; Niu, W.; Wang, Q.; Zheng, L. Interactive effects of temperature and body weight on oxygen consumption rate, ammonia excretion rate and asphyxiation point of Hippocampus erectus. *Fish. Sci. Technol. Inf.* **2022**, *49*, 67–72.

34. Zhu, S.; Long, X.; Turchini, G.M.; Deng, D.; Cheng, Y.; Wu, X. Towards defining optimal dietary protein levels for male and female sub-adult Chinese mitten crab, *Eriocheir sinensis* reared in earthen ponds: Performances, nutrient composition and metabolism, antioxidant capacity and immunity. *Aquaculture* **2021**, *536*, 736442. [CrossRef]

35. Nur Syafaat, M.; Mohammad, S.; Nor Azra, M.; Ma, H.; Abol-Munafi, A.B. Effect of water temperature on survival, growth and molting cycle during early crablet instar of mud crab, Scylla paramamosain (Estampador, 1950). *Thalass. Int. J. Mar. Sci.* **2020**, *36*, 543–551. [CrossRef]

36. Phuong, N.T.; Ha, N.T.K.; Nguyen, T.E.; Do Thi, T.H. Effects of different temperatures on the growth, survival and digestive enzyme activities of mud crab *Scylla paramamosain* at juvenile stage. *Aquac. Aquar. Conserv. Legis.* **2021**, *14*, 2741–2750.

37. Warren, C.E.; Davis, G.E. Laboratory Studies on the Feeding, Bioenergetics, and Growth of Fish. 1967. Available online: https:// ir.library.oregonstate.edu/concern/administrative_report_or_publications/x920g185p?locale=en (accessed on 19 August 2022).

38. Volkoff, H.; Rønnestad, I. Effects of temperature on feeding and digestive processes in fish. *Temperature* **2020**, *7*, 307–320. [CrossRef] [PubMed]

39. Imsland, A.K.; Foss, A.; Gunnarsson, S.; Berntssen, M.H.; FitzGerald, R.; Bonga, S.W.; Stefansson, S.O. The interaction of temperature and salinity on growth and food conversion in juvenile turbot (*Scophthalmus maximus*). *Aquaculture* **2001**, *198*, 353–367. [CrossRef]

40. Hewitt, D.R.; Duncan, P.F. Effect of high water temperature on the survival, moulting and food consumption of Penaeus (*Marsupenaeus*) *japonicus* (Bate, 1888). *Aquac. Res.* **2001**, *32*, 305–313. [CrossRef]

41. Handeland, S.O.; Imsland, A.K.; Stefansson, S.O. The effect of temperature and fish size on growth, feed intake, food conversion efficiency and stomach evacuation rate of Atlantic salmon post-smolts. *Aquaculture* **2008**, *283*, 36–42. [CrossRef]

42. Zhou, L.; Li, S.; Wang, Z.; Li, F.; Xiang, J. An eclosion hormone-like gene participates in the molting process of Palaemonid shrimp *Exopalaemon carinicauda*. *Dev. Genes Evol.* **2017**, *227*, 189–199. [CrossRef]

43. Tian, H.; Yang, C.; Yu, Y.; Yang, W.; Lu, N.; Wang, H.; Liu, F.; Wang, A.; Xu, X. Dietary cholesterol level affects growth, molting performance and ecdysteroid signal transduction in *Procambarus clarkii*. *Aquaculture* **2020**, *523*, 735198. [CrossRef]

44. Liu, T.; Xu, H.; Han, T.; Wang, J.; Yin, F.; Wang, C. Effect of dietary egg yolk lecithin levels on survival, growth, lipid metabolism, and antioxidant capacity of early juvenile green mud crab *Scylla paramamosain*. *Aquaculture* **2021**, *540*, 736706. [CrossRef]

45. Yuan, Q.; Qian, J.; Ren, Y.; Zhang, T.; Li, Z.; Liu, J. Effects of stocking density and water temperature on survival and growth of the juvenile Chinese mitten crab, *Eriocheir sinensis*, reared under laboratory conditions. *Aquaculture* **2018**, *495*, 631–636. [CrossRef]

46. Cai, Y.; Summerfelt, R.C. Effects of temperature and size on oxygen consumption and ammonia excretion by walleye. *Aquaculture* **1992**, *104*, 127–138. [CrossRef]

47. Clarke, A.; Fraser, K.P.P. Why does metabolism scale with temperature? *Funct. Ecol.* **2004**, *18*, 243–251. [CrossRef]

48. Mayzaud, P.; Conover, R.J. O:N atomic ratio as a tool to describe zooplankton metabolism. *Mar. Ecol. Prog. Ser.* **1988**, *45*, 289–302. [CrossRef]

49. Brito, R.; Chimal, M.E.; Gaxiola, G.; Rosas, C. Growth, metabolic rate, and digestive enzyme activity in the white shrimp *Litopenaeus setiferus* early postlarvae fed different diets. *J. Exp. Mar. Biol. Ecol.* **2000**, *255*, 21–36. [CrossRef]

50. Wendelaar Bonga, S.E. The stress response in fish. *Physiol. Rev.* **1997**, *77*, 591–625. [CrossRef]

51. Mommsen, T.P.; Vijayan, M.M.; Moon, T.W. Cortisol in teleosts: Dynamics, mechanisms of action, and metabolic regulation. *Rev. Fish Biol. Fish.* **1999**, *9*, 211–268. [CrossRef]

52. Jia, Y.; Fan, Y.S.; Wang, Y.Z.; Chai, Y.H.; Gao, J.W.; Dou, Y.; Peng, S. Physiological and energy metabolism responses of chinese Loach *Paramisgurnus dabryanus* (Dabry de Thiersant, 1872) to waterless preservation during transport. *Turk. J. Fish. Aquat. Sci.* **2018**, *19*, 279–287.

53. Schreck, C.B. Stress and fish reproduction: The roles of allostasis and hormesis. *Gen. Comp. Endocrinol.* **2010**, *165*, 549–556. [CrossRef]

54. Pörtner, H.O. Climate variations and the physiological basis of temperature dependent biogeography: Systemic to molecular hierarchy of thermal tolerance in animals. *Comp. Biochem. Physiol. Part A Mol. Integr. Physiol.* **2002**, *132*, 739–761. [CrossRef]

55. Pörtner, H.-O. Oxygen- and capacity-limitation of thermal tolerance: A matrix for integrating climate-related stressor effects in marine ecosystems. *J. Exp. Biol.* **2010**, *213*, 881–893. [CrossRef] [PubMed]

Article

Interactions between Cultivated *Gracilariopsis lemaneiformis* and Floating *Sargassum horneri* under Controlled Laboratory Conditions

Hanmo Song, Yan Liu *, Jingyu Li, Qingli Gong and Xu Gao *

Key Laboratory of Mariculture (Ministry of Education), Fisheries College, Ocean University of China, Qingdao 266003, China

* Correspondence: qd_liuyan@ouc.edu.cn (Y.L.); gaoxu@ouc.edu.cn (X.G.); Tel.: +86-532-8203-2377 (Y.L. & X.G.)

Abstract: The golden tide dominated by *Sargassum* has become a frequently-occurring marine ecological event that may constitute major biotic threats to seaweed aquaculture. In this study, the interaction between cultivated *Gracilariopsis lemaneiformis* (GL) and floating *Sargassum horneri* (SH) was investigated by physiological and biochemical measurements under mono-culture and co-culture with different biomass density ratios of 2:1 (2GL:1SH), 1:1 (1GL:1SH), and 1:2 (1GL:2SH). The relative growth rate, net photosynthetic rate, and NO_3-N uptake rate of *G. lemaneiformis* were significantly greater at the biomass density ratio of 2:1 than at mono-culture. However, these physiological parameters and biochemical composition contents (chlorophyll *a* and soluble protein) of *G. lemaneiformis* decreased significantly with increasing biomass of *S. horneri*. Meanwhile, these physiological and biochemical parameters of *S. horneri* were greater in all co-culture models than at mono-culture. They decreased significantly with decreasing biomass of *G. lemaneiformis*. These results indicate that the occurrence of floating *S. horneri* with low biomass can stimulate the growth of *G. lemaneiformis*, whereas its outbreak may significantly reduce the production and quality of *G. lemaneiformis*. *G. lemaneiformis* cultivation may be beneficial to the increased biomass of floating *S. horneri*.

Keywords: biochemical composition; golden tide; growth; NO_3-N uptake; photosynthesis; seaweed cultivation

Citation: Song, H.; Liu, Y.; Li, J.; Gong, Q.; Gao, X. Interactions between Cultivated *Gracilariopsis lemaneiformis* and Floating *Sargassum horneri* under Controlled Laboratory Conditions. *Water* **2022**, *14*, 2664. https://doi.org/10.3390/w14172664

Academic Editors: Xiangli Tian and Li Li

Received: 20 July 2022
Accepted: 26 August 2022
Published: 28 August 2022

Publisher's Note: MDPI stays neutral with regard to jurisdictional claims in published maps and institutional affiliations.

1. Introduction

In recent years, multiple large-scale outbreaks of golden tide caused by *Sargassum* have occurred along the coastal waters of the Pacific and Atlantic Ocean [1–4]. This phenomenon may be closely correlated with marine ecological deterioration, including seawater eutrophication and warming [5,6]. Along the northwest Pacific coast, *Sargassum horneri* is a predominant golden-tide-forming seaweed that originally forms extensive underwater forests [7,8]. Sessile populations of *S. horneri* have continuously decreased in this region [9], whereas a prominent drifting biomass has been frequently detected in the coastal regions of China in the last decade [3,10]. Golden tides have caused considerable disruption to coastal ecosystems, including the death of aquatic organisms by hypoxia, resource competition with native species, and even shifts in biological community structures [11,12]. Furthermore, it has brought considerable damage to local economic industries such as tourism, traditional fisheries, and mariculture [13–15].

Positive and negative interactions among plants are widespread in both terrestrial and aquatic ecosystems, and have complicated and changeable effects on population dynamics, community structure, and biodiversity patterns [16–18]. In coastal regions, a common interspecific interaction pattern is for marine plants with overlapping niches to compete for diverse limited environmental resources [19–21]. Epiphytic *Ulva lactuca* has been proven to severely interfere with mariculture of *Neoporphyra haitanensis*, as they compete for carbon and nitrogen resources [22]. Additionally, the growth of *Neopyropia yezoensis* was apparently restrained by *Ulva prolifera* proliferation because of competition for space, nutrients, and light for spore attachments [23,24]. Allelopathy is another pattern of interspecific interaction among marine algae, which can stimulate or inhibit each other through the release of biochemical metabolites [25–27]. Yuan et al. [28] showed that aqueous extract from *Sargassum fusiforme* promoted the growth of *Karenia mikimotoi* at 0.2 g L^{-1}, while its growth was inhibited at 1.6 g L^{-1}. Similarly, Patil et al. [29] documented that the growth of *Skeletonema costatum* was not significantly decreased by culture filtrates of *Pyropia haitanensis* at 0.625–10 g FW L^{-1}, while it was greatly inhibited at 15 and 20 g FW L^{-1}. These results imply that interaction through allelopathy is closely associated with the biomass density ratio among species. However, thus far there has been little information produced regarding the influence of the biomass density ratio on the interaction between marine macroalgae.

Gracilariopsis lemaneiformis is a common intertidal red seaweed that is naturally distributed across temperate regions worldwide [30–32]. In China, *G. lemaneiformis* is extensively cultivated due to its widespread applications in agar production, the food industry, and abalone aquaculture [33–35]. It is the seaweed species with the third highest production rate in terms of mass, with 368,967 tons of dry weight produced in 2020 [36]. Additionally, the cultivation of this species is beneficial to coastal ecosystems in many ways, including alleviation of harmful microalgal blooms, increasing the dissolved oxygen in the seawater, and maintaining coastal ecological balance [37]. In recent years, the deterioration of the marine environment caused by climate change and human activities has posed a severe threat to the mariculture production of seaweeds [38–40]. Due to the economic and ecological importance of *G. lemaneiformis*, a considerable number of studies about the impacts of diverse abiotic factors on its physiological and biochemical performance have been conducted [41–44]. Nevertheless, the effects of biotic stresses such as macroalgal blooms on *G. lemaneiformis* have rarely been reported.

In the current study, we performed a short-term culture experiment to examine the interactive influences between cultivated *G. lemaneiformis* and floating *S. horneri*. Changes in growth, photosynthesis, NO_3-N uptake, and biochemical compositions were estimated. Our results provide significant information to improve the mariculture management of this valuable species and to address marine ecological hazards.

2. Materials and Methods

2.1. Sampling and Maintenance

G. lemaneiformis thalli were sampled from farmed populations on Lidao Island (36°26′ N, 122°56′ E), China, in June 2020. Meanwhile, *S. horneri* thalli were sampled from floating populations in the cultivation area of *G. lemaneiformis*. Healthy samples were chosen and completely rinsed using sterilized seawater to remove epiphytes and detritus. Algal fragments of *G. lemaneiformis* (6 cm in length) and *S. horneri* (3 cm in length) were respectively excised from the apical position of branches for the experiments. Then, they were incubated in tanks containing 6 L of 25% PESI medium [45]. These fragments were kept for three days at 20 °C, 90 µmol photons m^{-2} s^{-1}, and a 12:12 L:D cycle in order to minimize the impacts of excision.

2.2. Culture Experiment

A twelve-day culture experiment was performed under mono-culture conditions of *G. lemaneiformis* and *S. horneri* and co-culture at different biomass density ratios (BDRs). For the mono-culture, the initial biomass density was set at 6 g for both species. For the co-culture, the initial BDR was set at 2:1 (4 g *G. lemaneiformis* cultured with 2 g *S. horneri*), 1:1 (3 g *G. lemaneiformis* cultured with 3 g *S. horneri*), and 1:2 (2 g *G. lemaneiformis* cultured with 4 g *S. horneri*). A total of five treatments were set up, and each treatment was conducted in three replicates. In this experiment, 20 °C, 90 µmol photons m^{-2} s^{-1}, and a 12:12 L:D cycle were maintained. This experiment used fifteen tanks, with each tank containing 6 L of sterilized seawater enriched with 25% PESI medium. Gentle aeration was conducted during this experiment. The medium was changed every three days. The fresh weights (FWs) of all fragments before and after incubation were determined after each fragment was blotted dry. The relative growth rate (RGR; % day^{-1}) was calculated by Equation (1):

$$RGR = 100 \times (\ln W_t - \ln W_0) \, / \, t \qquad (1)$$

where W_0 is the initial FW, W_t is the final FW, and t is the culture time in days.

2.3. Measurement of Photosynthesis

Following the culture experiment, we determined the net photosynthetic rates of *G. lemaneiformis* and *S. horneri* using a manual FireStingO$_2$II oxygen meter (Pyro Science GmbH, Aachen, Germany). For each species, 0.33 g (FW) of samples from each treatment were moved to the oxygen electrode cuvette containing 330 mL of 25% PESI medium. Next, the medium was continuously stirred to ensure homogenous oxygen diffusion. The culture conditions were the same as for the experiment above. The oxygen increase in the seawater was regarded as the net photosynthetic rate after an increase in the light density. Before measurements, these samples were placed into the cuvette for 5 min acclimation. The oxygen concentration in the medium was documented every 1 min for 15 min.

2.4. Measurement of NO$_3$-N Uptake Rate

Following the culture experiment, 0.2 g (FW) of *G. lemaneiformis* and/or *S. horneri* were randomly selected from each tank for NO$_3$-N uptake measurements. They were transferred into conical flasks containing 200 mL of sterilized seawater enriched with 25% PESI medium and gently shaken for 2 h using a horizontal oscillator. The light and temperature conditions were the same as those described for the culture experiment. For each treatment, the media before and after the culture experiment were separately collected and the NO$_3$-N concentrations were determined by the cadmium column reduction method and molybdenum blue method with ultraviolet absorption spectrophotometer, respectively [46,47]. The NO$_3$-N uptake rate was calculated by Equation (2):

$$U_N = (C_0 - C_t) \times V \, / \, (T \times W) \qquad (2)$$

where U_N is the NO$_3$-N uptake rate, C_0 is the initial concentration of NO$_3$-N (mg L^{-1}), C_t is the final concentration of NO$_3$-N (mg L^{-1}) following the experiment, V is the medium volume (mL), T is the experiment period, and W is the FW of samples (g).

2.5. Measurement of Chlorophyll a (Chl a)

0.1 g (FW) of *G. lemaneiformis* was used to extract the Chl *a* in each replicate of all the experimental treatments. These samples were ground using liquid nitrogen, then 3 mL of phosphate buffer solution (0.1 M, pH = 6.8) was added. After transferring the samples into tubes, they were centrifuged for 30 min at 4000 rpm at 4 °C. Subsequently, 8 mL of dimethylformamide (DMF) was added to the precipitates obtained following centrifugation and maintained at 4 °C for 1 d. Extracts were then centrifuged for 10 min at 6000 rpm at 4 °C. The supernatant was collected and the absorption was determined at 750, 664, 647, and 625 nm. The Chl *a* content (mg g^{-1}) was calculated by Equation (3):

$$\text{Chl } a = [12.65 \times (A_{664} - A_{750}) - 2.99 \times (A_{647} - A_{750}) - 0.04 \times (A_{625} - A_{750})] \times V_e \, / \, (I \times W \times 1000) \qquad (3)$$

where $A_{750-625}$ is the absorption of extracts at 750, 664, 647, and 625 nm, V_e is the DMF volume (mL), I is the optical path in the cuvette (cm), and W is the FW of the samples (g).

0.25 g (FW) of *S. horneri* was used to extract the Chl *a* in each replicate of all the experimental treatments. These samples were placed into 2 mL of dimethyl sulfoxide for 5 min, and the supernatant absorption was determined at 665, 631, 582, and 480 nm. Subsequently, the same samples were placed into 3 mL of acetone for 2 h. The supernatant was moved to a 10 mL tube, then 1 mL of methanol and 1 mL of distilled water were added.

The supernatant absorption was determined at 664, 631, 581, and 470 nm. The Chl *a* content (mg g^{-1}) was calculated by Equation (4):

$$\text{Chl } a = [(A_{665} / 72.8) \times V_1 + (A_{664} / 73.6) \times V_2] / (I \times W) \tag{4}$$

where A_{665} and A_{664} are the absorption rates of the extracts at 665 and 664 nm, respectively, V_1 is the volume of the extract at the first extraction process (mL), V_2 is the volume of extract at the second extraction process (mL), I is the optical path in the cuvette (cm), and W is the FW of the samples (g).

2.6. Measurements of Soluble Protein and Carbohydrate

For all the experimental treatments, 0.1 g (FW) of samples from each replicate were homogenized with a pestle and mortar and 0.9 mL of phosphate buffer solution (0.1 M, pH = 7.4). The extract was centrifuged for 30 min at 12,000 rpm at 4 °C. The supernatant absorption at 595 nm was recorded using an ultraviolet spectrophotometer. The soluble proteins of samples were evaluated using Coomassie Brilliant Blue G-250 dye (Nanjing Jiancheng Bioengineering Institute, Nanjing, China) and bovine albumin [48].

An amount (0.1 g FW) of the samples from each replicate were ground in 2 mL of distilled water and diluted to 10 mL after adding 2 mL of $MgCO_3$ suspension liquid. The mixture was centrifuged for 5 min at 4000 rpm at 4 °C. Then, 1 mL of supernatant was moved to a tube and diluted to 2 mL. After that, 8 mL of anthrone reagent was added. The mixture was bathed in boiled water for 10 min and cooled to room temperature. The absorption at 620 nm was recorded and the mixture was standardized. The soluble carbohydrate content was measured according to [49].

2.7. Statistical Analysis

For both species, one-way ANOVA was applied to analyze the RGR, net photosynthetic rate, NO_3-N uptake rate, and the contents of biochemical compositions among different culture models (mono- or co-cultures with three BDRs). Prior to these analyses, the variances of the data were subjected to homogeneity tests. Duncan's multiple range test was used if significant differences were detected ($p < 0.05$). Statistical analyses were carried out using SPSS 26.0 software.

3. Results

3.1. Growth

The RGRs of *G. lemaneiformis* were significantly affected by culture model (Figure 1A and Table 1). The RGR at BDR of 2:1 was significantly greater than that at mono-culture. However, the RGR decreased significantly with increasing biomass of *S. horneri*, from BDR of 2:1 to 1:2. The RGRs of *S. horneri* differed significantly among culture models (Figure 1B and Table 1). The RGRs of all co-culture treatments were greater than that at mono-culture, and significantly higher values were found at BDRs of 2:1 and 1:1. Additionally, the RGR at BDR of 2:1 was significantly greater than that at BDR of 1:1.

Table 1. One-way ANOVA testing of the effects of biomass density ratio on RGR, net photosynthetic rate, NO_3-N uptake rate, Chl *a*, soluble protein, and carbohydrate of *Gracilariopsis lemaneiformis* and *Sargassum horneri*.

Factors	*Gracilariopsis lemaneiformis*			*Sargassum horneri*		
	df	F	*p*	df	F	*p*
RGR	3	112.721	<0.001	3	16.595	<0.001
Net photosynthetic rate	3	29.975	<0.001	3	6.836	<0.05
NO_3-N uptake rate	3	52.763	<0.001	3	32.004	<0.001
Chl *a*	3	4.510	<0.05	3	4.157	<0.05
Soluble protein	3	6.398	<0.05	3	11.253	<0.01
Soluble carbohydrate	3	39.980	<0.001	3	33.784	<0.001

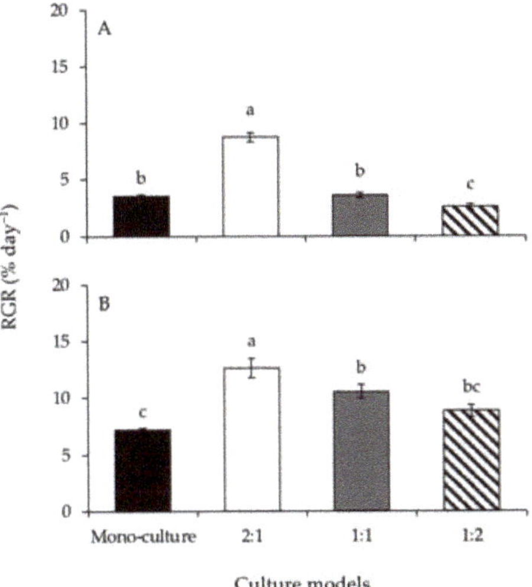

Figure 1. The RGRs of *Gracilariopsis lemaneiformis* (**A**) and *Sargassum horneri* (**B**) cultured for twelve days at mono-culture and co-culture with biomass density ratios of 2:1, 1:1, and 1:2. Different lowercase letters represent significant differences ($p < 0.05$) among culture models.

3.2. Photosynthesis

The net photosynthetic rates of *G. lemaneiformis* differed significantly among culture models (Figure 2A and Table 1). The net photosynthetic rate at BDR of 2:1 was significantly higher than those of the other culture models. Additionally, the net photosynthetic rate at BDR of 1:2 was significantly lower than that at mono-culture. The net photosynthetic rates of *S. horneri* were significantly affected by culture model (Figure 2B and Table 1). The net photosynthetic rate at BDR of 2:1 was significantly higher than that at mono-culture. However, the net photosynthetic rates decreased significantly with increasing biomass of *S. horneri*, from BDR of 2:1 to 1:2.

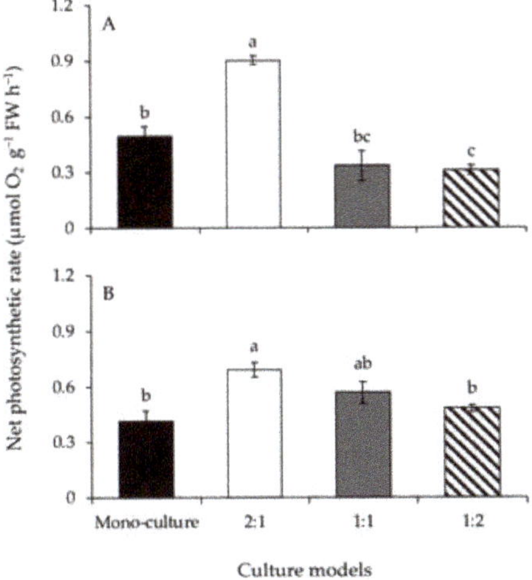

Figure 2. The net photosynthetic rates of *Gracilariopsis lemaneiformis* (**A**) and *Sargassum horneri* (**B**) cultured for twelve days at mono-culture and co-culture with biomass density ratios of 2:1, 1:1, and 1:2. Different lowercase letters represent significant differences ($p < 0.05$) among culture models.

3.3. NO$_3$-N Uptake Rate

A significant effect of the culture model on the NO$_3$-N uptake rates of *G. lemaneiformis* was detected (Figure 3A and Table 1). The NO$_3$-N uptake rate at BDR of 2:1 was significantly greater than those of the other culture models. The NO$_3$-N uptake rates of *S. horneri* were significantly different among culture models (Figure 3B and Table 1). The NO$_3$-N uptake rates of the co-culture treatments were significantly greater than that at mono-culture. The NO$_3$-N uptake rates decreased significantly with increasing biomass of *S. horneri*, from BDR of 2:1 to 1:2.

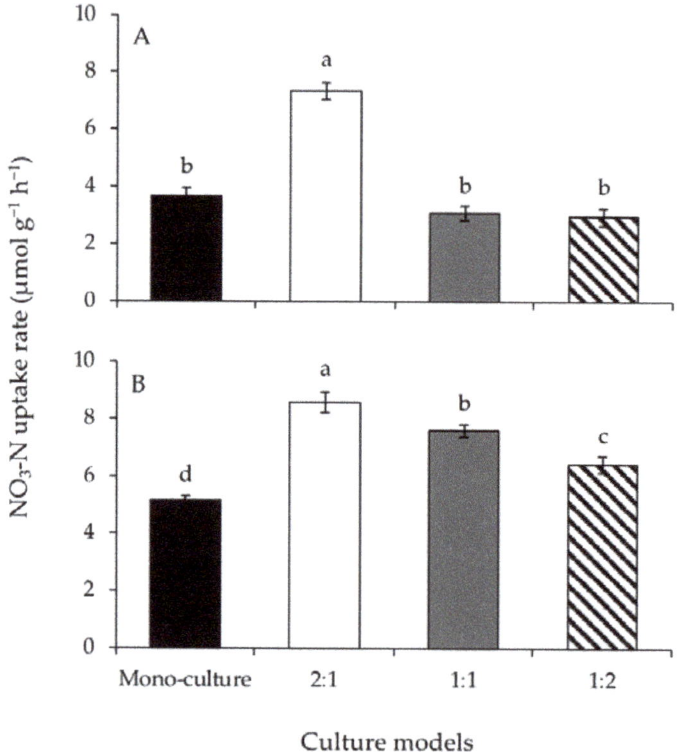

Figure 3. The NO$_3$-N uptake rates of *Gracilariopsis lemaneiformis* (**A**) and *Sargassum horneri* (**B**) cultured for twelve days at mono-culture and co-culture with biomass density ratios of 2:1, 1:1, and 1:2. Different lowercase letters represent significant differences ($p < 0.05$) among culture models.

3.4. Chl a

The Chl *a* contents of *G. lemaneiformis* differed significantly among culture models (Figure 4A and Table 1). The Chl *a* content at BDR of 1:2 was significantly lower than at mono-culture and BDR of 2:1. Similarly, the Chl *a* contents of *S. horneri* differed significantly among culture models (Figure 4B and Table 1). The Chl *a* content at BDR of 2:1 was significantly greater than at mono-culture and at BDR of 1:2.

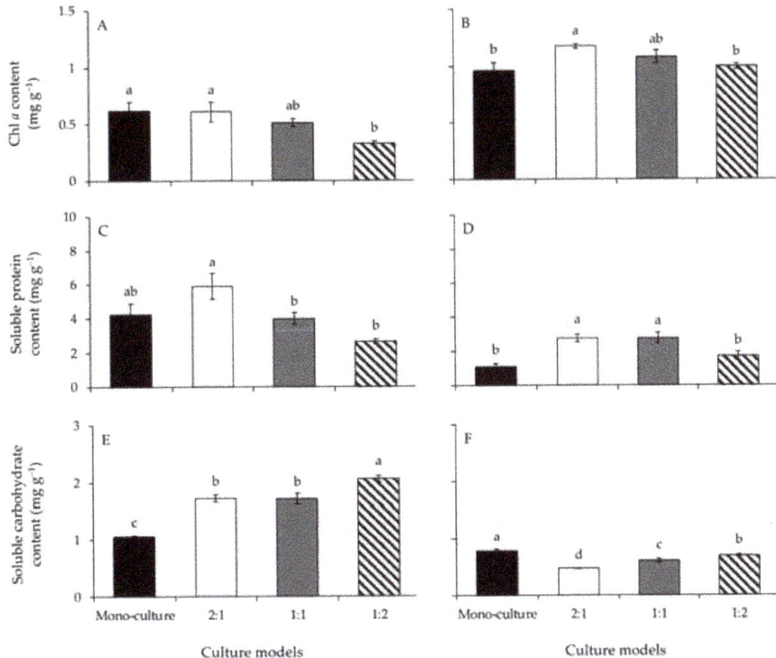

Figure 4. The contents of Chl *a*, soluble protein, and soluble carbohydrate of *Gracilariopsis lemaneiformis* (**A,C,E**) and *Sargassum horneri* (**B,D,F**) cultured for twelve days at mono-culture and co-culture with biomass density ratios of 2:1, 1:1, and 1:2. Different lowercase letters represent significant differences ($p < 0.05$) among culture models.

3.5. Soluble Protein and Carbohydrate

The one-way ANOVA results showed a significant difference among culture models on the soluble protein contents of *G. lemaneiformis* (Figure 4C and Table 1). The soluble protein content at BDR of 2:1 was significantly greater than at BDRs of 1:1 and 1:2. The soluble protein contents of *S. horneri* differed significantly among culture models (Figure 4D and Table 1). The soluble protein contents at BDRs of 2:1 and 1:1 were significantly greater than at mono-culture and BDR of 1:2.

A significant effect of the culture model on the soluble carbohydrate contents of *G. lemaneiformis* was detected (Figure 4E and Table 1). Compared to the mono-culture, the soluble carbohydrate content had significantly higher values at co-culture with different BDRs. Moreover, the soluble carbohydrate content at BDR of 1:2 was significantly greater than at BDRs of 2:1 and 1:1. Similarly, the soluble carbohydrate contents of *S. horneri* varied significantly among culture models (Figure 4F and Table 1). The soluble carbohydrate content was significantly higher at mono-culture than co-culture with different BDRs. The soluble carbohydrate contents increased significantly with increasing biomass of *S. horneri*, from BDR of 2:1 to 1:2.

4. Discussion

Diverse positive and negative interspecific interactions exist among plants in natural communities [50]. The principal mechanisms of these interactions, such as resource competition or allelopathy, are often species-specific and environment-dependent [51,52]. In this study, the RGR of *G. lemaneiformis* was significantly enhanced under co-culture with BDR of 2:1 compared to under mono-culture, indicating that the presence of *S. horneri* with a relatively lower biomass had a stimulating impact on *G. lemaneiformis*. A similar finding was documented for *G. lemaneiformis* and epiphytic *U. prolifera*, showing that the RGR of *U. prolifera* was greatly increased under co-culture with BDR of 1:1 [53]. This 'hormesis' effect may have resulted from the division and expansion of cells through the enzymatic or non-enzymatic processes following low-dose phytotoxin stimulation, thereby further increasing the growth of thalli [54,55]. Additionally, this may be due to an overcompensation response following the initial damage caused by low concentrations of phytotoxins [56,57]. Therefore, we speculated that the secretion of allelochemicals played a more critical role in interspecific interactions than resource competition when *G. lemaneiformis* suffered from a low biomass value of *S. horneri*. Moreover, allelopathy presents complex mechanisms that can influence a variety of physiological and biochemical processes [58]. In this study, we observed a significant increase in the net photosynthetic rate, NO_3-N uptake rate, and soluble protein content of *G. lemaneiformis* under co-culture with BDR of 2:1. Similarly, Pan et al. [53] showed that, compared to mono-culture, the Chl *a* content and photosynthetic rate of *G. lemaneiformis* were significantly enhanced under co-culture with 0.5 g L^{-1} of *U. prolifera*. Further experiments are required to clarify the correlative mechanisms of these findings.

Compared to the stimulating effect of *S. horneri* on *G. lemaneiformis* with a relatively lower biomass, the RGR at BDR of 1:2 was significantly lower than at mono-culture. Similarly, Xie et al. [59] demonstrated that the initial population density rate greatly influences the interaction of *Heterosigma akashiwo* (H) and *Prorocentrum donghaiense* (P). The growth of *H. akashiwo* was restrained by *P. donghaiense* at the inoculation proportion of 1H:4P, whereas *H. akashiwo* possessed a higher growth rate at the inoculation proportion of 4H:1P. Both nutrient competition and allelopathy have been demonstrated to play a critical role in interspecific interaction between these two microalgal species. In addition, the net photosynthetic rate, Chl *a* content, and soluble protein content of *G. lemaneiformis* significantly decrease at BDR of 1:2. A similar result was presented in [24], where the growth, photosynthetic activity, and biochemical metabolism of *N. yezoensis* were significantly inhibited by epiphytic *U. prolifera*, mainly as a result of nutrition and light competition. On the other hand, high concentrations of allelochemicals may lead to cell expansion and rupture, while cells can maintain osmotic pressure and intracellular homeostasis by accumulating soluble carbohydrates [60–62]. In the current study, the soluble carbohydrate content of *G. lemaneiformis* presented a significant increase at BDR of 1:2, indicating that *S. horneri* may have exerted a strong allelopathic effect on *G. lemaneiformis* under this condition. Therefore, the inhibitory effect of *S. horneri* with a relatively higher biomass on *G. lemaneiformis* may be associated with a combination of resource competition and allelopathy.

Compared to the mono-culture, the RGR, net photosynthetic rate, NO_3-N uptake rate, Chl *a* content, and soluble protein content of *S. horneri* were enhanced under co-culture conditions with *G. lemaneiformis* at different BDRs. Macroalgal species with a stronger growth capacity usually have advantages in interspecific competition scenarios [51,63]. Our results showed that *S. horneri* (7.23%) has a better growth rate than *G. lemaneiformis* (3.53%) under mono-culture conditions. The outstanding growth performance of *S. horneri* may be correlated with its adaptability to frequent marine environment changes during the long-distance floating processes [64]. Additionally, *U. lactuca* showed greater growth and maximal photosynthetic ability than *G. lemaneiformis* under competitive conditions, benefiting from its larger frond surface area and higher nutrient uptake rate [65]. All these results indicate that *S. horneri* has a dominant role in its competition with *G. lemaneiformis* because of its physiological and morphological advantages. In addition, we observed that all physiological and biochemical parameters of *S. horneri* increased significantly with increasing biomass of *G. lemaneiformis* under co-culture conditions. Wang et al. [66] reported that co-culture with a small quantity of *G. lemaneiformis* thalli was significantly beneficial to the growth of the red tide species *H. akashiwo*. Similarly, *G. lemaneiformis* can stimulate the growth of *S. costatum* by secreting various types of allelochemicals [67]. Therefore, we suggest that certain allelochemicals produced by *G. lemaneiformis* may partially contribute to the increased growth rate of *S. horneri*.

The golden tide caused by *Sargassum* species has become a major ecological threat to the coastal environment and aquaculture production. According to the present data, although the occurrence of small quantities of *S. horneri* can stimulate the growth of *G. lemaneiformis*, its outbreak has a greatly harmful impact on the growth and biochemical metabolism of *G. lemaneiformis*, resulting in reduced yield and quality. Inversely, the growth of *S. horneri* is enhanced by the occurrence of *G. lemaneiformis*, regardless of BDR. This observation suggests that *G. lemaneiformis* cultivation has no inhibitory effect on golden tide, rather favoring its further expansion. Hence, *S. horneri* should be removed or intercepted in time to minimize economic and ecological losses. Due to the limited availability of data in this study, more studies involving related physiological and molecular mechanisms are required to evaluate the coping strategies of *G. lemaneiformis* against golden tide outbreaks.

Author Contributions: Investigation, data curation, visualization, writing—original draft, H.S.; conceptualization, methodology, project administration, funding acquisition, writing—review and editing, Y.L. and X.G.; formal analysis, resources, J.L.; validation, supervision, Q.G. All authors have read and agreed to the published version of the manuscript.

Funding: This work was financially supported by the National Key R&D Program of China (No. 2020YFD0900201) and the Fundamental Research Funds for the Central Universities of China (No. 842212015).

Institutional Review Board Statement: Not applicable.

Informed Consent Statement: Not applicable.

Data Availability Statement: The data in this study are available from the corresponding author upon reasonable request.

Acknowledgments: We are grateful to Qiaohan Wang of Ocean University of China for experimental assistance.

Conflicts of Interest: The authors declare no conflict of interest.

References

1. Komatsu, T.; Tatsukawa, K.; Filippi, J.B.; Sagawa, T.; Matsunaga, D.; Mikami, A.; Ishida, K.; Ajisaka, T.; Tanaka, K.; Aoki, M.; et al. Distribution of drifting seaweeds in eastern East China Sea. *J. Mar. Syst.* **2007**, *67*, 245–252. [CrossRef]
2. Schell, J.M.; Goodwin, D.S.; Siuda, A.N.S. Recent *Sargassum* inundation events in the Caribbean: Shipboard observations reveal dominance of a previously rare form. *Oceanography* **2015**, *28*, 8–10. [CrossRef]
3. Xing, Q.G.; Guo, R.H.; Wu, L.L.; An, D.Y.; Cong, M.; Qin, S.; Li, X.R. High-resolution satellite observations of a new hazard of golden tides caused by floating *Sargassum* in winter in the Yellow Sea. *IEEE Geosci. Remote Sens. Lett.* **2017**, *14*, 1815–1819. [CrossRef]
4. García-Sánchez, M.; Graham, C.; Vera, E.; Escalante-Mancera, E.; Álvarez-Filip, L.; van Tussenbroek, B.I. Temporal changes in the composition and biomass of beached pelagic *Sargassum* species in the Mexican Caribbean. *Aquat. Bot.* **2020**, *167*, 103275. [CrossRef]
5. Lapointe, B.E. A comparison of nutrient-limited productivity in *Sargassum natans* from neritic vs. oceanic waters of the western North Atlantic Ocean. *Limnol. Oceanogr.* **1995**, *40*, 625–633. [CrossRef]
6. Huffard, C.L.; von Thun, S.; Sherman, A.D.; Sealey, K.; Smith, K.L. Pelagic *Sargassum* community change over a 40-year period: Temporal and spatial variability. *Mar. Biol.* **2014**, *161*, 2735–2751. [CrossRef]
7. Hu, Z.M.; Uwai, S.; Yu, S.H.; Komatsu, T.; Ajisaka, T.; Duan, D.L. Phylogeographic heterogeneity of the brown macroalga *Sargassum horneri* (Fucaceae) in the northwestern Pacific in relation to late *Pleistocene glaciation* and tectonic configurations. *Mol. Ecol.* **2011**, *20*, 3894–3909. [CrossRef]
8. Xu, M.; Sakamoto, S.; Komatsu, T. Attachment strength of the subtidal seaweed *Sargassum horneri* (Turner) C. Agardh varies among development stages and depths. *J. Appl. Phycol.* **2016**, *28*, 3679–3687. [CrossRef]
9. Sun, J.Z.; Chen, W.D.; Zhuang, D.G.; Zheng, H.Y.; Li, L.; Pang, S.J. In situ ecological studies of the subtidal brown alga *Sargasssum horneri* at Nanji Island of China. *S. China Fish. Sci.* **2008**, *4*, 58–63, (In Chinese with English abstract).
10. Qi, L.; Hu, C.M.; Wang, M.Q.; Shang, S.L.; Wilson, C. Floating algae blooms in the East China Sea. *Geophys. Res. Lett.* **2017**, *44*, 11501–11509. [CrossRef]
11. Smetacek, V.; Zingone, A. Green and golden seaweed tides on the rise. *Nature* **2013**, *504*, 84–88. [CrossRef] [PubMed]
12. Cruz-Rivera, E.; Flores-Díaz, M.; Hawkins, A. A fish kill coincident with dense *Sargassum* accumulation in a tropical bay. *Bull. Mar. Sci.* **2015**, *91*, 455–456. [CrossRef]

13. Solarin, B.B.; Bolaji, D.A.; Fakayode, O.S.; Akinnigbagbe, R.O. Impacts of an invasive seaweed *Sargassum hystrix var. fluitans* (Børgesen 1914) on the fisheries and other economic implications for the Nigerian coastal waters. *IOSR J. Agric. Vet. Sci.* **2014**, *7*, 1–6.

14. Milledge, J.J.; Harvey, P.J. Golden tides: Problem or golden opportunity? The valorisation of *Sargassum* from beach inundations. *J. Mar. Sci. Eng.* **2016**, *4*, 60. [CrossRef]

15. Wang, X.M.; Shan, T.F.; Pang, S.J.; Su, L. Assessment of the genetic relationship between the recently established benthic population and the adjacent floating populations of *Sargassum horneri* (Phaeophyceae) in Dalian of China by newly developed trinucleotide microsatellite markers. *J. Appl. Phycol.* **2019**, *31*, 3989–3996. [CrossRef]

16. Dickie, I.A.; Schnitzer, S.A.; Reich, P.B.; Hobbie, S.E. Spatially disjunct effects of co-occurring competition and facilitation. *Ecol. Lett.* **2005**, *8*, 1191–1200. [CrossRef] [PubMed]

17. Narwani, A.; Bentlage, B.; Alexandrou, M.A.; Fritschie, K.J.; Delwiche, C.; Oakley, T.H.; Cardinale, B.J. Ecological interactions and coexistence are predicted by gene expression similarity in freshwater green algae. *J. Ecol.* **2017**, *105*, 580–591. [CrossRef]

18. Cameron, H.; Coulson, T.; Marshall, D.J. Size and density mediate transitions between competition and facilitation. *Ecol. Lett.* **2019**, *22*, 1879–1888. [CrossRef]

19. Choi, H.G.; Norton, T.A. Competition and facilitation between germlings of *Ascophyllum nodosum* and *Fucus vesiculosus*. *Mar. Biol.* **2005**, *147*, 525–532. [CrossRef]

20. White, L.F.; Shurin, J.B. Density dependent effects of an exotic marine macroalga on native community diversity. *J. Exp. Mar. Biol. Ecol.* **2011**, *405*, 111–119. [CrossRef]

21. Gavina, M.K.A.; Tahara, T.; Tainaka, K.; Ito, H.; Morita, S.; Ichinose, G.; Okabe, T.; Togashi, T.; Nagatani, T.; Yoshimura, J. Multi-species coexistence in Lotka–Volterra competitive systems with crowding effects. *Sci. Rep.* **2018**, *8*, 1198. [CrossRef] [PubMed]

22. Chen, B.; Lin, L.; Ma, Z.; Zhang, T.; Chen, W.; Zou, D. Carbon and nitrogen accumulation and interspecific competition in two algae species, *Pyropia haitanensis* and *Ulva lactuca*, under ocean acidification conditions. *Aquac. Int.* **2019**, *27*, 721–733. [CrossRef]

23. Li, X.S.; Xu, J.T.; He, P.M. Comparative research on inorganic carbon acquisition by the macroalgae *Ulva prolifera* (Chlorophyta) and *Pyropia yezoensis* (Rhodophyta). *J. Appl. Phycol.* **2016**, *28*, 491–497. [CrossRef]

24. Sun, J.Y.; Bao, M.L.; Xu, T.P.; Li, F.T.; Wu, H.L.; Li, X.S.; Xu, J.T. Elevated CO_2 influences competition for growth, photosynthetic performance and biochemical composition in *Neopyropia yezoensis* and *Ulva prolifera*. *Algal Res.* **2021**, *56*, 102313. [CrossRef]

25. Sukenik, A.; Eshkol, R.; Livne, A.; Hadas, O.; Rom, M.; Tchernov, D.; Vardi, A.; Kaplan, A. Inhibition of growth and photosynthesis of the dinoflagellate *Peridinium gatunense* by *Microcystis* sp. (Cyanobacteria): A novel allelopathic mechanism. *Limnol. Oceanogr.* **2002**, *47*, 1656–1663. [CrossRef]

26. Wang, Y.; Yu, Z.M.; Song, X.X.; Tang, X.X.; Zhang, S.D. Effects of macroalgae *Ulva pertusa* (Chlorophyta) and *Gracilaria lemaneiformis* (Rhodophyta) on growth of four species of bloom-forming dinoflagellates. *Aquat. Bot.* **2007**, *86*, 139–147. [CrossRef]

27. He, D.; Liu, J.; Hao, Q.; Ran, L.H.; Zhou, B.; Tang, X.X. Interspecific competition and allelopathic interaction between *Karenia mikimotoi* and *Dunaliella salina* in laboratory culture. *Chin. J. Oceanol. Limnol.* **2016**, *34*, 301–313. [CrossRef]

28. Yuan, C.Y.; Zhao, Y.; Yang, S.; Cui, Q.M. Allelopathic effect of *Sargassum fusiforme* on growth of *Karenia mikimotoi*. *Appl. Mech. Mater.* **2014**, *522–524*, 721–724. [CrossRef]

29. Patil, V.; Abate, R.; Yang, Y.F.; Zhang, J.W.; Lin, H.N.; Chen, C.P.; Liang, J.R.; Sun, L.; Li, X.S.; Gao, Y.H. Allelopathic effect of *Pyropia haitanensis* (Rhodophyta) on the bloom-forming *Skeletonema costatum* (Bacillariophyta). *J. Appl. Phycol.* **2020**, *32*, 1275–1286. [CrossRef]

30. Steentoft, M.; Irvine, L.M.; Farnham, W.F. Two terete species of *Gracilaria* and *Gracilariopsis* (Gracilariales, Rhodophyta) in Britain. *Phycologia* **1995**, *34*, 113–127. [CrossRef]

31. Yoshida, T. *Marine Algae of Japan*; Uchida Rokakuho Publishing Company: Tokyo, Japan, 1998. (In Japanese)

32. Iyer, R.; De Clerck, O.; Bolton, J.J.; Coyne, V.E. Morphological and taxonomic studies of *Gracilaria* and *Gracilariopsis* species (Gracilariales, Rhodophyta) from South Africa. *S. Afr. J. Bot.* **2004**, *70*, 521–539. [CrossRef]

33. Qi, Z.H.; Liu, H.M.; Li, B.; Mao, Y.Z.; Jiang, Z.J.; Zhang, J.H.; Fang, J.G. Suitability of two seaweeds, *Gracilaria lemaneiformis* and *Sargassum pallidum*, as feed for the abalone *Haliotis discus hannai* Ino. *Aquaculture* **2010**, *300*, 189–193. [CrossRef]

34. Liu, X.J.; Zhang, Q.F.; Huan, Z.Y.; Zhong, M.Q.; Chen, W.Z.; Du, H. Identification and characterization of glutamine synthetase isozymes in *Gracilaria lemaneiformis*. *Aquat. Bot.* **2018**, *146*, 23–30. [CrossRef]

35. Yuan, S.L.; Duan, Z.H.; Lu, Y.N.; Ma, X.L.; Wang, S. Optimization of decolorization process in agar production from *Gracilaria lemaneiformis* and evaluation of ntioxidant activities of the extract rich in natural pigments. *3 Biotech* **2018**, *8*, 8. [CrossRef] [PubMed]

36. Fishery Bureau of Agriculture Ministry of China. *Chinese Fishery Statistical Yearbook*; China Agriculture Press: Beijing, China, 2021; pp. 23–28. (In Chinese)

37. Yang, Y.F.; Chai, Z.Y.; Wang, Q.; Chen, W.Z.; He, Z.L.; Jiang, S.J. Cultivation of seaweed *Gracilaria* in Chinese coastal waters and its contribution to environmental improvements. *Algal Res.* **2015**, *9*, 236–244. [CrossRef]

38. Kim, J.K.; Yarish, C.; Hwang, E.K.; Park, M.; Kim, Y. Seaweed aquaculture: Cultivation technologies, challenges and its ecosystem services. *Algae* **2017**, *32*, 1–13. [CrossRef]

39. Gao, X.; Kim, J.H.; Park, S.K.; Yu, O.H.; Kim, Y.S.; Choi, H.G. Diverse responses of sporophytic photochemical efficiency and gametophytic growth for two edible kelps, *Saccharina japonica* and *Undaria pinnatifida*, to ocean acidification and warming. *Mar. Pollut. Bull.* **2019**, *142*, 315–320. [CrossRef]

40. Li, S.F.; Liu, Y.; Gong, Q.L.; Gao, X.; Li, J.Y. Physiological and ultrastructural responses of the brown seaweed *Undaria pinnatifida* to triphenyltin chloride (TPTCL) stress. *Mar. Pollut. Bull.* **2020**, *153*, 110978. [CrossRef]

41. Liu, C.X.; Zou, D.H.; Yang, Y.F.; Chen, B.B.; Jiang, H. Temperature responses of pigment contents, chlorophyll fluorescence characteristics, and antioxidant defenses in *Gracilariopsis lemaneiformis* (Gracilariales, Rhodophyta) under different CO_2 levels. *J. Appl. Phycol.* **2017**, *29*, 983–991. [CrossRef]

42. Wang, Y.; Feng, Y.Q.; Liu, X.J.; Zhong, M.Q.; Chen, W.Z.; Wang, F.; Du, H. Response of *Gracilaria lemaneiformis* to nitrogen deprivation. *Algal Res.* **2018**, *34*, 82–96. [CrossRef]

43. Yang, Y.L.; Li, W.; Li, Y.H.; Xu, N.J. Photophysiological responses of the marine macroalga *Gracilariopsis lemaneiformis* to ocean acidification and warming. *Mar. Environ. Res.* **2020**, *163*, 105204. [CrossRef] [PubMed]

44. Xue, J.R.; Pang, T.; Liu, J.G. Changes in the temperature tolerance profile of *Gracilariopsis lemaneiformis* from the perspective of photosynthesis, respiration, and biochemical markers after many years of vegetative propagation. *J. Appl. Phycol.* **2022**, *34*, 1045–1058. [CrossRef]

45. Tatewaki, M. Formation of a crustaceous sporophyte with unilocular sporangia in *Scytosiphon lomentaria*. *Phycologia* **1966**, *6*, 62–66. [CrossRef]

46. Li, S.; Liu, D. Improvement of spectrophotometric determination of phosphorus in water by phospho–molybdenum blue. *Environ. Prot. Chem. Ind.* **2006**, *26*, 78–80, (In Chinese with English abstract).

47. Sun, X.Y.; Hong, L.C.; Ye, H.M. Experiment determining nitrate nitrogen in water samples by on-line cadmium column reduction-flow injection method. *Water Resour. Prot.* **2010**, *26*, 75–77, (In Chinese with English abstract).

48. Kochert, G. Protein determination by dye binding. In *Handbook of Phycological Methods. Physiological and Biochemical Methods*; Hellebust, J.A., Craigie, J.S., Eds.; Cambridge University Press: Cambridge, UK, 1978; pp. 91–93.

49. Tang, S.Q.; Gong, M.G.; Chen, D.C. Study of methods for determination of carbohydrates in sea water 2. The anthrone method for total amount of particulate carbohydrate. *J. Shandong Coll. Oceanol.* **1985**, *15*, 47–54. (In Chinese)

50. Hirsch, A.M.; Bauer, W.D.; Bird, D.M.; Cullimore, J.; Tyler, B.; Yoder, J.I. Molecular signals and receptors: Controlling rhizosphere interactions between plants and other organisms. *Ecology* **2003**, *84*, 858–868. [CrossRef]

51. Xu, D.; Gao, Z.Q.; Zhang, X.F.; Fan, X.; Wang, Y.T.; Li, D.M.; Wang, W.; Zhuang, Z.M.; Ye, N.H. Allelopathic interactions between the opportunistic species *Ulva prolifera* and the native macroalga *Gracilaria lichvoides*. *PLoS ONE* **2012**, *7*, e33648. [CrossRef]

52. Bieberich, J.; Lauerer, M.; Drachsler, M.; Heinrichs, J.; Müller, S.; Feldhaar, H. Species- and developmental stage-specific effects of allelopathy and competition of invasive *Impatiens glandulifera* on cooccurring plants. *PLoS ONE* **2018**, *13*, e0205843. [CrossRef]

53. Pan, Z.L.; Yu, Y.Y.; Chen, Y.L.; Yu, C.Z.; Xu, N.J.; Li, Y.H. Combined effects of biomass density and low-nighttime temperature on the competition for growth and physiological performance of *Gracilariopsis lemaneiformis* and *Ulva prolifera*. *Algal Res.* **2022**, *62*, 102638. [CrossRef]

54. Bais, H.P.; Venkatachalam, L.; Biedrzycki, M.L. Stimulation or inhibition: Conflicting evidence for (±)-catechin's role as a chemical facilitator and disease protecting agent. *Plant Signal. Behav.* **2010**, *5*, 239–246. [CrossRef] [PubMed]

55. Chen, X.J.; Huang, G.H.; Fu, H.Y.; An, C.J.; Yao, Y.; Cheng, G.H.; Suo, M.Q. Allelopathy inhibitory effects of *Hydrodictyon reticulatum* on *Chlorella pyrenoidosa* under co-culture and liquor-cultured conditions. *Water* **2017**, *9*, 416. [CrossRef]

56. Belz, R.G. Stimulation versus inhibition–bioactivity of parthenin, a phytochemical from *Parthenium hysterophorus* L. *Dose-Response* **2008**, *6*, 80–96. [CrossRef]

57. Belz, R.G. Investigating a potential auxin–related mode of hermetic/inhibitory action of the phytotoxin parthenin. *J. Chem. Ecol.* **2016**, *42*, 71–83. [CrossRef] [PubMed]

58. Budzałek, G.; Śliwińska-Wilczewska, S.; Wiśniewska, K.; Wochna, A.; Bubak, I.; Latała, A.; Wiktor, J.M. Macroalgal defense against competitors and herbivores. *Int. J. Mol. Sci.* **2021**, *22*, 7865. [CrossRef]

59. Xie, Z.H.; Xiao, H.; Cai, H.J.; Wang, R.J.; Tang, X.X. Influence of UV-B irradiation on the interspecific growth interaction between *Heterosigma akashiwo* and *Prorocentrum donghaiense*. *Int. Rev. Hydrobiol.* **2006**, *91*, 555–573. [CrossRef]

60. Nagayama, K.; Shibata, T.; Fujimoto, K.; Honjo, T.; Nakamura, T. Algicidal effect of phlorotannins from the brown alga *Ecklonia kurome* on red tide microalgae. *Aquaculture* **2003**, *218*, 601–611. [CrossRef]

61. Accoroni, S.; Percopo, I.; Cerino, F.; Romagnoli, T.; Pichierri, S.; Perrone, C.; Totti, C. Allelopathic interactions between the HAB dinoflagellate *Ostreopsis* cf. ovata and macroalgae. *Harmful Algae* **2015**, *49*, 147–155. [CrossRef]

62. Zhang, L.; Liu, B.Y.; Ge, F.J.; Liu, Q.; Zhang, Y.Y.; Zhou, Q.H.; Xu, D.; Wu, Z.B. Interspecific competition for nutrients between submerged macrophytes (*Vallisneria natans, Ceratophyllum demersum*) and filamentous green algae (*Cladophora oligoclona*) in a co-culture system. *Pol. J. Environ. Stud.* **2019**, *28*, 1483–1494. [CrossRef]

63. Zhang, S.D.; Yu, Z.M.; Song, X.X.; Song, F.; Wang, Y. Competition about nutrients between *Gracilaria lemaneiformis* and *Prorocentrum donghaiense*. *Acta Ecol. Sin.* **2005**, *25*, 2676–2680, (In Chinese with English abstract).

64. Liu, Z.Y.; Sun, P.; Qin, S.; Li, J.J.; Zhuang, L.C.; Song, W.L.; Bi, Y.X.; Zhong, Z.H. Comparative analysis of growth and photosynthetic characteristics in benthic and floating *Sargassum horneri*. *Chin. J. Ecol.* **2021**, *40*, 76–83, (In Chinese with English abstract).

65. Chen, B.B.; Zou, D.H.; Jiang, H. Elevated CO_2 exacerbates competition for growth and photosynthesis between *Gracilaria lemaneiformis* and *Ulva lactuca*. *Aquaculture* **2015**, *443*, 49–55. [CrossRef]

66. Wang, Y.; Zhou, B.; Tang, X. Effects of two species of macroalgae—*Ulva pertusa* and *Gracilaria lemaneiformis*—On growth of *Heterosigma akashiwo* (Raphidophyceae). *J. Appl. Phycol.* **2009**, *21*, 375–385. [CrossRef]
67. Lu, H.M.; Xie, H.H.; Gong, Y.X.; Wang, Q.; Yang, Y.F. Secondary metabolites from the seaweed *Gracilaria lemaneiformis* and their allelopathic effects on *Skeletonema costatum*. *Biochem. Syst. Ecol.* **2011**, *39*, 397–400. [CrossRef]

water

Article

Effects of Protein Level on the Production and Growth Performance of Juvenile Chinese Mitten Crab (*Eriocheir sinensis*) and Environmental Parameters in Paddy Fields

Yilin Yu [1], Jiwu Wan [2], Xiaochen Liang [1], Yuquan Wang [1], Xueshen Liu [3], Jie Mei [1], Na Sun [4] and Xiaodong Li [1,4,*]

[1] College of Animal Science and Veterinary Medicine, Shenyang Agricultural University, Shenyang 110866, China; yuyilin0722@163.com (Y.Y.); q1051079182@163.com (X.L.); q785241512@163.com (Y.W.); meijie08114010@gmail.com (J.M.)
[2] Jilin Fisheries Technology Promotion Station, Changchun 130012, China; jilintmj@sina.com
[3] College of Fisheries and Life Sciences, Dalian Ocean University, Dalian 116023, China; lxs912453956@gmail.com
[4] Panjin Guanghe Crab Industry Co., Ltd., Panjin 124200, China; sunna0911@163.com
* Correspondence: lixiaodong@syau.cn; Tel.: +86-137-000-729-18

Abstract: Rice–crab co-culture systems represent integrated agriculture–aquaculture systems developed in China over the last 30 years. The rice–crab co-culture area comprised approximately 1.386×10^5 hm^2 in 2019. However, there is no specific feed designed for Chinese mitten crab (*Eriocheir sinensis*) cultured in this system until now. In this study, we investigated feed formulae for the nutritional requirements of Chinese mitten crab in this mode. The control group was not fed with any artificial feed (Co), and the experimental groups were fed with three different feeds of 15% (T15), 30% (T30), or 45% (T45) protein content, respectively. Growth performance variations in *E. sinensis* were investigated along with water quality, phytoplankton, zooplankton, aquatic vascular plants, and benthic animals in the paddy fields to determine the effect of crabs and their diet on the paddy ecosystem. Dietary protein levels had no significant effect on water quality. The biomass and species of phytoplankton, zooplankton, aquatic vascular plants, and zoobenthos in the paddy field were affected by crabs and their diet. Morphological parameters of crabs were significantly more pronounced in the high-protein group than in the other groups. However, the T45 diet negatively affected production by increasing feed costs, causing precocious puberty and inducing water eutrophication. In conclusion, adding a 15% protein compound feed can meet the nutritional needs of crabs, reduce culture costs, and improve water quality. The discharged water had low ammonia nitrogen and nitrite content and no eutrophication occurred, so the water could be recycled. These findings provide a scientific reference for supporting rice and fish co-cultivation.

Keywords: rice–crab co-culture; *Eriocheir sinensis*; dietary protein content; ecological environment

Citation: Yu, Y.; Wan, J.; Liang, X.; Wang, Y.; Liu, X.; Mei, J.; Sun, N.; Li, X. Effects of Protein Level on the Production and Growth Performance of Juvenile Chinese Mitten Crab (*Eriocheir sinensis*) and Environmental Parameters in Paddy Fields. *Water* **2022**, *14*, 1941. https://doi.org/10.3390/w14121941

Academic Editors: Abasiofiok Mark Ibekwe and Thomas Hein

Received: 2 April 2022
Accepted: 14 June 2022
Published: 16 June 2022

1. Introduction

The Chinese mitten crab (*Eriocheir sinensis*) is the most commonly farmed crab species in China. In 2020, the General Office of the Ministry of Agriculture and Rural Affairs proposed the implementation of "five major actions," including the promotion of ecological and healthy farming modes. The new rice–crab co-culture mode integrates the culturing of rice and crabs with ecological, economical, and social benefits [1]. The food chain in the ecosystem of this mode is quite complex, creating a more stable ecosystem than that in single-species aquaculture. Crabs are at the top of this food chain and feed on plankton, weeds, and benthic animals in rice fields, ensuring an efficient matter circulation and a smooth energy flow through the whole system [2]. Furthermore, this mode produces a double harvest of rice and crabs [3].

The ecological environment in rice–crab co-culture may be affected by several factors. For instance, high-density culture adversely affects phytoplankton and benthic animals [4].

Zhang et al. [5] demonstrated that the phytoplankton biomass in crab culture was significantly higher than that in rice culture or rice co-culture. The daily activities of crabs change the physical and chemical environment of water bodies, which indirectly affects the plankton community structure [6]. These studies showed that the feeding behavior of crabs on plankton and benthic animals influenced the rice–crab co-culture ecosystem.

The quality and output of the mode are not the only important factors; it is also necessary to minimize expenses. Feed costs are the main expense in aquaculture. Almost all crab artificial feeds use fish meal (FM) as the main protein source [2]. However, there is a limited supply of FM, and it is costly [7], necessitating the use of expensive compound feeds. The optimum crude protein for the growth of juvenile crabs is 347.8 g/kg under the indoor individual Chinese mitten crab system [8]. Xu et al. found that a certain amount of fish meal replaced by soybean meal effectively reduced the cost of feed and had no effect on the growth performance, related enzyme activities and genes expression of Chinese mitten crabs [9]. These experiments showed that the protein in the feed played a key role in the growth performance of crabs and cost of feed.

After more than 30 years of development, China's rice–crab co-culture area comprised approximately 1.386×10^5 hm^2 in 2019, which accounted for 5.94% of the national rice and fishery planting area and produced a yield of 6.18×10^4 tons [10]. However, although paddy fields are rich in natural nutrients in rice–crab co-culture systems, there is a lack of details on the nutritional requirements, feed costs, and environmental response mechanisms of crabs, especially regarding the interaction between crabs and the ecosystem in paddy fields. In this study, we investigated the effect of three different protein levels in compound feeds on crabs, plankton, aquatic vascular plants, and benthic animals in a rice–crab co-culture system. We aimed to comprehensively analyze the protein requirements of juvenile Chinese mitten crabs under a rice–crab co-culture system. The findings from this research will provide a reference for the optimization of the feeding strategy in the rice–crab co-culture system.

2. Materials and Methods

2.1. Experimental Animals and Experimental Paddy Field Management

The experimental animals were obtained from Panjin Photosynthetic Crab Co., Ltd. (Panjin, China) and raised in the nursery pond at their facility. Crabs with complete appendages and of uniform size were randomly selected for our experiment. The experimental paddy field was routinely managed, and a base fertilizer was applied once, prior to rice planting.

2.2. Experimental Design

The paddy-field crab–culture experiment was conducted in the paddy field at Panjin Photosynthetic Crab Industry Co., Ltd. from 25 May to 8 October 2020 (Figure 1; E 121°50′38.73″–121°50′41.85″, N 40°54′1.07″). Water samples were collected nine times, including a sample in May prior to the initiation of the experiment and to any rice field being stocked with crabs. Subsequently, samples were collected twice a month throughout the experiment. The experimental design involved a total of 12 enclosures (6 m × 6.7 m). The enclosures were divided into three experimental groups and a control group, with three repetitions for each group. Crabs were stocked in each enclosure (the macrophthalmia size was 160/g at 30,000/hm^2), including the control group, which was not supplied with any artificial food.

The feed used in the experimental group was designed by the research group with FM and soybean meal as the main protein sources, and fish oil was used as the main fat source (Table 1). Three types of isolipid feeds with different protein contents were formulated by simultaneously increasing the FM and soybean meal content (FM: soybean meal = 2). The feed protein levels were 15%, 30%, and 45%. The various solid raw materials were accurately weighed according to the required formula ratio and, then, were fully mixed according to the principle of step-by-step enlargement. Subsequently, the artificial feed was pulverized through a 100-mesh sieve, the oil was added, and all ingredients were stirred to an even consistency. Finally, water was added (30%), and the feed was mixed again. A double helix A pellet mill (DES-TS1280, Jinan Dingrun Machinery Equipment Co., Ltd., Jinan, China) was used to press the feed into 3 mm diameter pellets. The pellets were naturally air-dried and packaged and sealed in plastic bags. The bags were stored in a refrigerator at −20 °C. Crabs were fed at a rate of 10% of their body weight. A detailed list of contents of each experimental feed is shown in Table 1, and the feed costs are presented in Table A1.

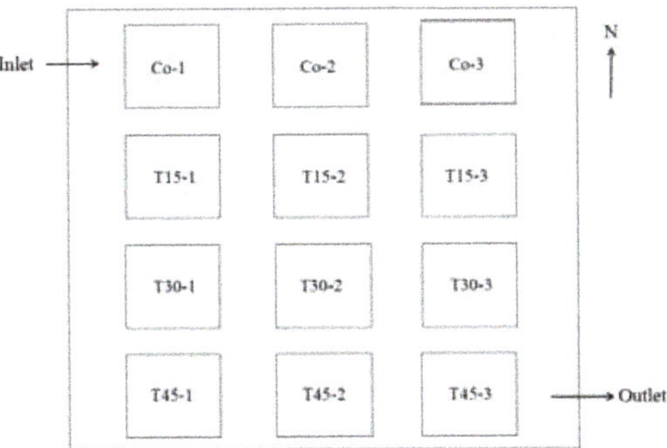

Figure 1. Experimental design field of the crab–rice field. Inlet is where the enclosure received water; the outlet is where the enclosure drained water. Co is the control group with no artificial feed supplied, and T15, T30, and T45 represent the treatment groups fed with experimental feeds of 15%, 30%, and 45% protein content, respectively.

Table 1. Detailed list of contents of each experimental feed.

Ingredients/%	Content		
	T15	**T30**	**T45**
Fish meal	6	27	47
Soybean meal	3	13.5	23.5
Beer yeast	3	3	3
Wheatmeal	72.59	43.09	15.09
Fish oil	7.5	5.5	3.5
Lecitin	0.5	0.5	0.5
Mineral premix	2	2	2
Vitamin premix	2	2	2
Squid paste	1	1	1
Glycine betaine	0.5	0.5	0.5
Choline chloride	0.2	0.2	0.2
Calcium dihydrogen phosphate	1.5	1.5	1.5
Chromium oxide	0.1	0.1	0.1
Ethoxyquin	0.01	0.01	0.01
Calcium propionate	0.1	0.1	0.1
Total	100	100	100
Crude protein	15.00	30.49	45.25
Crude lipid	9.13	9.27	9.31
Total energy/KJ·g^{-1}	18.09	16.57	17.17

2.3. Test Methods

2.3.1. Monitoring Water Quality in the Paddy Fields

During the experiment, physical and chemical water quality indicators were recorded and monitored in each enclosure. The indicators included temperature (oxygen dissolving instrument, YSI550-A, Vasey Instrument Company, Exton, PA, USA), pH (pH meter, PHB-1,

Shanghai Thunder Magnetic, Shanghai, China), salinity (Pen salinity meter, AR-8012, Xima Instrument Co., Ltd., Dongguan, China), dissolved oxygen (oxygen dissolving instrument, YSI550-A, Vasey Instrument Company), ammonium nitrogen (visible spectrophotometer, V-1100, Shanghai Meitong Instrument Co., Ltd., Shanghai, China), and nitrite nitrogen [11].

2.3.2. Growth Performance and Yield of Crabs

The megalopae of crabs were collected and weighed to ensure all enclosures contained the same number of crabs (160/g) on 25 May. The megalopae were cultured in the enclosures for 46 days until they reached the feeding phase. The experiment was initiated after the crabs were measured. The average body length, width and height was 1.08 ± 0.08, 1.16 ± 0.09 and 0.53 ± 0.05 cm, respectively, and average body weight was 0.63 ± 0.13 g. During the experiment, the body length, width, height, and weight of the crabs were measured on 10 July, 28 July, 17 August, 8 September, and 8 October. All crabs were humanely harvested at the end of the experiment. Crabs were caught using a plastic bucket inserted into a hole dug in the bottom of the enclosure. The frequency of collection depended on the number of crabs. All specimens were counted, measured, and weighed. Precocious puberty was assessed by comparing the abdomen, junction, villi, gonad, color, and crab patterns with those of the representative crab specimens [12]. Growth performance indicators were calculated using the following formulae:

$$\text{Survival rate}/\% = n_t/n_0 \times 100\%,$$

$$\text{Weight gain rate}/\% = (m_t - m_0)/m_0 \times 100\%,$$

$$\text{Specific growth rate}/\%/d = (\ln m_t - \ln m_0)/t \times 100\%,$$

$$\text{Total output}/\text{g·m}^{-2} = W/S,$$

$$\text{Net output}/\text{g·m}^{-2} = (W - W_0)/S,$$

where n_0 represents the initial number of crabs, n_t represents the final number of crabs, m_t represents the final average body weight, m_0 represents the initial average body weight, t represents the total number of days of the experiment, W represents the final total weight of crabs in an enclosure, W_0 represents the initial total weight of crabs in an enclosure, and S represents the area of the enclosure ($6 \text{ m} \times 6.7 \text{ m} = 40.2 \text{ m}^2$).

2.3.3. Qualitative and Quantitative Analysis of Phytoplankton

The sampling and measurement methods used to assess the phytoplankton were based on those of Zhang [6]. Briefly, 1 L of water was collected by five-point sampling at each point and mixed in a bucket. A lugol solution (10–15 mL) was then evenly mixed into the water. After 48 h, the sample was concentrated by siphonage, fixed at 100 mL volume, and then put into an iodometric bottle for qualitative analysis. The qualitative and quantitative analyses followed Li et al. [13], and Zhao [14], respectively. The specific gravity of phytoplankton is approximately 1. Therefore, the volume was directly converted into wet weight, and the phytoplankton biomass was calculated (Table A2).

2.3.4. Qualitative and Quantitative Analyses of Zooplankton

The sampling and measurement of the zooplankton were based on methods of Zhang [6]. The qualitative and quantitative methods followed those described in Section 2.3.3, and the zooplankton biomass was calculated (Table A3).

2.3.5. Qualitative and Quantitative Analyses of Aquatic Vascular Plants

The aquatic vascular plants were sampled by selecting two points that were consistent for each enclosure. A 30 cm $\times$ 30 cm iron frame was used to divide the sampling area. The plants (except rice) were uprooted, species were identified, and plant wet weight was determined.

2.3.6. Qualitative and Quantitative Analysis of Benthic Animals

The quantification of the benthic animals was conducted at the same sampling points as those mentioned in Section 2.3.5 at a depth of approximately 10 cm using a self-made barrel dredger [15]. The benthic animals were screened using a sieve with an aperture of 0.2–2 mm and then wet-weighed, identified, and counted with precision.

2.4. Statistical Analysis

The experimental data were collated using Excel. The homogeneity of variance test and one-way ANOVA were performed using SPSS 24.0. Any significant differences between groups were further analyzed using Duncan's multiple comparison tests. The results were expressed as the mean ± standard deviation. In all analyses, a probability value less than 0.05 was considered significant ($p < 0.05$).

Dominance (Y) was calculated according to the formula:

$$Y = ni/N \times fi$$

where Y is the degree of dominance, ni is the number of individuals of species i, N is the total number of individuals, and fi is the frequency of occurrence of species i at five sampling points within an enclosure.

The Shannon–Wiener diversity index (H′) was calculated using the formula:

$$H' = -\sum [(ni/N) \times \ln(ni/N)],$$

where ni is the number of individuals of species i, and N is the total number of individuals of the species.

3. Results

3.1. Growth Performance and Yield of Crabs

The morphological parameters of the crabs in the high- and low-protein diet groups varied significantly at each measurement (Figure 2). At the end of the experiment, the carapace length, width, and height of both T30 and T45 groups were significantly higher than those of the Co group ($p < 0.05$), and the carapace length and width were significantly higher than those in the T15 group ($p < 0.05$). The final body weight and weight gain rate of the crabs in the T45 group were significantly higher than those of the Co group ($p < 0.05$). The growth rate of the crabs in the Co group was significantly lower than that of the T30 and T45 groups ($p < 0.05$).

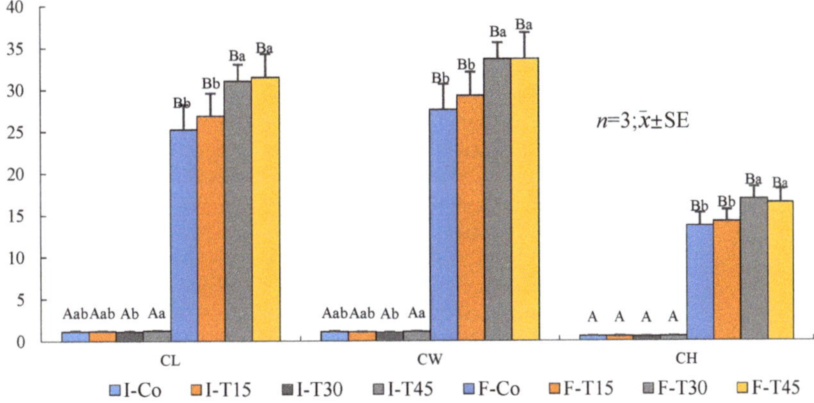

Figure 2. Comparison of morphological parameters of crabs among different protein-content groups. I, the initial size in the beginning of experiment; F, the final size at end of experiment; CL, carapace

length; CW, carapace width; CH, carapace height. Different letters indicate the significant difference among treatment groups ($p < 0.05$).

The final body weight of the crabs ranged from 8.30 g to 17.28 g (Table 2). The final body weight of the crabs that were fed diets increased significantly with the increase of protein content ($p < 0.05$). The body weight increase rate varied from 9641.86% to 20,181.73% and significantly increased with the increase in dietary protein content ($p < 0.05$). The specific growth rate of the crabs varied from 3.30%/d to 3.85%/d and significantly increased as the dietary protein content increased ($p < 0.05$).

Table 2. Effects of different dietary protein content levels on the growth performance and yield ($n = 3$; $\bar{x} \pm$ SD) of juvenile Chinese mitten crabs.

Growth Performance and Yield	Groups			
	Co	T15	T30	T45
Initial body weight/g·ind^{-1}	0.09 ± 0.02	0.09 ± 0.01	0.09 ± 0.01	0.09 ± 0.02
Final body weight/g·ind^{-1}	8.30 ± 3.19 [a]	13.10 ± 1.59 [ab]	14.35 ± 1.87 [ab]	17.28 ± 6.23 [b]
Survival rate/%	17.93 ± 8.25	12.51 ± 3.38	8.46 ± 2.76	10.38 ± 3.43
Weight gain rate/%	9641.86 ± 3738.90 [a]	$15,277.93 \pm 1860.99$ [ab]	$16,737.02 \pm 2194.882$ [ab]	$20,181.73 \pm 7316.62$ [b]
Specific growth rate/%·d^{-1}	3.30 ± 0.33 [a]	3.67 ± 0.09 [ab]	3.74 ± 0.09 [b]	3.85 ± 0.26 [b]
Total output/g·m^{-2}	48.38 ± 10.14	58.62 ± 10.05	43.86 ± 12.27	60.41 ± 1.18
Net output/g·m^{-2}	42.62 ± 10.14	52.86 ± 10.05	38.10 ± 12.27	54.64 ± 1.18

Note: Values in each row with different superscripts are significantly different ($p < 0.05$).

The total and net yields of crabs varied from 43.86 g/m^2 to 60.41 g/m^2 and 38.10 g/m^2 to 54.64 g/m^2, respectively. The highest total and net yields were for crabs in the T45 group, followed by those of the T15 and Co groups; the lowest was for those of the T30 group. There was no significant variation in the total and net yield of crabs in the different experimental groups ($p > 0.05$). Towards the end of the experiment, approximately 10% of crabs in the T45 group experienced precocious puberty.

3.2. Water Quality

The physical and chemical parameters of water quality in each enclosure during the experimental period was assessed (Figure 3). Generally, the parameters of water quality in each enclosure were consistent with the fishery water quality standard of the People's Republic of China (GB 11607-89). During the experiment, the water temperature ranged from 24.5 °C to 28.2 °C, and the salinity from 0.493‰ to 1.743‰. No significant difference was found in dissolved oxygen, ammonia, and nitrate levels in all enclosures (Figure 3).

3.3. Phytoplankton in the Paddy Fields and Variations with Time
3.3.1. Phytoplankton Biodiversity

A total of 54 species of phytoplankton from seven phyla were detected in four treatments during the experiment (Table A4). Seven phyla were present in all treatments (Figure 4). Bacillariophyta was the dominant group, with 19 species present, and accounted for 35.19% of the phytoplankton species observed. Chlorophyta was the phylum with the second highest number of species (18 species) and accounted for 33.33% of the total number of species. Other groups included Cyanophyta (9 species; 16.67%) and Euglenophyta (5 species; 9.26%). The phyla Cryptophyta, Chrysophyta, and Pyrrophyta were represented by one species each and accounted for 1.85% of the total species.

The rank of phytoplankton biomass of each group, from nine sampling events, was: Chlorophyta (32.48 mg/L, accounting for 26.96% of the total biomass), Eugelenophyta (29.19 mg/L; 24.23%), Chrysophyta (26.85 mg/L; 22.29%), Cyanobacteria (0.26 mg/L; 15.15%), Bacillariophyta (13.68 mg/L; 11.36%), and Cryptophyta (3.06 mg/L; 2.54%). Pyrrophyta was detected only once at a very low proportion.

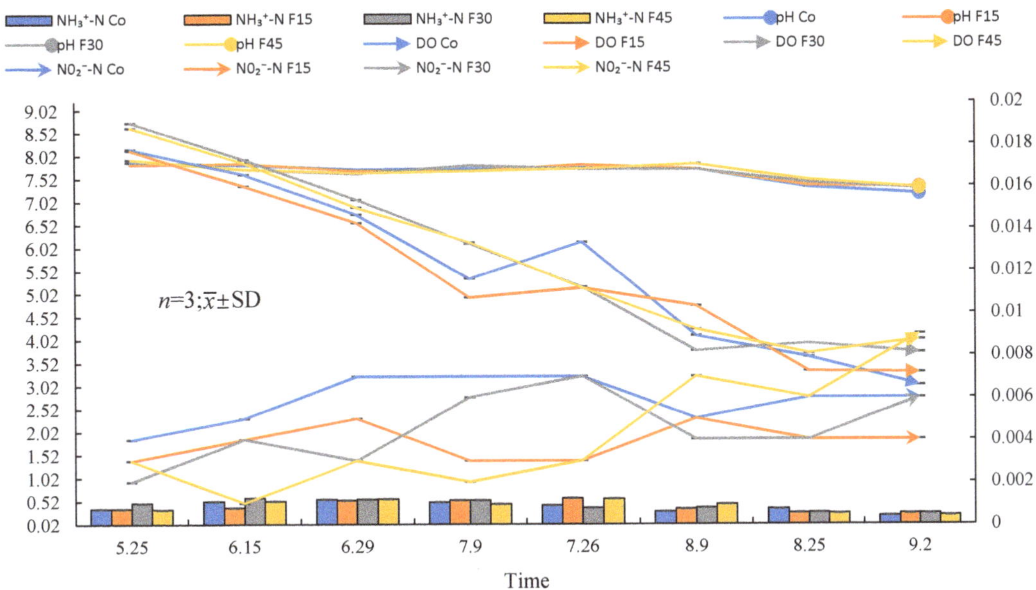

Figure 3. Water quality parameters of each enclosure during the experiment. The y-axis of $NO_2{}^{-}$-N is the secondary coordinate axis (right), and others correspond to the left primary coordinate axis.

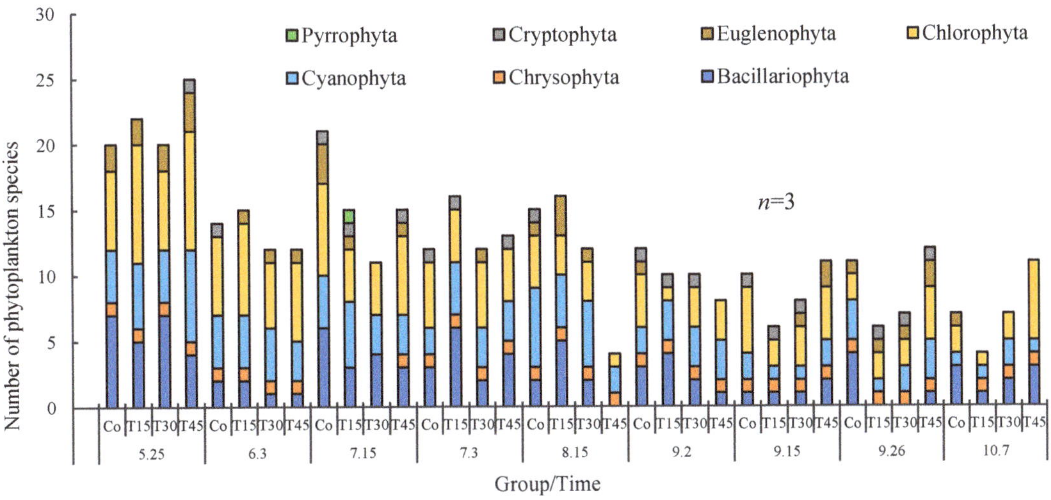

Figure 4. Species of phytoplankton observed in different treatment groups during the experiment.

The phytoplankton diversity in each group was analyzed using the Shannon–Wiener diversity index (Figure 5), with an overall average of 1.07. The diversity index of the Co group was 0.59 to 1.43, with an average of 1.03. The diversity index of the T15 group was 0.46 to 1.62, with an average of 0.95. The diversity index of the T30 group was 0.54 to 1.65, with an average of 1.02. The diversity index of the T45 group was 0.93 to 1.62, with an average of 1.26.

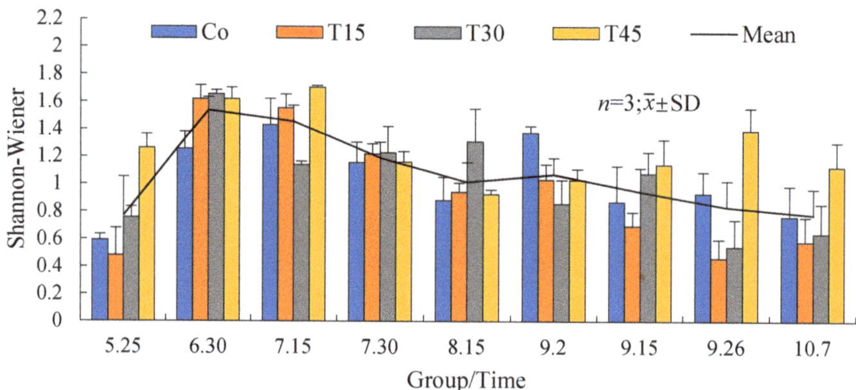

Figure 5. Shannon–Wiener diversity distribution indexes in different treatment groups during the experiment.

3.3.2. Variation Trends in Phytoplankton Biomass over Time

The phytoplankton biomass of different treatment groups over times was analyzed statistically. The overall average biomass of each experimental group showed a downward trend with time. However, an abnormal increase occurred for those in Co group on 15 August (Figure 6). On 25 May, the biomass in T45 was significantly higher than that in Co ($p < 0.05$). On 30 June, the biomass in T15 was significantly lower than that in Co and T45, while those in Co and T15 was significantly higher than that in T45 on 15 August ($p < 0.05$). The phytoplankton biomass in Co group was the lowest among all groups on 15 October ($p < 0.05$).

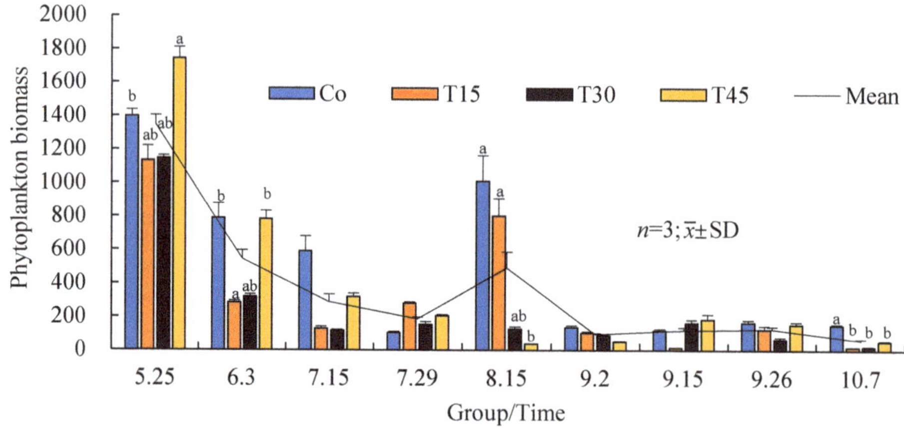

Figure 6. Phytoplankton biomass in different treatment groups during the experiment. Different lowercase letters in each group represent significant differences ($p < 0.05$).

3.3.3. Succession and Population Changes in the Dominant Phytoplankton Species

The dominant phytoplankton species were calculated, and the results are presented in Figure 7. If Y > 0.02, then the species was considered dominant. After the quantification and calculation analysis, there were five phyla of dominant zooplankton. The species composition and dominance of each phytoplankton species varied within different sampling time intervals in each treatment group (Table A5).

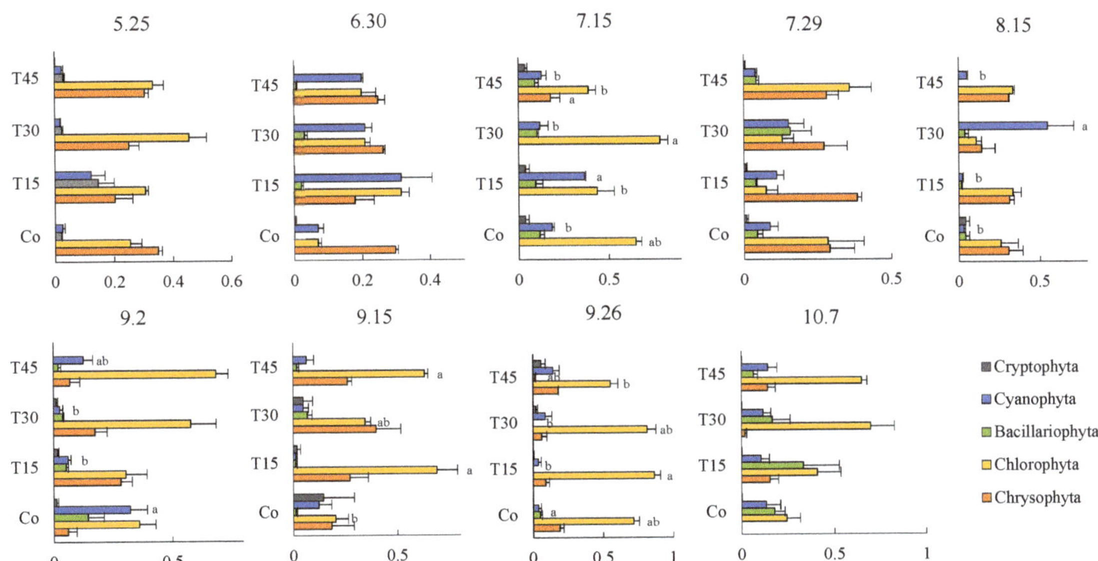

Figure 7. Dominance phytoplankton at phyla level in different treatment groups during the experiment.

Chromulina pygmaea and *Chlorella pyrenoidosa* were highly dominant species in different treatment groups during the experiment. The diversity of dominant phytoplankton species in the Co group increased over time. Comparatively, the dominant species of each feeding group were relatively simple. In the later stage of the experiment, the degree of dominance of *Chroococcus* in the T30 and T45 groups was higher than that in the Co and T15 groups (Figure 8). On 2 September, the degree of dominance of *Chroococcus* in T45 was significantly higher than in the other groups ($p < 0.05$).

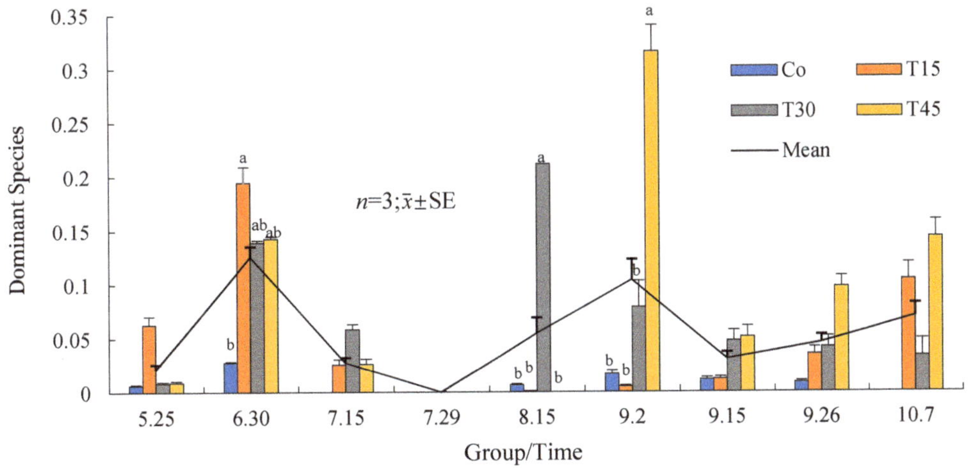

Figure 8. Dominance of *Chroococcus* sp. in different treatment groups during the experiment.

3.4. Species, Quantities, and Changes in Zooplankton

3.4.1. Zooplankton Species Diversity

A total of 50 species of zooplankton were detected during the experiment (Figure 9). Protozoa had the highest number of species, with 23 species detected, and accounted for

46% of the total number of species. Rotifera had 15 species detected, accounting for 30% of the total species. Cladocera had eight species, accounting for 16% of the total species. Copepoda had four species, accounting for 8% of the total species.

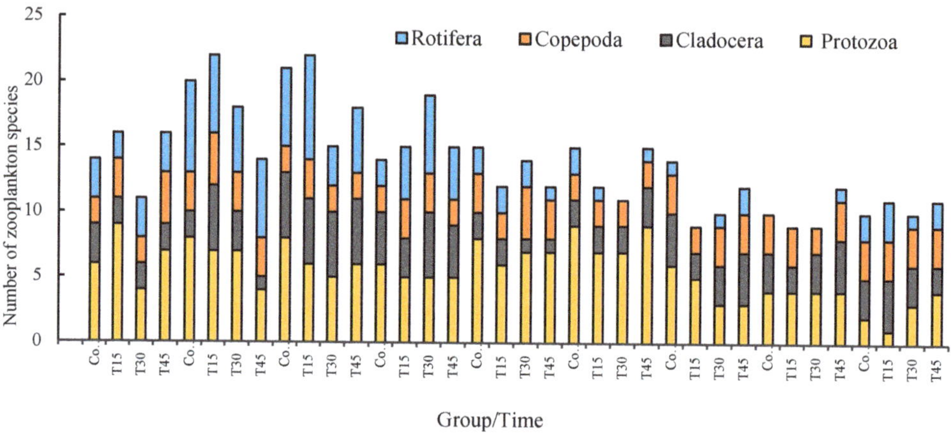

Figure 9. Species of zooplankton in different treatment groups during the experiment.

The zooplankton species in different treatment groups during the experiment are listed in Table A6. The average biomass of Copepoda was the largest at 31.61 mg/L and accounted for 50.00% of the overall zooplankton biomass. That of Cladocera was 29.29 mg/L and accounted for 46.33% of the total biomass. These two groups accounted for 96.32% of the total average biomass. The average biomass of Protozoa was 1.76 mg/L, accounting for 2.79% of the total. The zooplankton group with the lowest average biomass was the rotifers, with only 0.56 mg/L, accounting for 0.89% of the total average biomass.

The zooplankton diversity in each group was analyzed using the Shannon–Wiener diversity index (Figure 10). The overall average value was 1.69. The diversity index was from 1.19 to 2.16 for the Co group (1.74 average), 1.15 to 2.31 for the T15 group (1.67 average), 1.00 to 2.18 for the T30 group (1.61 average), and 1.27 to 2.08 for the T45 group (1.73 average), respectively.

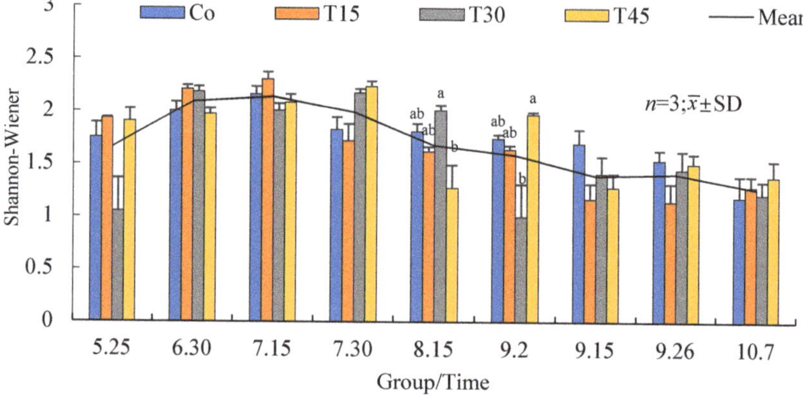

Figure 10. Shannon–Wiener diversity indexes of the zooplankton in different treatment groups during the experiment.

3.4.2. Variation Trends of the Zooplankton Biomass over Time

The total average zooplankton biomass in the paddy field fluctuated over time. There was an upward trend from the beginning of the experiment to 15 July, which then decreased to 15 August and increased to September 15 before decreasing again (Figure 11). The biomass of zooplankton in T45 was significantly lower than those in T15 and T30 on 30 June ($p < 0.05$), while the biomass in T45 was significantly higher than those in other groups on 29 July ($p < 0.05$).

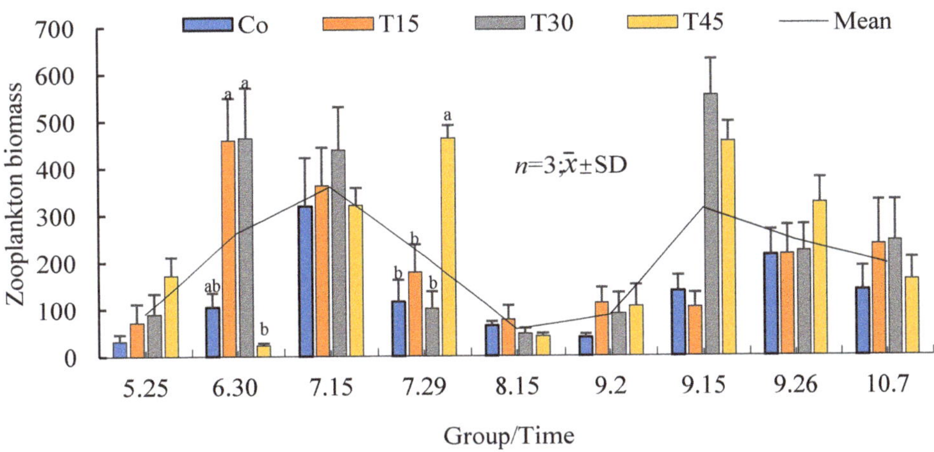

Figure 11. Zooplankton biomass in different treatment groups during the experiment.

3.4.3. Succession of Dominant Zooplankton Species and Community Changes

The dominant zooplankton species in the paddy fields during the experiment are shown in Table A7 and Figure 12. When dominance value (Y) > 0.02, the species was considered dominant. The dominant species in different treatment groups consisted of 32 species belonging to 4 zooplankton groups (i.e., Rotifera, Copepoda, Cladocera and Protozoa), respectively. The biomasses of the dominant zooplankton species were relatively low in the four treatment groups on 25 May and 7 October. The number of dominant species in T15 was the lowest on 15 August. The numbers of dominant species were 6 to 10 in Co group, 4 to 11 in T15 group, 4 to 9 in T30 group, and 6 to 9 in T45 group, respectively, from 30 June to 26 September.

3.5. Species, Quantities, and Changes in Aquatic Vascular Plants

The species diversity and biomass of the aquatic vascular plants in the enclosures of each treatment group are shown in Figure 13. Seven aquatic vascular plants were detected in the four treatment groups, i.e., *Vallisneria spiralis*, *Monochoria vaginalis*, *Sparganium stenophyllum*, *Spirodela polyrhiza*, *Potamogeton* sp., *Elodea nuttallii*, and *Scirpus validus*.

Sparganium stenophyllum was the most dominant species, with an average biomass of 535.71 g/m^2, accounting for 52.02% of the total aquatic vascular plant biomass. The following dominant species was *S. validus* with 287.12 g/m^2, accounting for 27.88% of the total biomass. Meanwhile, the biomasses for *M. vaginalis*, *V. spiralis*, *E. nuttallii*, *S. polyrhiza* and *Potamogeton* sp. were 82.79, 79.55, 24.09, 15.41 and 5.12 g/m^2, and accounted for 8.04%, 7.72%, 2.34%, 1.50% and 0.50% of the total biomass, respectively.

The changes of the submerged-plant biomass in different treatment groups over time are shown in Figure 14. The statistical analysis revealed significant variations in the biomass of submerged plants in each treatment group with different times. The overall submerged-plant biomass rapidly decreased to 30 June before rapidly increasing to 25 July and then increased gradually to 29 July and 15 August. The biomass in the T45 group was significantly higher than the other groups ($p < 0.05$). Afterwards, submerged-plant biomass rapidly decreased again. No submerged plant was found in all treatment groups from 15 September to 7 October.

The biomass of the emergent plants in different treatment groups over time are shown in Figure 15. The statistical analysis revealed significant variations in the biomass of emerged plants at different times. Generally, the biomass increased throughout the experiment and only decreased in the last sample collection. The emergent plants biomass in the Co group was significantly higher than that in the T45 group on July 15 and July 29, while the biomass in the T15 and T30 groups was significantly higher than that in the T45 group on July 29 ($p < 0.05$).

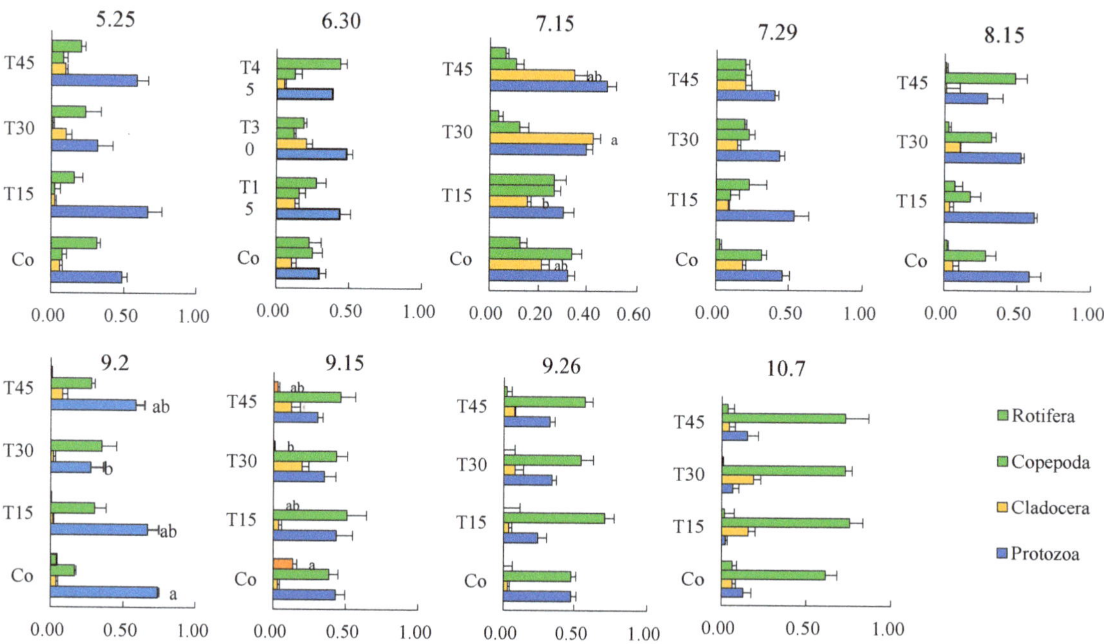

Figure 12. Composition of the predominant zooplankton species in different treatment groups during the experiment.

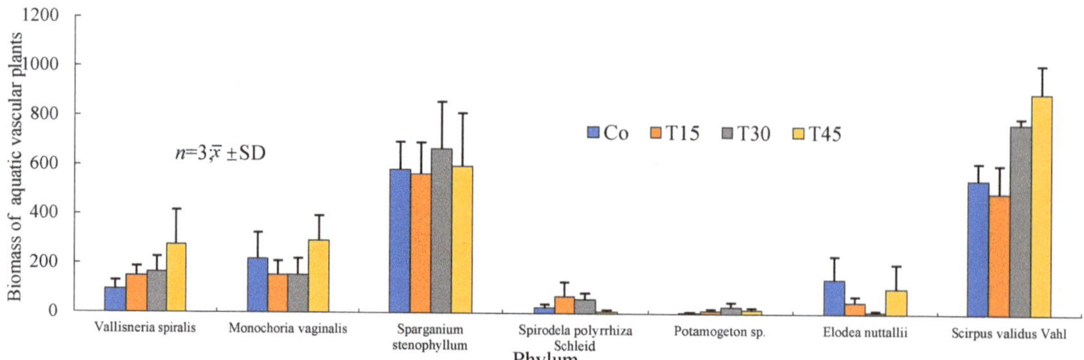

Figure 13. Species and biomass of aquatic vascular plants in different treatment groups during the experiment.

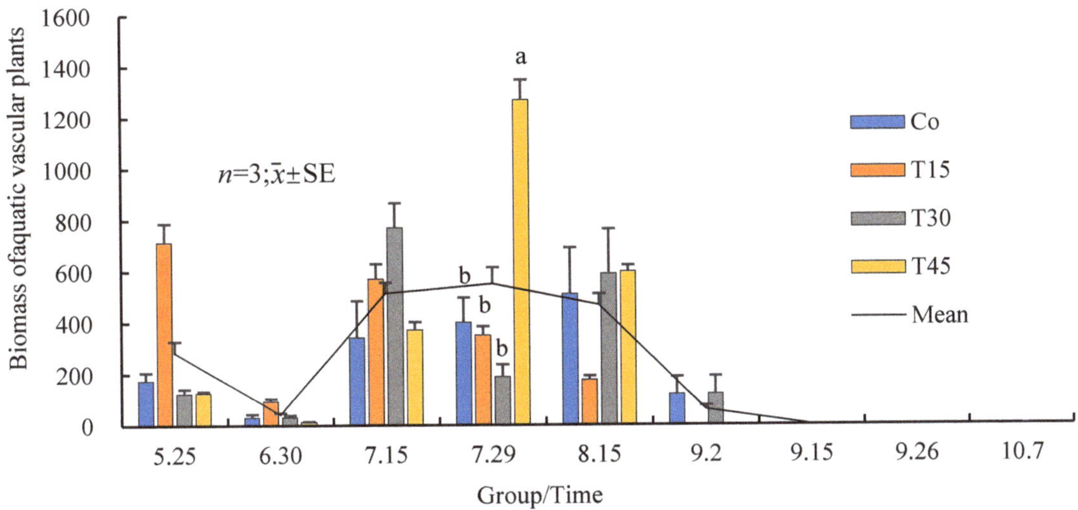

Figure 14. Biomass of the submerged plants in different treatment groups during the experiment.

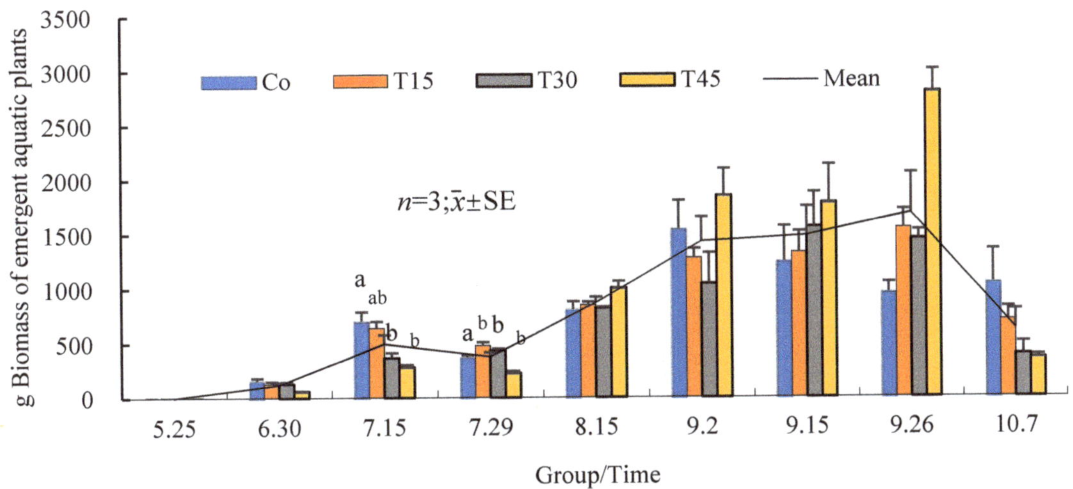

Figure 15. Biomass of the emergent aquatic plants in different treatment groups during the experiment.

3.6. Variations in the Benthic Animal Species and Quantities with Time

The diversity and biomass of benthic animals in different treatment groups are shown in Table A8 and Figure 16. Five taxa were found in the nine samples from the four treatment groups, i.e., *Gyraulus* sp., *Euconulus* sp., *Limnodrilus* sp., *Branchiura* sp., and Insecta.

The biomass of benthic animals in different treatment groups changed over time. Generally, they initially increased from 23 May to 29 July, then decreased rapidly to 15 August, followed by a gradual decline through September until the end of the experiment (Figure 17). On 2 September, the biomass of the benthic animals in the Co group was significantly higher than that in the T45 and T30 groups ($p < 0.05$). No significant difference was found among various treatment groups at the same sampling time.

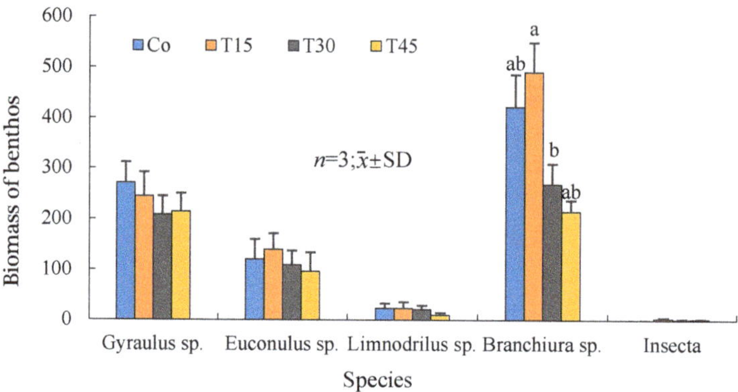

Figure 16. Species and biomass of benthos in different treatment groups during the experiment.

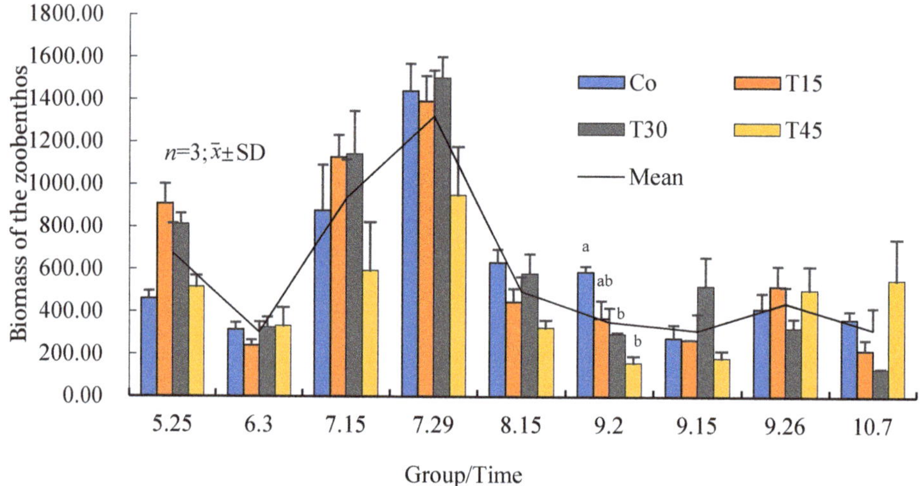

Figure 17. Biomass of the zoobenthos in different treatment groups during the experiment.

4. Discussion

4.1. Effects of Diets with Different Protein Levels on the Growth Performance and Yield of Juvenile Crabs

Protein is one of the most important nutritional components in the diet of crabs, and the level required varies depending on growth stages [16]. The five separate measurements of morphological parameters of the crabs revealed that the high-protein compound feed resulted in significantly higher crab carapace length, width, and height and body weight compared with those in the low-protein group. Feeding with a high-protein compound feed had a significantly positive effect on the weight gain rate, final body weight, and specific growth rate of juvenile crabs. These results support the findings of Zhang [6]. However, the T30 group in this experiment may have escaped and/or had "milky disease" [17]. The rate of disease varied according to the original health status of the crabs and may have resulted in the observed differences in survival rates; however, there was no significant difference in the crab yield between the different diet treatments. These results do not correspond with the growth rate results. Therefore, the contribution of natural food to crab growth may be underestimated in the rice–crab co-culture mode.

Integrated agricultural and aquaculture systems can effectively contribute to green and sustainable agricultural development and ensure food security [18]. In 2017, the

"General Principles of Technical Specifications for Rice and Fishing Integrated Planting and Culture," issued by the Chinese Agriculture and Rural Affairs Bureau, highlighted the fact that animals raised in aquaculture should make full use of natural bait present in the environment (in this case the rice paddies), reducing the use of fish feed. The results of this experiment strongly support this statement. The lower temperature in the paddies is more favorable to the growth of crabs compared with the temperature in monoculture crab systems, which can reduce crab sexual precocity [19].

In this experiment, the sexual precocity rate in the T45 group was approximately 10%, which may be due to the high protein content. Chen et al. [20] demonstrated that when the protein content in the feed is too high, excess protein is converted into fat and stored in the hepatopancreas, resulting in sexual precocity in crabs. Sexual precocity during the breeding process reduces culture efficiency. Studies have shown that the survival rate of adult crabs cultured with precocious crab species in the second year is already extremely low [21]. Therefore, it is important to prevent the sexual precocity of crabs in production.

4.2. Changes in Physical and Chemical Properties of Paddy Water Environment

There were no obvious changes in water temperature, pH, salinity, ammonia nitrogen, or nitrite nitrogen over the course of this experiment. There were also no significant differences between the experimental groups. The ammonia nitrogen and nitrite content of the water was low. However, there was a downward trend in dissolved oxygen levels in the water, which may have been caused by a variety of factors. The daily photosynthesis of plants is the main source of dissolved oxygen in water [22]. Animal respiration in the water releases large amounts of organic matter. Additional organic matter is produced during decay (after death) and after feeding, which leads to an increase in the respiration in water and sediments, and is also the main destination of dissolved oxygen [23]. The crabs were placed in the experimental enclosures on 29 May. On the same day, the dissolved oxygen in the water began to decrease, indicating that crab respiration was the main factor causing the low dissolved oxygen in the rice–crab co-culture. As the experiment progressed, the shading effect of large vascular plants (including the rice) led to a decrease in phytoplankton photosynthesis. This is also one of the reasons for the decrease in dissolved oxygen levels. In the later stages of the experiment, the plankton species and biomass and the biomass of zooplankton increased, while the biomass of the phytoplankton decreased. Consequently, there were more aerobic biological factors and less oxygen-producing organisms in the environment, resulting in a decrease in the dissolved oxygen levels in the water. The levels of dissolved oxygen ranged from 3.04 to 8.75 mg/L, which is lower than the normal dissolved oxygen requirements of crabs (5 mg/L). In low dissolved oxygen conditions, crabs tend to escape. Crabs also crawl to the shore in the later culture stage. The dissolved oxygen content of the water in rice–crab co-culture is lower than that in conventional rice fields [24,25]. This may cause a stress response in the crabs and is, therefore, one of the shortcomings of breeding crabs in rice paddies.

4.3. Effects of Diets with Different Protein Levels on the Phytoplankton in Paddy Fields

As primary producers, phytoplankton also act as a natural food source for crabs in rice–crab co-culture systems. Through the experiment, the aquatic organisms and water quality factors affected and correlated with each other. Species diversity is a basic characteristic of biological communities and is an important indicator of a healthy system [26]. In this experiment, the overall average Shannon–Wiener diversity index of the phytoplankton in the paddy field was 1.07, indicating that the phytoplankton diversity in the paddy field environment was not extremely diverse but superior to that in polluted waters. Studies have shown the Shannon–Wiener diversity index for phytoplankton in the rice–crab co-culture mode is higher than that in conventional rice fields [5,27]. This is due to the reduction in the use of chemical fertilizers and pesticides in rice–fish symbiosis [27,28]. Thus, biodiversity in the paddy fields has been well protected [29,30].

The rice growth and the subsequent shading effect of the rice reduced the light-receiving area of the water in the paddy field. This weakened the photosynthesis by phytoplankton. Furthermore, the physical and chemical factors involved in water quality and organisms in the water environment interact with each other [30,31]. Individual phytoplankton are small and varied, and different species can affect the environment in diverse ways. Adaptability of the species differs, and most can intuitively reflect the changes in water physicochemical factors after environmental changes. Rice will absorb nutrients and ions in the paddy field, effectively regulating the physical properties and chemicals in the environment, and can inhibit the absorption of nutrients by phytoplankton. Therefore, it was expected that the total average biomass of phytoplankton would show a decreasing trend with time. When the phytoplankton productivity is low, the consumption of phytoplankton by zooplankton is an important factor affecting phytoplankton growth [32]. For example, the sample collected on 15 August revealed that the phytoplankton biomass had abnormally increased. When the data were combined with the analysis of changes in zooplankton biomass, the zooplankton biomass had decreased significantly at that time. Reduced grazing pressure on phytoplankton results in abnormally elevated phytoplankton biomass. In the natural environment, there are many reasons for an increase in phytoplankton biomass. For example, an increase in nutrient concentration can lead to similar results. The water quality monitoring results showed that the nutrient index in the water was not significantly different from that in other periods; therefore, there was no increased nutrient concentration.

The results of this experiment revealed that the diversity of the dominant phytoplankton species in the control group presented an increasing trend. However, it did not vary among the feeding groups. This may be because the addition of exogenous nutrients in the feed led to the eutrophication of the water body, resulting in a decrease in biodiversity. Water eutrophication destroys the ecosystem balance and can even lead to the collapse of the entire aquatic system [33]. When the nutrients in the water increased, Cyanophyta phytoplankton gradually became the dominant species at the expense of other species, indicating that Cyanophyta are indicator organisms for water eutrophication [34]. In this experiment, the dominance of *Chroococcus* in the T45 group was significantly higher than that of the other three groups on 2 September, after which it decreased with no significant difference, indicating that high protein levels could cause water eutrophication. However, in a paddy field environment, water eutrophication is not a concern owing to the self-purification function of rice.

4.4. Effects of Different Protein Levels in the Crab Diet on Zooplankton in Paddy Fields

In the rice–crab co-culture environment, zooplankton can feed on phytoplankton, which is also the main food of crabs. Zooplankton is an important link between energy flow and material cycling in the ecosystem [35]. Zooplankton species and community structure are affected by environmental factors. A total of 51 species of zooplankton were identified in this experiment, and the Shannon–Wiener diversity index was 1.73. Previous research has shown that the Shannon–Wiener diversity index of Cladocera and Rotifers in a water environment under rice–crab co-culture is higher than that of conventional rice fields [36]. In addition, owing to the purification effect of rice, zooplankton in crab paddy fields is highly diverse.

The results of this experiment showed that the zooplankton biomass initially increased and then decreased before increasing again. Combined with the analysis of the changes in phytoplankton in the paddy field, the biomass of the phytoplankton was relatively high in the early stage of the experiment. Then, the zooplankton fed on the phytoplankton and grew rapidly; its biomass also increased. As crabs grew, they preyed on the zooplankton, and the zooplankton biomass decreased. Horn et al. [37] tracked zooplankton body length and found that not only the maximum body length of zooplankton decreases, but the average body length and length frequency distribution of zooplankton also shifted to that of smaller individuals under predation pressure. The results showed that predation pressure on zooplankton by crabs led to the miniaturization of zooplankton. In the later stage of this experiment, the miniaturization of zooplankton, combined with the larger size and mouthparts of crabs, reduced the crab predation on zooplankton, so the biomass of zooplankton increased.

The dominant zooplankton species in the early stages of the experiment, on 25 May and 30 June, were rotifers, especially *Polyarthra trigla*, which is consistent with Zhang's [6] results. Rotifers, Cladocera, and copepods all competitively feed on phytoplankton in paddy fields. According to Gilbert [38], when there is competition between Cladocera and rotifers, Cladocera has an advantage. Therefore, the existence of Cladocera affects the diversity and quantity of rotifers. As the experiment progressed, some Cladocera species gradually became dominant. In the later stage of this experiment, the dominant species of zooplankton in the crab paddy field were small and mainly existed in the state of copepod nauplii, with the dominant species being mainly protozoa.

4.5. Effects of Crabs on Aquatic Vascular Plants in Paddy Fields

Large plants control zooplankton, provide habitats for fish that feed on zooplankton, and provide shelter for phytoplankton [39]. In the early stages of this experiment, the main submerged plant in the paddy field environment was *S. polyrhiza*. The growth of the submerged plants, including *V. spiralis* and *Potamogeton* sp., increased, and the biomass of submerged plants also increased. Shading also increases with the rapid growth of emergent plants [40]. Crabs began feeding on the submerged plants, which reduced their growth until no submerged plants were detected from 15 September. On 7 October, the emergent plants decayed and died, decreasing the biomass.

Farmers have traditionally used chemical weeding machines to remove large weeds from rice fields. Over time, weed resistance and herbicide damage have become increasingly serious problems [41], since crabs are omnivorous and feed on large plants, such as aquatic vascular plants [42]. Even if there is an excess of animal food in the environment, crabs will still consume aquatic plants, especially submerged plants [43]. However, crabs rarely feed on emergent plants, which allows the emergent plants to grow and absorb fertilizer from the crab pond sediment [44]. Lv et al. [45] found that the fresh and dry weights of weeds in the experimental group without crabs being provided artificial feed were significantly lower than those in the crab feeding group. Other studies have also shown that weed control by rice crabs is more effective than traditional weed control methods used in rice production [19,46].

4.6. Changes in Benthic Animals in Rice–Crab Co-Culture

Benthic animals are the main food source for crabs [47]. Xu et al. [48] found that crabs affect habitat structure in two ways: feeding and reducing competition with aquatic plants by preying on attached organisms, thereby promoting the growth of aquatic plants. At the beginning of the experiment, the local benthic animals were in the culture period, and the biomass showed an upward trend, which is consistent with Li et al. [46]. As crabs grew, the predation of benthic animals by crabs increased, and the biomass of the benthic animals decreased. The stress of predation by crabs caused *Branchiura* sp., *Limnodrilus* sp., and other benthic animals that reproduce via burrowing, to increase in numbers, causing an overall increase in the benthic animal biomass.

In this experiment, the crabs fed with high-protein compound feed were larger and had a stronger predation ability on benthic animals than the representative crabs. On 2 September, the biomass of benthic animals increased with the level of dietary protein. The biomass of benthic animals in the control group was significantly higher than that of the three experimental groups. Xu et al. [48] found that the stocking of crabs reduced the benthic animal diversity in the environment based on the lakes where crabs were stocked, outside the lake enclosure, in natural lake waters, and in lakes where fish were stocked. Benthic species diversity decreased significantly, and the production volume and density decreased by more than 60% compared with those of the control water body. The results vary from the results in our experiment, which may be caused by the different environments, sampling times, and stocking densities of the crabs. First, lake and paddy field environments are quite different. Second, unlike Xu et al.'s [48] experiment, which took four samples twice a year for two years, we monitored the dynamics of zoobenthos nine times over a period of nearly five months (from May to October). By comparing the culture densities, Xu et al. [48] showed that the over farming of crabs causes the high variations. Conversely, our experiment used normal crab culture density.

5. Conclusions

In this study, we evaluated how different levels of protein in crab feed could affect the performance of crabs and the environment in rice–crab co-culture in paddy fields. The results showed that feed with 15% protein level compound diet can not only meet the nutritional requirements of crabs but also reduce the cost of cultivation and improve the water quality of the paddy field. The discharged water had low ammonia nitrogen and nitrite content, and no eutrophication was observed. Consequently, the water could be recycled. These findings provide a scientific basis for feed formulation for juvenile crabs in rice–crab co-culture.

Author Contributions: Conceptualization, X.L. (Xiaodong Li), J.W. and Y.Y.; data curation, Y.Y.; formal analysis, Y.Y. and X.L. (Xiaochen Liang); funding acquisition, X.L. (Xiaodong Li); investigation, X.L. (Xiaochen Liang), Y.W., X.L. (Xueshen Liu), J.M. and Y.Y.; methodology, N.S. and Y.Y.; project administration, Y.Y. and N.S.; resources, X.L. (Xiaodong Li) and N.S.; supervision, X.L. (Xiaodong Li) and N.S.; validation, X.L. (Xiaodong Li); writing—original draft preparation, Y.Y.; writing—review and editing, X.L. (Xiaodong Li). All authors have read and agreed to the published version of the manuscript.

Funding: This research was funded by Liaoning Province Key R&D Planning Project (2019JH2/10200006), National Key R&D Program of China (2018YFD0901702), Liaoning Province Key R&D Guidance Program (201802120), Shenyang Agricultural University High Level Introduction of talent Program (880416005), and Liaoning Province "The Open Competition Mechanism to Select the Best Candidates" Project, Grant (2021JH1/10400040). The funders had no role in the design of the study; in the collection, analyses, or interpretation of data; in the writing of the manuscript, or in the decision to publish the results.

Institutional Review Board Statement: Our study did not involve endangered or protected species. In China, breeding and catching Chinese mitten crabs, *Eriocheir sinensis*, in rice fields does not require specific permits. All efforts were made to minimize animal suffering and discomfort. The animal study protocol was approved by the Animal Ethics Committee of Shenyang Agriculture University.

Data Availability Statement: The data presented in this study are not publicly available but are available upon request from the corresponding author.

Acknowledgments: The author would like to acknowledge my mentor Li for imparting knowledge to me and helping me revise the article carefully. We would also like to thank all employees of the Panjin Guanghe Crab Industry Co., Ltd. for patiently helping us and providing us with an excellent test environment. We thank teacher Hu, who is a teacher and a friend, for helping me to revise the article and overcome many difficulties. We are very grateful to Tian for his suggestions on the revision of the article and teaching us a lot of knowledge. Thanks to Jiang for helping me contact the company to polish the article. Thanks to my employees at the company's R and D center for their experimental help and guidance. The author would also like to acknowledge Zuo and Liu from Dalian Ocean University for their great help with the formula and production of the feed required for the experiment. Thanks also to Liu and Zheng who gave me practical guidance and help with the culture. I would also like to thank MSA Bi who provided guidance on testing techniques. Thanks to my junior Liang for assisting me in data processing and analysis.

Conflicts of Interest: The authors declare no conflict of interest.

Appendix A

Table A1. The cost of each experimental feed.

Unit Price/RMB yuan/ton	Cost/RMB yuan/ton		
	T15	**T30**	**T45**
7360	442	1987	3459
3380	101	456	794
3920	118	118	118
2623	1904	1130	396
15,000	1125	825	525
11,000	55	55	55
19,600	392	392	392
133,000	2660	2660	2660
30,000	300	300	300
44,000	220	220	220
65,600	131	131	131
7200	108	108	108
178,000	178	178	178
1,000,000	100	100	100
22,900	23	23	23
	7857	8684	9459

Table A2. Phytoplankton in each enclosure with the average wet weight.

Species	Average Wet Weight of Cells/mg
Bacillariophyta	
Nitzschia sp.	0.003
Cyclotella meneghiniana	0.00125
Nitzschia frustulum	0.006
Nitzschia acicularis	0.005
Synedra sp.	0.005
Chaetoceros sp.	0.0014
Navicula amphibola	0.0017
Navicula placentula	0.006
Melosira sp.	0.006
Coscinodisous sp.	0.02
Fragilaria sp.	0.001
Gyrosigma sp.	0.03
Mastogloia sp.	0.00325
Navicula directa	0.03
Amphora exigua	0.0017
Pleurosigma sp.	0.047
Eunotogramma sp.	0.001
Navicula sp.	0.001
Chtysophyta	
Chromulina pygmaea Nygaard	0.00065
Cyanophyta	
Chroococcus sp.	0.0001
Spirulina sp.	0.0077
Merismopedia sinica	0.00025
Oscillatoria sp.	0.01
Microcystis sp.	0.0016
Nostoc sp.	0.00025
Anabaena sp.	0.0005
Merismopedia sp.	0.00003
Phormidium sp.	0.002
Chlorophyta	
Oocystis borgei	0.004
Schroederi krosch	0.003

Table A2. *Cont.*

Species	Average Wet Weight of Cells/mg
Closterium sp.	0.08
Actinastrum lag sp.	0.001
Chlorella pyrenoidesa	0.00015
Selenastrum bibraianum	0.001
Chlamydomonas sp.	0.01
Pandorina morum	0.04
Dictyosphaerium sp.	0.001
Eudorina elegans	0.02
Tetraedron trilobulatum	0.003
Crucigenia sp.	0.001
Kirchneriellalunaris lunatis	0.001
Platymonas sp.	0.012
Spirogyra sp.	0.02
Ankistrodesmus convolutus	0.002
Pediastrum sp.	0.01
Sceneclesmus sp.	0.002
Euglenophyta	
Euglena sp.	0.04
Euglena oxyuris	0.15
Phacus sp.	0.06
Euglena viridis	0.04
Euglena pisciformis	0.15
Cryptophyta	
Cryptomonas sp.	0.01
Phrrophyta	
Gymmodinium sp.	0.008

Table A3. Phytoplankton in each enclosure with the average wet weight.

Species	Average Wet Weight of Cells/mg
Protozoa	
Sarcomastigophora	
Saccamoeba sp.	0.00003
Difflugia sp.	0.00003
Difflugia oblonga	0.00024
Tintinnidium fliuviatile	0.00024
Arcella vulgaris	0.00003
Globigerinoides sp.	0.00002
Ciliophora	
Coleps sp.	0.00003
Strobilidium sp.	0.00003
Litonotus sp.	0.00003
Tintinnopsis sp.	0.00003
Lembadion sp.	0.000016
Zoothamnium sp.	0.0000017
Euplotes sp.	0.000016
Pseudoprorodon sp.	0.00005
Didinium nasufum	0.00045
Actinobolina sp.	0.00003
Prorodon ovum	0.00005
Pleuronema sp.	0.0000017
Stentor polymorphrus	0.000002
Vorticella sp.	0.000014
Colpoda sp.	0.0000017
Trachelius sp.	0.000007
Rhabdostyla sp.	0.00005
Cladocera	
Daphnia carinata	0.2

Table A3. *Cont.*

Species	Average Wet Weight of Cells/mg
Daphnia magna	0.01
Moinidae brachiata	0.1
Moinidae rectirostris	0.01
Moinidae macrocopa	0.05
Chydoroidea quadrangula	0.01
Chydoroidea sphaericus	0.03
Chydoroidea longirostris	0.03
Copepoda	
Cyclops sp.	0.03
Copepodid	0.003
Calanoida sp.	0.312
Copepod nauplius	0.003
Rotifera	
Brachionus calyciflorus	0.0025
Brachionus ureus	0.00024
Brachionus quadridentatus	0.00055
Polyarthra trigla	0.000331
Polyarthra sp.	0.0025
Lecanidae inermis	0.026
Asplachna brightwelli	0.0005
Brachionus diversicornis	0.0005
Rotaria citrine	0.00028
Filinia sp.	0.0003
Lepadella ovalis	0.0003
Pedalia mira	0.000027
Keratella cochlearis	0.0003
Keratella valga	0.00024
Euchlanis pellucida	0.0025

Table A4. Species and biomass of the phytoplankton in each enclosure and grouped by crab diet ($n = 3$).

Species	Biomass/mg·L^{-1}			
	Co	T15	T30	T45
Bacillariophyta	16.64 ± 6.83	15.29 ± 6.16	13.46 ± 5.92	9.85 ± 5.59
Nitzschia sp.	1.33 ± 0.79	0.78 ± 0.44	0.78 ± 0.44	0.44 ± 0.36
Cyclotella meneghiniana	0.14 ± 0.24	0.23 ± 0.25	0.05 ± 0.08	0.00 ± 0.00
Nitzschia frustulum	1.11 ± 1.09	1.56 ± 1.5	2.00 ± 1.54	1.11 ± 1.09
Nitzschia acicularis	0.00 ± 0.00	0.00 ± 0.00	0.00 ± 0.00	0.00 ± 0.00
Synedra sp.	4.07 ± 3.90	1.67 ± 1.01	1.11 ± 0.69	0.19 ± 0.31
Chaetoceros sp.	0.00 ± 0.00	0.00 ± 0.00	0.05 ± 0.09	0.05 ± 0.09
Navicula amphibola	0.25 ± 0.25	0.06 ± 0.11	0.06 ± 0.11	0.00 ± 0.00
Navicula placentula	0.67 ± 0.83	0.22 ± 0.38	0.00 ± 0.00	0.00 ± 0.00
Melosira sp.	0.44 ± 0.52	0.00 ± 0.00	0.00 ± 0.00	0.89 ± 1.51
Coscinodisous sp.	0.00 ± 0.00	0.00 ± 0.00	0.00 ± 0.00	0.00 ± 0.00
Fragilaria sp.	0.11 ± 0.14	0.04 ± 0.06	0.00 ± 0.00	0.00 ± 0.00
Gyrosigma sp.	1.11 ± 1.89	0.00 ± 0.00	0.00 ± 0.00	0.00 ± 0.00
Mastogloia sp.	0.00 ± 0.00	0.00 ± 0.00	0.00 ± 0.00	0.24 ± 0.41
Navicula directa	3.89 ± 3.22	4.44 ± 3.55	7.78 ± 5.83	5.56 ± 5.47
Amphora exigua	0.06 ± 0.11	0.00 ± 0.00	0.00 ± 0.00	0.00 ± 0.00
Pleurosigma sp.	0.00 ± 0.00	5.22 ± 4.92	0.00 ± 0.00	0.00 ± 0.00
Eunotogramma sp.	0.78 ± 1.26	0.00 ± 0.00	0.00 ± 0.00	0.00 ± 0.00
Navicula sp.	1.63 ± 1.1	1.07 ± 0.77	1.63 ± 0.93	2.00 ± 0.96
Chrysophyta	47.47 ± 35.67	24.46 ± 19.63	15.74 ± 11.73	19.72 ± 9.42
Chromulina pygmaea Nygaard	47.47 ± 35.67	24.46 ± 19.63	15.74 ± 11.73	19.72 ± 9.42
Cyanophyta	16.1 ± 5.87	15.23 ± 5.37	11.37 ± 4.13	17.87 ± 6.72
Chroococcus sp.	0.25 ± 0.16	0.28 ± 0.14	0.28 ± 0.15	0.76 ± 0.55
Spirulina sp.	1.03 ± 1.33	0.57 ± 0.67	0.57 ± 0.67	0.29 ± 0.48

Table A4. *Cont.*

Species	Biomass/mg·L^{-1}			
	Co	**T15**	**T30**	**T45**
Merismopedia sinica	0.05 ± 0.06	0.02 ± 0.03	0.01 ± 0.02	0.12 ± 0.19
Oscillatoria sp.	10.00 ± 3.85	10.00 ± 4.06	7.78 ± 2.92	11.11 ± 4.66
Microcystis sp.	0.53 ± 0.48	1.13 ± 0.87	0.12 ± 0.14	0.47 ± 0.41
Nostoc sp.	0.17 ± 0.25	0.00 ± 0.00	0.00 ± 0.00	0.00 ± 0.00
Anabaena sp.	0.15 ± 0.17	0.06 ± 0.09	0.02 ± 0.03	0.00 ± 0.00
Merismopedia sp.	0.00 ± 0.01	0.00 ± 0.00	0.00 ± 0.00	0.01 ± 0.00
Phormidium sp.	3.93 ± 2.11	3.19 ± 1.34	2.59 ± 1.28	5.11 ± 3.35
Chlorophyta	42.28 ± 23.32	24.23 ± 12.36	21.59 ± 10.26	41.79 ± 23.33
Oocystis borgei	0.15 ± 0.25	0.3 ± 0.35	0.59 ± 0.79	0.3 ± 0.35
Schroederi krosch	0.32 ± 0.41	0.56 ± 0.47	0.00 ± 0.00	0.22 ± 0.26
Closterium sp.	2.96 ± 5.04	5.93 ± 6.98	2.96 ± 5.04	5.93 ± 6.98
Actinastrum lag sp.	0.00 ± 0.00	0.04 ± 0.06	0.04 ± 0.06	0.04 ± 0.06
Chlorella pyrenoidesa	7.25 ± 5.09	4.71 ± 2.34	5.56 ± 3.48	5.12 ± 2.42
Selenastrum bibraianum	0.63 ± 0.42	0.48 ± 0.26	0.78 ± 0.62	0.89 ± 0.83
Chlamydomonas sp.	14 ± 19.46	8.15 ± 7.70	1.11 ± 1.39	19.20 ± 2.62
Pandorina morum	4.44 ± 5.54	1.48 ± 2.52	7.41 ± 5.18	3.07 ± 0.15
Dictyosphaerium sp.	0.00 ± 0.00	0.00 ± 0.00	0.04 ± 0.06	0.04 ± 0.06
Eudorina elegans	8.89 ± 8.57	1.48 ± 1.75	1.48 ± 1.75	2.59 ± 2.81
Tetraedron trilobulatum	0.22 ± 0.26	0.22 ± 0.26	0.00 ± 0.00	0.33 ± 0.42
Crucigenia sp.	0.00 ± 0.00	0.00 ± 0.00	0.37 ± 0.63	0.37 ± 0.33
Kirchneriellalunaris lunatis	0.00 ± 0.00	0.00 ± 0.00	0.00 ± 0.00	0.04 ± 0.06
Platymonas sp.	0.89 ± 1.05	0.00 ± 0.00	0.00 ± 0.00	0.00 ± 0.00
Spirogyra sp.	0.00 ± 0.00	0.00 ± 0.00	0.00 ± 0.00	0.74 ± 1.26
Ankistrodesmus convolutus	0.67 ± 0.68	0.22 ± 0.28	0.15 ± 0.25	0.33 ± 0.35
Pediastrum sp.	0.37 ± 0.63	0.00 ± 0.00	0.00 ± 0.00	0.00 ± 0.00
Sceneclesmus sp.	1.41 ± 0.72	0.67 ± 0.51	1.11 ± 0.61	2.52 ± 1.87
Euglenophyta	136.60 ± 22.87	104.40 ± 1.94	77.78 ± 20.02	183.40 ± 30.75
Euglena sp.	33.33 ± 20.79	33.33 ± 24.85	33.33 ± 24.85	61.11 ± 12.33
Euglena oxyuris	16.67 ± 15.71	22.22 ± 22.38	22.22 ± 22.38	22.30 ± 17.75
Phacus sp.	6.67 ± 11.33	4.44 ± 5.24	0.00 ± 0.00	0.00 ± 0.00
Euglena viridis	68.89 ± 29.37	44.44 ± 19.04	22.22 ± 4.44	88.89 ± 22.84
Euglena pisciformis	11.11 ± 18.89	0.00 ± 0.00	0.00 ± 0.00	11.11 ± 18.89
Cryptophyta	5.19 ± 3.19	2.22 ± 1.39	1.11 ± 1.05	3.70 ± 3.03
Cryptomonas sp.	5.19 ± 3.19	2.22 ± 1.39	1.11 ± 1.05	3.7 ± 3.03
Pyrrophyta	0.00 ± 0.00	0.30 ± 0.00.5	0.00 ± 0.00	0.00 ± 0.00
Gymmodinium sp.	0.00 ± 0.00	0.30 ± 0.5	0.00 ± 0.00	0.00 ± 0.00
Total biomass	160.7	108.4	85.49	127.1

Note: 0.00 mg·L^{-1} is provided when the average biomass is less than 0.005 mg·L^{-1} or not detected.

Table A5. Composition of the predominant species of phytoplankton in the enclosures of each treatment group (*n* = 3).

Date	Group	Dominant Species (Degree of Dominance)
05-25	Co	*Chromulina pygmaea* Nygaard (0.34) *Chlorella pyrenoidesa* (0.26)
	T15	*Navicula* sp. (0.03) *Chromulina pygmaea* Nygaard (0.21) *Chlorella pyrenoidesa* (0.17)
	T30	*Chromulina pygmaea* Nygaard (0.27) *Chlorella pyrenoidesa* (0.41)
	T45	*Navicula* sp. (0.02) *Chromulina pygmaea* Nygaard (0.29) *Chlorella pyrenoidesa* (0.33)
06-30	Co	*Chromulina pygmaea* Nygaard (0.31) *Chlorella pyrenoidesa* (0.32)
	T15	*Chromulina pygmaea* Nygaard (0.17) *Chroococcus* sp. (0.14) *Phormidium* sp. (0.03) *Chlorella pyrenoidesa* (0.13)
	T30	*Navicula* sp. (0.02) *Chromulina pygmaea* Nygaard (0.27) *Chroococcus* sp. (0.14) *Phormidium* sp. (0.03) *Chlorella pyrenoidesa* (0.17) *Selenastrum bibraianum* (0.02)
	T45	*Chromulina pygmaea* Nygaard (0.26) *Chroococcus* sp. (0.15) *Phormidium* sp. (0.05) *Chlorella pyrenoidesa* (0.19) *Chlamydomonas* sp. (0.02) *Sceneclesmus* sp. (0.02)
07-15	Co	*Navicula* sp. (0.04) *Oscillatoria* sp. (0.05) *Microcystis* sp. (0.02) *Phormidium* sp. (0.06) *Chlorella pyrenoidesa* (0.53)
	T15	*Oscillatoria* sp. (0.11) *Microcystis* sp. (0.09) *Phormidium* sp. (0.07) *Chlorella pyrenoidesa* (0.30)

Table A5. *Cont.*

Date	Group	Dominant Species (Degree of Dominance)
07-15	T30	*Navicula* sp. (0.03) *Chroococcus* sp. (0.05) *Phormidium* sp. (0.04) *Chlorella pyrenoidesa* (0.69)
	T45	*Navicula* sp. (0.07) *Chromulina pygmaea* Nygaard (0.12) *Oscillatoria* sp. (0.05) *Phormidium* sp. (0.06) *Chlorella pyrenoidesa* (0.35)
07-29	Co	*Chromulina pygmaea* Nygaard (0.17) *Oscillatoria* sp. (0.03) *Chlorella pyrenoidesa* (0.06) *Ankistrodesmus convolutus* (0.08)
	T15	*Chromulina pygmaea* Nygaard (0.38) *Oscillatoria* sp. (0.03) *Phormidium* sp. (0.05) *Chlorella pyrenoidesa* (0.03)
	T30	*Navicula directa* (0.04) *Navicula* sp. (0.03) *Chromulina pygmaea* Nygaard (0.19) *Oscillatoria* sp. (0.08)
	T45	*Chromulina pygmaea* Nygaard (0.30) *Chlorella pyrenoidesa* (0.31)
08-15	Co	*Chromulina pygmaea* Nygaard (0.30) *Chlorella pyrenoidesa* (0.07)
	T15	*Chromulina pygmaea* Nygaard (0.38) *Chlorella pyrenoidesa* (0.22)
	T30	*Chromulina pygmaea* Nygaard (0.11) *Chroococcus* sp. (0.05) *Oscillatoria* sp. (0.03) *Phormidium* sp. (0.04) *Chlorella pyrenoidesa* (0.04)
	T45	*Chromulina pygmaea* Nygaard (0.31) *Oscillatoria* sp. (0.04) *Chlorella pyrenoidesa* (0.33)
09-02	Co	*Nitzschia* sp. (0.04) *Chromulina pygmaea* Nygaard (0.05) *Chroococcus* sp. (0.15) *Chlorella pyrenoidesa* (0.11) *Sceneclesmus* sp. (0.06)
	T15	*Chromulina pygmaea* Nygaard (0.25) *Chlorella pyrenoidesa* (0.25)
	T30	*Chromulina pygmaea* Nygaard (0.14) *Chlorella pyrenoidesa* (0.48)
	T45	*Chromulina pygmaea* Nygaard (0.03) *Chroococcus* sp. (0.04) *Oscillatoria* sp. (0.03) *Chlorella pyrenoidesa* (0.64)
09-15	Co	*Chromulina pygmaea* Nygaard (0.17) *Schroederi krosch* (0.02) *Pandorina morum* (0.02) *Eudorina elegans* (0.03) *Cryptomonas* sp. (0.13)
	T15	*Chromulina pygmaea* Nygaard (0.21) *Chlorella pyrenoidesa* (0.61)
	T30	*Synedra* sp. (0.03) *Chromulina pygmaea* Nygaard (0.35) *Chlorella pyrenoidesa* (0.19)
	T45	*Chromulina pygmaea* Nygaard (0.24) *Chlorella pyrenoidesa* (0.53) *Sceneclesmus* sp. (0.04)
09-26	Co	*Chromulina pygmaea* Nygaard (0.17) *Chlorella pyrenoidesa* (0.68)
	T15	*Chromulina pygmaea* Nygaard (0.05) *Chlorella pyrenoidesa* (0.87)
	T30	*Chlorella pyrenoidesa* (0.51) *Crucigenia* sp. (0.04)
	T45	*Chromulina pygmaea* Nygaard (0.18) *Chroococcus* sp. (0.07) *Chlorella pyrenoidesa* (0.31) *Crucigenia* sp. (0.06)
10-07	Co	*Synedra* sp. (0.02) *Navicula amphibola* (0.02) *Anabaena* sp. (0.10) *Selenastrum bibraianum* (0.02) *Eudorina elegans* (0.10)
	T15	*Chromulina pygmaea* Nygaard (0.15) *Chroococcus* sp. (0.10) *Chlorella pyrenoidesa* (0.41)
	T30	*Chlorella pyrenoidesa* (0.56)
	T45	*Navicula* sp. (0.02) *Chromulina pygmaea* Nygaard (0.11) *Chroococcus* sp. (0.08) *Chlorella pyrenoidesa* (0.55)

Table A6. Biomass and species of zooplankton in the enclosures of each treatment group ($n = 3$).

Species	Biomass/mg/L			
	Co	T15	T30	T45
Protozoa	1.75 ± 0.47	2.34 ± 1.02	1.36 ± 0.48	1.63 ± 0.47
Sarcomastigophora	1.52 ± 0.47	2.09 ± 1.01	1.25 ± 0.47	1.47 ± 0.48
Saccamoeba sp.	0.04 ± 0.03	0.05 ± 0.06	0.03 ± 0.02	0.02 ± 0.01
Difflugia sp.	0.13 ± 0.05	0.15 ± 0.08	0.13 ± 0.05	0.11 ± 0.05
Difflugia oblonga	0.37 ± 0.23	0.82 ± 0.61	0.41 ± 0.30	0.44 ± 0.23
Tintinnidium fliuviatile	0.97 ± 0.39	1.07 ± 0.49	0.68 ± 0.38	0.00 ± 0.37
Arcella vulgaris	0.00 ± 0.00	0.00 ± 0.00	0.00 ± 0.00	0.00 ± 0.00
Globigerinoides sp.	0.00 ± 0.00	0.00 ± 0.00	0.00 ± 0.00	0.00 ± 0.00
Ciliophora	0.24 ± 0.1	0.25 ± 0.21	0.11 ± 0.04	0.15 ± 0.05
Coleps sp.	0.08 ± 0.06	0.07 ± 0.06	0.04 ± 0.02	0.05 ± 0.03
Strobilidium sp.	0.03 ± 0.02	0.08 ± 0.09	0.04 ± 0.02	0.06 ± 0.03
Litonotus sp.	0.00 ± 0.00	0.00 ± 0.00	0.00 ± 0.01	0.01 ± 0.01
Tintinnopsis sp.	0.01 ± 0.01	0.05 ± 0.06	0.01 ± 0.02	0.00 ± 0.00
Lembadion sp.	0.06 ± 0.05	0.03 ± 0.02	0.01 ± 0.01	0.02 ± 0.01
Zoothamnium sp.	0.00 ± 0.00	0.00 ± 0.00	0.00 ± 0.00	0.00 ± 0.00
Euplotes sp.	0.00 ± 0.00	0.01 ± 0.01	0.00 ± 0.00	0.00 ± 0.00
Pseudoprorodon sp.	0.00 ± 0.00	0.00 ± 0.00	0.00 ± 0.01	0.01 ± 0.01
Didinium nasufum	0.03 ± 0.06	0.00 ± 0.00	0.00 ± 0.00	0.00 ± 0.00

Table A6. *Cont.*

Species	Biomass/mg/L			
	Co	**T15**	**T30**	**T45**
Actinobolina sp.	0.00 ± 0.00	0.00 ± 0.00	0.00 ± 0.00	0.00 ± 0.00
Prorodon ovum	0.01 ± 0.01	0.00 ± 0.00	0.00 ± 0.00	0.00 ± 0.00
Pleuronema sp.	0.00 ± 0.00	0.00 ± 0.00	0.00 ± 0.00	0.00 ± 0.00
Stentor polymorphrus	0.00 ± 0.00	0.00 ± 0.00	0.00 ± 0.00	0.00 ± 0.00
Vorticella sp.	0.00 ± 0.00	0.00 ± 0.00	0.00 ± 0.00	0.00 ± 0.00
Colpoda sp.	0.00 ± 0.00	0.00 ± 0.00	0.00 ± 0.00	0.00 ± 0.00
Trachelius sp.	0.00 ± 0.00	0.01 ± 0.00	0.00 ± 0.00	0.01 ± 0.01
Rhabdostyla sp.	0.00 ± 0.00	0.00 ± 0.00	0.00 ± 0.00	0.00 ± 0.00
Cladocera	22.16 ± 4.57	25.30 ± 5.77	29.19 ± 5.35	40.52 ± 6.71
Daphnia carinata	1.48 ± 2.52	8.89 ± 2.74	2.30 ± 0.44	0.74 ± 1.26
Daphnia magna	6.67 ± 1.33	0.00 ± 0.00	6.67 ± 1.33	16.67 ± 18.32
Moinidae brachiata	2.30 ± 0.78	1.52 ± 0.66	3.56 ± 1.73	2.52 ± 1.32
Moinidae rectirostris	5.93 ± 1.4	5.93 ± 0.21	4.44 ± 3.31	9.63 ± 3.05
Moinidae macrocopa	1.19 ± 0.42	0.52 ± 0.29	0.96 ± 0.12	0.44 ± 0.19
Chydoroidea quadrangula	2.59 ± 0.62	7.04 ± 0.46	9.26 ± 0.84	7.78 ± 0.61
Chydoroidea sphaericus	0.00 ± 0.00	0.07 ± 0.13	0.00 ± 0.00	0.07 ± 0.13
Chydoroidea longirostris	2.00 ± 1.89	1.33 ± 1.37	2.00 ± 2.31	2.67 ± 2.39
Copepoda	27.00 ± 5.19	29.65 ± 2.88	39.10 ± 3.27	30.70 ± 1.25
Cyclops sp.	8.44 ± 2.51	9.67 ± 0.22	10.89 ± 1.8	13.11 ± 2.83
Copepodid	0.60 ± 0.50	0.30 ± 0.17	0.30 ± 0.17	0.22 ± 0.14
Calanoida sp.	13.87 ± 1.78	16.18 ± 12.13	25.42 ± 0.86	13.87 ± 1.78
Copepod nauplius	4.09 ± 0.02	3.50 ± 0.04	2.49 ± 0.29	3.50 ± 0.87
Rotifera	0.70 ± 0.70	0.48 ± 0.29	0.21 ± 0.09	0.86 ± 0.72
Brachionus calyciflorus	0.07 ± 0.10	0.15 ± 0.16	0.04 ± 0.04	0.05 ± 0.06
Brachionus ureus	0.03 ± 0.02	0.10 ± 0.09	0.03 ± 0.03	0.05 ± 0.03
Brachionus quadridentatus	0.00 ± 0.00	0.00 ± 0.00	0.00 ± 0.00	0.00 ± 0.00
Polyarthra trigla	0.16 ± 0.09	0.16 ± 0.14	0.06 ± 0.04	0.11 ± 0.08
Polyarthra sp.	0.00 ± 0.00	0.00 ± 0.00	0.00 ± 0.00	0.00 ± 0.00
Lecanidae inermis	0.00 ± 0.00	0.00 ± 0.00	0.02 ± 0.03	0.00 ± 0.00
Asplachna brightwelli	0.39 ± 0.65	0.00 ± 0.00	0.00 ± 0.00	0.58 ± 0.72
Brachionus diversicornis	0.01 ± 0.01	0.01 ± 0.02	0.00 ± 0.00	0.02 ± 0.04
Rotaria citrine	0.01 ± 0.01	0.03 ± 0.02	0.02 ± 0.02	0.04 ± 0.03
Filinia sp.	0.01 ± 0.01	0.00 ± 0.00	0.01 ± 0.01	0.00 ± 0.00
Lepadella ovalis	0.01 ± 0.01	0.02 ± 0.02	0.01 ± 0.01	0.01 ± 0.01
Pedalia mira	0.00 ± 0.00	0.00 ± 0.00	0.00 ± 0.00	0.00 ± 0.00
Keratella cochlearis	0.00 ± 0.00	0.00 ± 0.00	0.00 ± 0.00	0.00 ± 0.00
Keratella valga	0.00 ± 0.00	0.00 ± 0.00	0.00 ± 0.00	0.00 ± 0.00
Euchlanis pellucida	0.00 ± 0.00	0.00 ± 0.00	0.02 ± 0.03	0.00 ± 0.00
Total biomass	51.59	57.77	69.85	73.71

Note: 0.00 mg·L^{-1} is the term used when the average biomass is less than 0.005 mg·L^{-1} or is not detected.

Table A7. Composition of the predominant zooplankton species in the enclosures of each treatment group ($n = 3$).

Date	Group	Dominant Species (Degree of Dominance)
05-25	Co	*Strobilidium* sp. (0.26) *Tintinnopsis* sp. (0.03) *Difflugia* sp. (0.06) *Saccamoeba* sp. (0.05) *Copepod nauplius* (0.06) *Brachionus calyciflorus* (0.05) *Brachionus ureus* (0.23)
	T15	*Coleps* sp. (0.09) *Strobilidium* sp. (0.25) *Tintinnopsis* sp. (0.15) *Saccamoeba* sp. (0.05) *Brachionus calyciflorus* (0.04) *Brachionus ureus* (0.12)
	T30	*Strobilidium* sp. (0.18) *Daphnia carinata* (0.13) *Brachionus ureus* (0.18)
	T45	*Coleps* sp. (0.07) *Strobilidium* sp. (0.25) *Tintinnidium fliuviatile* (0.04) *Daphnia magna* (0.06) *Brachionus ureus* (0.10)
06-30	Co	*Tintinnidium fliuviatile* (0.13) *Moinidae brachiata* (0.06) *Cyclops* sp. (0.02) *Copepod nauplius* (0.22) *Polyarthra trigla* (0.13) *Brachionus diversicornis* (0.02) *Rotaria citrine* (0.02)
	T15	*Strobilidium* sp. (0.02) *Difflugia* sp. (0.18) *Tintinnidium fliuviatile* (0.12) *Zoothamnium* sp. (0.02) *Stentor polymorphrus* (0.02) *Moinidae brachiata* (0.06) *Chydoroidea quadrangula* (0.04) *Copepod nauplius* (0.12) *Brachionus ureus* (0.02) *Polyarthra trigla* (0.17) *Rotaria citrine* (0.03)

Table A7. *Cont.*

Date	Group	Dominant Species (Degree of Dominance)
06-30	T30	*Tintinnopsis* sp. (0.03) *Difflugia* sp. (0.20) *Tintinnidium fliuviatile* (0.17) *Moinidae brachiata* (0.07) *Moinidae macrocopa* (0.03) *Chydoroidea quadrangula* (0.04) *Copepod nauplius* (0.08) *Brachionus ureus* (0.08) *Polyarthra trigla* (0.09)
	T45	*Strobilidium* sp. (0.12) *Difflugia* sp. (0.12) *Tintinnidium fliuviatile* (0.13) *Moinidae brachiata* (0.05) *Copepod nauplius* (0.11) *Brachionus ureus* (0.09) *Polyarthra trigla* (0.21) *Brachionus diversicornis* (0.02)
07-15	Co	*Strobilidium* sp. (0.03) *Difflugia* sp. (0.09) *Tintinnidium fliuviatile* (0.13) *Moinidae brachiata* (0.08) *Moinidae macrocopa* (0.05) *Cyclops* sp. (0.03) *Copepod nauplius* (0.30) *Polyarthra trigla* (0.03)
	T15	*Tintinnopsis* sp. (0.03) *Difflugia* sp. (0.10) *Tintinnidium fliuviatile* (0.09) *Moinidae brachiata* (0.06) *Cyclops* sp. (0.03) *Chydoroidea quadrangula* (0.03) *Copepod nauplius* (0.19) *Polyarthra trigla* (0.13) *Lepadella ovalis* (0.08)
	T30	*Difflugia* sp. (0.21) *Tintinnidium fliuviatile* (0.15) *Moinidae brachiata* (0.11) *Cyclops* sp. (0.04) *Moinidae macrocopa* (0.07) *Chydoroidea quadrangula* (0.08) *Chydoroidea longirostris* (0.06) *Copepod nauplius* (0.05)
	T45	*Difflugia* sp. (0.15) *Difflugia oblonga* (0.04) *Tintinnidium fliuviatile* (0.20) *Cyclops* sp. (0.09) *Moinidae brachiata* (0.17) *Moinidae macrocopa* (0.03) *Chydoroidea quadrangula* (0.04) *Chydoroidea longirostris* (0.05)
07-29	Co	*Difflugia* sp. (0.29) *Tintinnidium fliuviatile* (0.07) *Moinidae brachiata* (0.09) *Cyclops* sp. (0.06) *Chydoroidea longirostris* (0.05) *Copepod nauplius* (0.23)
	T15	*Difflugia* sp. (0.25) *Difflugia oblonga* (0.08) *Tintinnidium fliuviatile* (0.14) *Cyclops* sp. (0.04) *Moinidae rectirostris* (0.03) *Chydoroidea longirostris* (0.03) *Copepod nauplius* (0.05) *Rotaria citrine* (0.03)
	T30	*Difflugia* sp. (0.19) *Tintinnidium fliuviatile* (0.05) *Saccamoeba* sp. (0.10) *Moinidae brachiata* (0.08) *Cyclops* sp. (0.04) *Copepod nauplius* (0.15) *Polyarthra trigla* (0.02) *Filinia* sp. (0.02)
	T45	*Rotaria citrine* (0.11) *Polyarthra trigla* (0.04) *Copepod nauplius* (0.11) *Chydoroidea quadrangula* (0.09) *Cyclops* sp. (0.09) *Moinidae brachiata* (0.05) *Tintinnidium fliuviatile* (0.03) *Difflugia oblonga* (0.09) *Difflugia* sp. (0.24)
08-15	Co	*Coleps* sp. (0.05) *Difflugia* sp. (0.05) *Tintinnidium fliuviatile* (0.03) *Pleuronema* sp. (0.03) *Trachelius* sp. (0.04) *Saccamoeba* sp. (0.04) *Moinidae brachiata* (0.04) *Moinidae macrocopa* (0.02) *Cyclops* sp. (0.02) *Copepod nauplius* (0.21)
	T15	*Copepod nauplius* (0.13) *Trachelius* sp. (0.22) *Difflugia* sp. (0.03) *Coleps* sp. (0.13)
	T30	*Coleps* sp. (0.15) *Strobilidium* sp. (0.03) *Difflugia* sp. (0.09) *Trachelius* sp. (0.06) *Rhabdostyla* sp. (0.03) *Moinidae brachiata* (0.11) *Cyclops* sp. (0.09) *Copepod nauplius* (0.15)
	T45	*Coleps* sp. (0.02) *Difflugia* sp. (0.05) *Trachelius* sp. (0.05) *Saccamoeba* sp. (0.02) *Cyclops* sp. (0.07) *Copepod nauplius* (0.23)
09-02	Co	*Coleps* sp. (0.03) *Difflugia* sp. (0.03) *Tintinnidium fliuviatile* (0.08) *Lembadion* sp. (0.07) *Zoothamnium* sp. (0.02) *Saccamoeba* sp. (0.02) *Cyclops* sp. (0.05) *Copepod nauplius* (0.12) *Polyarthra trigla* (0.03)
	T15	*Copepod nauplius* (0.20) *Cyclops* sp. (0.07) *Lembadion* sp. (0.20) *Difflugia oblonga* (0.15)
	T30	*Difflugia* sp. (0.03) *Difflugia oblonga* (0.07) *Cyclops* sp. (0.03) *Copepod nauplius* (0.30)
	T45	*Copepod nauplius* (0.12) *Moinidae rectirostris* (0.04) *Lembadion* sp. (0.08) *Tintinnidium fliuviatile* (0.05) *Difflugia oblonga* (0.03) *Strobilidium* sp. (0.08) *Coleps* sp. (0.05)
09-15	Co	*Coleps* sp. (0.08) *Difflugia oblonga* (0.05) *Tintinnidium fliuviatile* (0.05) *Zoothamnium* sp. (0.04) *Cyclops* sp. (0.02) *Copepod nauplius* (0.34) *Polyarthra trigla* (0.14)
	T15	*Copepod nauplius* (0.42) *Cyclops* sp. (0.06) *Moinidae brachiata* (0.03) *Difflugia oblonga* (0.07) *Coleps* sp. (0.06)
	T30	*Coleps* sp. (0.12) *Difflugia oblonga* (0.13) *Moinidae rectirostris* (0.02) *Cyclops* sp. (0.04) *Chydoroidea quadrangula* (0.09) *Copepod nauplius* (0.35)
	T45	*Copepod nauplius* (0.42) *Chydoroidea quadrangula* (0.06) *Zoothamnium* sp. (0.05) *Tintinnidium fliuviatile* (0.03)
09-26	Co	*Difflugia* sp. (0.07) *Difflugia oblonga* (0.03) *Lembadion* sp. (0.32) *Cyclops* sp. (0.07) *Calanoida* sp. (0.03) *Copepod nauplius* (0.36)
	T15	*Copepod nauplius* (0.58) *Calanoida* sp. (0.02) *Chydoroidea quadrangula* (0.02) *Cyclops* sp. (0.06) *Lembadion* sp. (0.08)
	T30	*Difflugia* sp. (0.03) *Difflugia oblonga* (0.06) *Lembadion* sp. (0.10) *Cyclops* sp. (0.08) *Chydoroidea quadrangula* (0.05) *Copepod nauplius* (0.43)
	T45	*Brachionus ureus* (0.02) *Copepod nauplius* (0.39) *Chydoroidea quadrangula* (0.03) *Cyclops* sp. (0.11) *Lembadion* sp. (0.03) *Difflugia oblonga* (0.09)
10-07	Co	*Coleps* sp. (0.06) *Moinidae rectirostris* (0.04) *Cyclops* sp. (0.05) *Copepod nauplius* (0.47) *Copepodid* (0.02) *Polyarthra trigla* (0.03)
	T15	*Moinidae brachiata* (0.02) *Moinidae rectirostris* (0.04) *Cyclops* sp. (0.08) *Copepod nauplius* (0.61)
	T30	*Coleps* sp. (0.03) *Cyclops* sp. (0.17) *Calanoida* sp. (0.06) *Copepod nauplius* (0.38)
	T45	*Coleps* sp. (0.05) *Moinidae brachiata* (0.03) *Cyclops* sp. (0.12) *Copepod nauplius* (0.52) *Asplachna brightwelli* (0.02)

Table A8. Species and biomass of benthos in the enclosures of each treatment group ($n = 3$; $x \pm$ SD).

Species	Biomass/g·m^{-2}			
	Co	T15	T30	T45
Annelida	205.54 ± 438.17	290.72 ± 465.51	71.53 ± 208.79	124.92 ± 184.82
Oligochaeta	205.54 ± 438.17	290.72 ± 465.51	71.53 ± 208.79	124.92 ± 184.82

Table A8. *Cont.*

Species	Biomass/g·m^{-2}			
	Co	**T15**	**T30**	**T45**
Limnodrilus sp.	18.14 ± 26.03	18.14 ± 32.55	11.91 ± 19.69	5.67 ± 10.33
Branchiura sp.	187.4 ± 424.71	272.58 ± 451.54	59.63 ± 193.87	119.25 ± 181.1
Mollusca	391.63 ± 174.88	383.92 ± 338.97	277.43 ± 153.15	300.6 ± 134.79
Gastropoda	391.63 ± 174.88	383.92 ± 338.97	277.43 ± 153.15	300.6 ± 134.79
Gyraulus sp.	271.18 ± 119.68	244.36 ± 141.38	208.6 ± 107.14	214.56 ± 106.47
Euconulus sp.	120.45 ± 115	139.57 ± 274.88	68.83 ± 83.12	86.04 ± 107
Arthropoda	0.00 ± 0.00	0.7 ± 3.55	0.35 ± 1.78	0.37 ± 1.78
Insecta	0.00 ± 0.00	0.7 ± 3.55	0.35 ± 1.78	0.37 ± 1.78
Ephydra sp.	0.00 ± 0.00	0.7 ± 3.55	0.35 ± 1.78	0.35 ± 1.78
Corixa substriata	0.00 ± 0.00	0.00 ± 0.00	0.00 ± 0.00	0.02 ± 0.1

References

1. Bao, J.; Jiang, H.; Li, X. Thirty years of rice-crab coculture in China—Research progress and prospects. *Rev. Aquac.* **2022**, *14*, 1597–1612. [CrossRef]
2. Jia, E.T.; Zheng, X.; Cheng, H.; Jie, L.; Zhang, D. Dietary fructooligosaccharide can mitigate the negative effects of immunity on Chinese mitten crab fed a high level of plant protein diet. *Fish Shellfish. Immunol.* **2019**, *84*, 100–107. [CrossRef] [PubMed]
3. Chen, F.; Zhang, Z. Ecological economic analysis of a rice-crab model. *Chin. J. Appl. Ecol.* **2002**, *13*, 323–326. (In Chinese)
4. Bashir, M.A.; Liu, J.; Geng, Y.; Wang, H.; Pan, J.; Zhang, D.; Rehim, A.; Aon, M.; Liu, H. Co-culture of rice and aquatic animals: An integrated system to achieve production and environmental sustainability. *J. Clean. Prod.* **2020**, *249*, 119310. [CrossRef]
5. Zhang, Q.; Lv, D.; Ma, X.; Wang, A.; Wang, W. The effects of different transplanting ways on the phytoplankton community in rice-crab culture system. *J. Food Agric. Environ.* **2014**, *12*, 199–204.
6. Zhang, C. In Partial Fulfillment of the Requirement for the Degree of Master in Hydrobiology. Master's Thesis, Dalian Ocean University, Dalian, China, 2015. (In Chinese).
7. Lunger, A.N.; Mclean, E.; Craig, S.R. The effects of organic protein supplementation upon growth, feed conversion and texture quality parameters of juvenile cobia (*Rachycentron canadum*). *Aquaculture* **2007**, *264*, 342–352. [CrossRef]
8. Lin, S.; Luo, L.; Ye, Y. Effects of dietary protein level on growth, feed utilization and digestive enzyme activity of the Chinese mitten crab, *Eriocheir sinensis*. *Aquac. Nutr.* **2010**, *16*, 290–298. [CrossRef]
9. Xu, C.; Wen, L.; Zhang, D.; Liu, J.; Zheng, X.; Zhang, C.; Yao, J.; Zhu, C.; Chi, C. Effects of partial fish meal replacement with two fermented soybean meals on the growth of and protein metabolism in the Chinese mitten crab (*Eriocheir sinensis*). *Aquac. Rep.* **2020**, *17*, 100328. [CrossRef]
10. Ministry of Agriculture and Rural Development of China. Development report of China's rice and fishery comprehensive planting and breeding industry. *China Fish.* **2020**, *10*, 12–19. (In Chinese)
11. Lei, Y. *Environmental Chemistry of Aquaculture Water (for Aquaculture Specialty)*, 1st ed.; China Agricultural Publishing House: Beijing, China, 2004; pp. 67–108. (In Chinese)
12. Fu, L.; Zhou, G.; Li, Y.; Lu, Q.; Pan, J. Research progress on precocious puberty of Chinese mitten crab. *Jiangsu Agric. Sci.* **2017**, *45*, 5. (In Chinese)
13. Li, H.; Sun, C.; Zhang, Q.; Zhang, P.; Bao, H.; Ren, H.; Yao, X.; Zhong, W.; Cui, K.; You, H.; et al. *Study on the Composition of Plankton Species in Tianjin Area*; Tianjin Aquatic Products Technology Promotion Statio: Tianjin, China, 2017. (In Chinese)
14. Zhao, W. *Aquatic Biology*, 1st ed.; China Agricultural Publishing House: Beijing, China, 2005; pp. 62–365. (In Chinese)
15. Li, Y.; Ding, J.; Xu, F. On the distribution and hatching of resting eggs of rotifer in fishponds. *Acta Hydrobiol. Sin.* **1985**, *9*, 20–31. (In Chinese)
16. Cheng, Y.X.; Du, N.S.; Lai, W. Lipid composition in hepatopancreas of Chinese mitten crab *Eriocheir sinensis* at different stages. *Acta Hydrobiol. Sin.* **1998**, *44*, 420–429.
17. Bao, J.; Jiang, H.; Shen, H.; Xing, Y.; Chen, Q. First description of milky disease in the Chinese mitten crab Eriocheir sinensis caused by the yeast *Metschnikowia bicuspidata*. *Aquaculture* **2021**, *532*, 735984. [CrossRef]
18. Prein, M. Integration of aquaculture into crop–animal systems in Asia. *Agric. Syst.* **2002**, *71*, 127–146. [CrossRef]
19. Hu, L.; Guo, L.; Zhao, L.; Shi, X.; Ren, W.; Zhang, J.; Tang, J.; Chen, X. Productivity and the complementary use of nitrogen in the coupled rice-crab system. *Agric. Syst.* **2020**, *178*, 102742. [CrossRef]
20. Chen, X.; Wang, J.; Xin, H.; Yue, W.; Shu, H. Tissue-expression profiles unveils the gene interaction of hepatopancreas, eyestalk, and ovary in precocious female Chinese mitten crab, *Eriocheir sinensis*. *BMC Genet.* **2018**, *20*, 12. [CrossRef]
21. Zhang, T.; Li, Z.; Cui, Y. Survival, Growth, Sex Ratio, and Maturity of the Chinese Mitten Crab (*Eriocheir sinensis*) Reared in a Chinese Pond. *J. Freshw. Ecol.* **2001**, *16*, 633–640. [CrossRef]
22. Guo, Q.; Li, W.; Liu, D.; Wu, W.; Liu, Y.; Wen, X.; Liao, Y. Seasonal characteristics of CO_2 fluxes in a rain-fed wheat field ecosystem at the Loess Plateau. *Span. J. Agric. Res.* **2013**, *11*, 980–988. [CrossRef]

23. Mb, A.; Mma, B.; Mk, C.; Ph, A.; Bta, B.; Lpa, B. Dissolved oxygen deficits in a shallow eutrophic aquatic ecosystem (fishpond)-sediment oxygen demand and water column respiration alternately drive the oxygen regime. *Sci. Total Environ.* **2020**, *766*, 142647.
24. Zhang, Y.J.; Wang, A.; Ma, X.; Wang, W.; Li, Y. Preliminary study on level changes of water quality in rice-crab culture. *Guangdong Agric. Sci.* **2013**, *40*, 16–19.
25. Wang, A.; Wang, W.; Ma, X.; Wang, Q.; Yu, Y.; Chen, W. Study on the changes of water nnvironmental factors in rice-crab culture system. *Hubei Agric. Sci.* **2011**, *50*, 6. (In Chinese)
26. Lobo-Araújo, L.W.; Toledo, M.; Efe, M.; Malhado, A.; Vital, M.; Toledo-lima, G.; Macario, P.; Santos, J.; Ladle, R. Bird communities in three forest types in the Pernambuco Centre of Endemism, Alagoas, Brazil. *Iheringia Série Zool.* **2013**, *103*, 85–96. [CrossRef]
27. Jian, X.; Hu, L.; Tang, J.; Wu, X.; Li, N. Ecological mechanisms underlying the sustainability of the agricultural heritage rice-fish coculture system. *Proc. Natl. Acad. Sci. USA* **2011**, *108*, E1381–E1387.
28. Frei, M.; Khan, M.; Razzak, M.; Hossain, M.; Dewan, S.; Becker, K. Effects of a mixed culture of common carp, *Cyprinus carpio* L.; Nile tilapia, *Oreochromis niloticus* (L.), on terrestrial arthropod population, benthic fauna, and weed biomass in rice fields in Bangladesh. *Biol. Control.* **2007**, *41*, 207–213. [CrossRef]
29. Altieri, M.A. Linking ecologists and traditional farmers in the search for sustainable agriculture. *Front. Ecol. Environ.* **2004**, *2*, 35–42. [CrossRef]
30. Ren, W.; Hu, L.; Guo, L.; Zhang, J.; Tang, L. Preservation of the genetic diversity of a local common carp in the agricultural heritage rice-fish system. *Proc. Natl. Acad. Sci. USA* **2018**, *115*, E546–E554. [CrossRef]
31. Lassen, M.F.; Bramm, M.; Richardson, K.; Yusoff, F.; Shariff, M. Phytoplankton community composition and size distribution in the Langat River estuary, Malaysia. *Estuaries* **2004**, *27*, 716–727. [CrossRef]
32. Sterner, R.W.; Chrzanowski, T.; Elser, J.; George, N. Sources of nitrogen and phosphorus supporting the growth of bacteria and phytoplankton in an oligotrophic Canadian shield lake. *Limnol. Oceanogr.* **1995**, *40*, 242–249. [CrossRef]
33. Odum, W.E. *Ecological Guidelines for Tropical Coastal Development*; IUCN Publications New Series. No. 42; IUCN: Gland, Switzerland, 1976.
34. Cabecinha, E.; Cortes, R.; Cabral, J.; Ferreira, T.; Lourenço, M.; Pardal, M. Multi-scale approach using phytoplankton as a first step towards the definition of the ecological status of reservoirs. *Ecol. Indic.* **2009**, *9*, 240–255. [CrossRef]
35. Sun, W.T.; Zhang, Q.; Xu-Zhou, M.; Wang, W.; Wang, A. A study on effects of different crab stocking density on water environment and rice yield. *J. Shanghai Ocean. Univ.* **2014**, *23*, 366–373.
36. Qingtian, L.I.; Wang, H.; Zheng, L.I. The biodiversity and dynamics of oribatid mites in rice paddies. *Syst. Appl. Acarol.* **2000**, *5*, 9.
37. Horn, W. Long-term development of the crustacean plankton in the Saidenbach Reservoir (Germany)–changes, causes, consequences. *Hydrobiologia* **2003**, *504*, 185–192. [CrossRef]
38. Gilbert, K. Quantitative comparison of food niches in some freshwater zooplankton. *Oecologia* **1987**, *72*, 331–340.
39. Perrow, M.R.; Jowitt, A.; Stansfield, J.; Phillips, G. *The Practical Importance of the Interactions between Fish, Zooplankton and Macrophytes in Shallow Lake Restoration*; Springer: Dordrecht, The Netherlands, 1999.
40. Wen, X.F.; Zhang, Y.; Zhen, L.; Jin, J.; Duan, T. Research on effection of the comprehensive cultivation mode of the shrimp, crab, submerged plant polyculture and the ecological floating bed. *Guangdong Agric. Sci.* **2012**, *24*, 138–142.
41. Ueji, M.; Inao, K. Rice paddy field herbicides and their effects on the environment and ecosystems. *Weed Biol. Manag.* **2001**, *1*, 71–79. [CrossRef]
42. Guo, H.; Tang, D.; Shi, X.; Wu, Q.; Wang, Z. Comparative transcriptome analysis reveals the expression and characterization of digestive enzyme genes in the hepatopancreas of the Chinese mitten crab. *Fish. Sci.* **2019**, *85*, 979–989. [CrossRef]
43. Sun, L.P.; Song, X.; Zhu, J.; Meng, X.; Zhang, L. Effects of submerged plants on growth performance and non-specific immunity of Chinese mitten crab (*Eriocheir sinensis*). *Freshw. Fish.* **2012**, *42*, 35–40.
44. Zeng, Q.; Jeppesen, E.; Gu, X.; Mao, Z.; Chen, H. Cannibalism and Habitat Selection of Cultured Chinese Mitten Crab: Effects of Submerged Aquatic Vegetation with Different Nutritional and Refuge Values. *Water* **2018**, *10*, 1542. [CrossRef]
45. Lv, D.; Wang, W.; Xu-Zhou, M.; Wang, Q.; Wang, A.; Chen, Z.; Tian, S. Ecological Prevention and Control of Weeds in Rice-crab Polycultured Field. *Hubei Agric. Sci.* **2011**, *50*, 1574–1578.
46. Li, X.; Dong, S.; Lei, Y.; Li, Y. The effect of stocking density of Chinese mitten crab *Eriocheir sinensis* on rice and crab seed yields in rice–crab culture systems. *Aquaculture* **2007**, *273*, 487–493. [CrossRef]
47. Hong, Y.; Jiang, C. Effects of stocking Chinese mitten crab on the zoobenthos and aquatic vascular plant in the east lake reservoir, Heilongjiang, China. *Acta Hydrobiol. Sin.* **2005**, *29*, 430–434, (In Chinese with English Abstract).
48. Xu, Q.; Wang, H.; Zhang, S. The impact of overstocking of mitten crab, Eriocheir sinensis, on lacustrine zoobentic community. *Acta Hydrobiol. Sin.* **2003**, *27*, 41–46, (In Chinese with English Abstract).

Article

The Effects of Different Carbon Sources on the Production Environment and Breeding Parameters of *Litopenaeus vannamei*

Yiming Xue [1,2], Li Li [1,2,*], Shuanglin Dong [1,2], Qinfeng Gao [1,2] and Xiangli Tian [1,2]

[1] Key Laboratory of Mariculture, Ministry of Education, Ocean University of China, Qingdao 266003, China; xueyiming@qdio.ac.cn (Y.X.); dongsl@ouc.edu.cn (S.D.); qfgao@ouc.edu.cn (Q.G.); xianglitian@ouc.edu.cn (X.T.)

[2] Function Laboratory for Marine Fisheries Science and Food Production Processes, Qingdao National Laboratory for Marine Science and Technology, Qingdao 266235, China

[*] Correspondence: l_li@ouc.edu.cn; Tel.: +86-532-82031590

Abstract: This study investigated the effect of different carbon sources on water quality, ammonia removal pathways, the bacterial community, and the production of *Litopenaeus vannamei* in outdoor culture tanks. Three systems were established: a clear water system (CW) and biofloc technology (BFT) systems with added molasses (M-BF) or poly (3-hydroxybutyric acid-co-3-hydrovaleric acid) (PHBV) (P-BF). The average pH, total alkalinity, total organic carbon, biofloc volume, chlorophyll a, nitrite, nitrate, total nitrogen, and nitrification rate were significantly different among the treatments. Microbial composition varied and different dominant taxa were identified in the treatments by linear discriminant analysis effect size. Redundancy analysis indicated that the water quality parameters affected the distribution of the microbial community. Moreover, the genus *Leucothrix* was closely related to the M-BF treatment. Chemoheterotrophy and aerobic chemoheterotrophy were the most abundant functions in all treatments. A comparison of functions using BugBase indicated that the relative abundance of several functions such as biofilm formation, stress tolerance and functions related to anaerobic processes increased in the M-BF treatment. The specific growth rate, growth rate, and survival rate of shrimp were significantly higher in the P-BF system than in the CW system and the feed conversion ratio in the BFT treatments was significantly lower than that in the CW system. Overall, adding carbon sources affected water quality, microbial community, and shrimp performance. The results show that PHBV is a good alternative to carbon sources.

Keywords: molasses; PHBV; water quality; nitrification rate; heterotrophic; bacteria community; bacterial function

Citation: Xue, Y.; Li, L.; Dong, S.; Gao, Q.; Tian, X. The Effects of Different Carbon Sources on the Production Environment and Breeding Parameters of *Litopenaeus vannamei*. *Water* **2021**, *13*, 3584. https://doi.org/10.3390/w13243584

Academic Editor: Christophe Piscart

Received: 4 November 2021
Accepted: 10 December 2021
Published: 14 December 2021

Publisher's Note: MDPI stays neutral with regard to jurisdictional claims in published maps and institutional affiliations.

1. Introduction

The Pacific white shrimp (*Litopenaeus vannamei*) is the most important commercially-traded species in shrimp aquaculture [1]. The culture systems for *L. vannamei* are becoming more intensive, resulting in ammonia and nitrite accumulation within grow-out systems [2,3]. Attention should be given to high ammonia concentrations because the stress can kill shrimp, fish, and other cultured animals [4–6]. Three processes involving ammonia removal from aquaculture systems are photoautotrophic uptake by algae, chemoautotrophic oxidation of ammonia to nitrate, and heterotrophic assimilation of ammonia directly to bacterial protein [7,8]. These three processes occur simultaneously in many aquaculture systems but seldom in equal importance, and changes in carbon sources may cause shifts in the dominant pathway [7].

The chemoautotrophic process is dominant in systems with a low total organic carbon/total nitrogen (C/N) ratio, such as a recirculating aquaculture system, [9]. Increasing the C/N ratio in biofloc technology (BFT) systems by adding organic carbon sources induce a shift in nitrogen use by the bacterial community from chemoautotrophy to heterotrophy [10]. The rate of nitrogen use by heterotrophic bacteria is 10 times faster than that of autotrophic nitrification [11]. Previous studies have demonstrated the efficiency of heterotrophic bacteria in controlling total ammonia nitrogen (TAN) in shrimp and tilapia BFT systems [5,8]. Although the growth of heterotrophic bacteria is stimulated in BFT systems, nitrification occurs in these systems, and the process is important for the long-term removal of TAN originating from the feed in intensive BFT systems [8,12,13]. However, it is unclear whether the nitrification rates differ in different carbon added systems, and whether the nitrification rate changes with the development of the bacterial community in a BFT system.

Molasses is a by-product of sugar production, that was suggested to be an appropriate carbon source to increase the C/N ratio in BFT systems. This water-soluble carbon source needs to be applied frequently with constant supervision to prevent overdosing and starvation of the bacterial floc, which increases the management effort. In recent years, insoluble biodegradable polymers (BDPs), such as poly (3-hydroxybutyric acid-co-3-hydrovaleric acid) (PHBV) have been used as carbon sources and biofilm carriers in aquaculture [14]. The relatively low cost of BDPs and the simple management requirements make them an attractive alternative carbon source for aquaculture [15]. However, the use of different carbon sources may lead to differences in water quality and changes in the bacterial community of a BFT system [6,16].

The conversion of nitrogen and waste recycling in BFT systems strongly depends on the capacity of microbes to assimilate and convert the nutrient waste [17]. The functions of biofloc, as a waste nutrient converting agent and a food source, are related to the microbial community. The changes in the bacterial community after adding carbons can affect the nutrient composition, which is essential for bacteria and cultured animals. Moreover, the relevance of host–microbiota–nutrient interactions is important in aquaculture [18]. Therefore, it is essential to explore the relationships between environmental factors and the microbiota [18].

The present study was designed to evaluate the effects of different carbon sources (molasses and PHBV) on water quality, nitrification rate and growth performance of *L. vannamei* in low water exchange grow-out systems. Moreover, the microbial community, the relationship between microbial structure and water environmental factors, as well as microbial function was investigated to develop a further understanding of the impact of different carbon sources on the microbial community in BFT systems.

2. Materials and Methods

2.1. Experimental Shrimp and Acclimation

The *L. vannamei* juveniles used in this experiment were supplied by Baorong Aquaculture Corp., Qingdao, China, and the experiment was conducted at the Ruizi Marine Food Product Co., Ltd., Qingdao City, China. The shrimp were acclimated for 10 days in 11 polyethylene fiber tanks (400 L) from which 50% of the water was renewed daily. The shrimp were fed a commercial feed (crude protein, 40%; moisture, 12%; ash, 16%; crude fat, 4% and NaCl, 3%) produced by the Yuehai Feed Co., Ltd., Guangzhou, China. Feed was offered twice daily (7:00 a.m. and 6:00 p.m.) based on a feeding amount equivalent to 5% of shrimp body weight.

2.2. Experimental Design

A 34-day trial was conducted. Three kinds of culture systems, each replicated three times (400-L of water in each circular tank), were established and 60 healthy shrimp (2.70 ± 0.02 g) were stocked into each tank. The treatments were: a clear water system (CW) and two BFT systems with added molasses (M-BF) or PHBV (P-BF). No molasses or PHBV was added to the CW tanks. Molasses was applied to the M-BF treatment at the ideal theoretical application amounts suggested by Avnimelech [10]. The molasses had a total organic carbon concentration of 26.5%, and it was diluted with water from the tanks and splashed uniformly over the tank surfaces twice daily. As suggested by Zhang, et al. [19], after 7 days of activation in the acclimation tanks, 633 g of PHBV was placed in a PVC tube with an inner diameter of 15 cm and an operating depth of 30 cm, and an air stone was placed inside the tube to sufficiently stir the PHBV. The PVC tube was wrapped with a filter screen, which was cleaned every five days.

The tanks were located within a greenhouse covered with clear plastic sheeting. One-fifth of the water in the CW treatment was replaced daily to ensure good water quality [13]. One-fifth of the water in the BFT tanks was replaced every 5 days [20]. The seawater used was pre-filtered using a sand filter. Each tank was continuously aerated with four air stones. The tanks were covered with shade netting to control sunlight. The feeding regime was the same as during the acclimation period.

2.3. Water Quality

Temperature, dissolved oxygen (DO), and pH were measured at 8:00 am daily using a portable DO meter (RDB20, Yuanmu, Shanghai, China) and a pH meter (pH-100, LICHEN, Shanghai, China). The salinity of the culture water was 31 g L^{-1} which was monitored using an optical salinity meter (LYT-610, Yulin, Shanghai, China). Water samples were collected from the tanks every 5 days. Total ammonia nitrogen (TAN), nitrite (NO_2^{-}-N), nitrate (NO_3^{-}-N), total nitrogen (TN), total alkalinity (TA) and chlorophyll (Chl) a were analyzed following the standard methods of the Chinese State Environmental Protection Agency [21]. Particulate organic carbon (POC) and dissolved organic carbon (DOC) were analyzed according to the method of Gilcreas [22]. Total organic carbon (TOC) was calculated as the sum of DOC and POC. Water samples (1000 mL) were transferred to Imhoff cones (1000-0010, Nalgene) at 3:00 pm every five days to determine biofloc volume. The volume of the biofloc plug accumulating on the bottom of the cone was determined 30 min after sedimentation.

2.4. Nitrification Rate

Water samples (500 mL) were collected on days 14, 19, 24, 29 and 34 to measure the nitrification rate following the methodology of Bratvold and Browdy [23]. Briefly, duplicate samples were enriched with ammonium chloride (2.5 mg L^{-1} of ammonia-N) and incubated in the absence or presence of 2.2 mg L^{-1} of the nitrification inhibitor N-allylthiourea (Sigma-Aldrich, St. Louis, MO, USA). The nitrification process is expressed as $NH_4^{+} + 2O_2 \rightarrow NO_3^{-} + 2H^{+} + H_2O$. The difference in the DO concentration of dissolved oxygen between the bottles with and without the inhibitor was the nitrification result. The nitrification rate (NR) (mg L^{-1} h^{-1}) was calculated using Equation (1):

$$NR = (DO_1 - DO_2)/h, \tag{1}$$

where (DO_1) = dissolved oxygen in the bottle with the inhibitor (mg L^{-1}); DO_2 = dissolved oxygen in the bottle without the inhibitor (mg L^{-1}); h = incubation period (h).

2.5. 16S rRNA Gene High Throughput Sequencing Analysis

At the end of the study, 1-L water sample was filtered with a 50-mm membrane (pore size 0.20 μm) and used for bacterial DNA extraction and analysis. Three replicates were conducted for each treatment. Total DNAs from these samples were extracted by the CTAB/SDS method [24] and were diluted to 1 ng/μL with sterile water and used for PCR template. 16S rRNA genes of distinct regions (16S V4/V5) were amplified using specific primers (515F: 5′-GTGCCAGCMGCCGCGG-3′ and 907R: 5′-CCGTCAATTCMTTTRAGTTT-3′). These PCR products were purified with a GeneJET Gel Extraction Kit (Thermo Scientific, Waltham, MA, USA) and prepared to construct libraries. Sequencing libraries were generated using NEB Next®Ultra™DNA Library Prep Kit for Illumina (NEB, Ipswich, MA, USA) following the manufacturer's instructions. The library quality was assessed on the Qubit@ 2.0 Fluorometer (Thermo Scientific, Waltham, MA, USA) and Agilent Bioanalyzer 2100 system (Agilent Technologies, Santa Clara, CA, USA). Then the library was sequenced on an Illumina MiSeq platform and 250 bp paired-end reads were generated. The raw amplicon sequencing data were deposited to NCBI Sequence Read Archive (accession number: PRJNA738270).

Paired-end reads were merged using FLASH (v. 1.2.11, http://ccb.jhu.edu/software/FLASH/, accessed on 29 April 2021), which was designed to merge paired-end reads when at least some of the reads overlap those generated from the opposite end of the same DNA fragments. The splicing sequences were called raw tags. Quality filtering on the raw tags were performed under specific filtering conditions to obtain the high-quality clean tags using Trimmomatic v. 0.33 [25]. The primer sequences were identified and removed using Cutadapt v. 1.9.1 [26]. The raw tags were compared with the reference database (Silva database, Release132, http://www.arb-silva.de/, 29 April 2021) using UCHIME algorithm v. 10 [27] to detect and remove chimera sequences [28]. The sequencing results were clustered into operation taxonomy units (OTUs) at a similarity threshold of 97% using USEARCH v. 10.0 [27]. The representative sequences for each OTU were made taxonomic annotations against the SILVA taxonomy database (Release132, http://www.arb-silva.de/, 29 April 2021) using Qiime 2 [29].

2.6. Shrimp Growth Performance and Survival

At the end of the experiment, all shrimp in the tanks were collected, counted, and weighed to calculate growth rate, feed conversion ratio (FCR), specific growth rate (SGR) and survival rate, which were determined as follows:

$$\text{Growth rate (g/wk)} = (\text{average final weight of shrimp} - \text{average initial weight of shrimp})/\text{culture weeks}; \quad (2)$$

$$\text{Feed conversion ratio (FCR)} = \text{total weight of feed intake (g)}/\text{total shrimp weight gain (g)}; \quad (3)$$

$$\text{Specific growth rate (SGR)} = (\ln (\text{initial wet weight}) - \ln (\text{final wet weight}))/\text{culture days} \times 100; \quad (4)$$

$$\text{Survival rate (\%)} = (\text{final shrimp number}/\text{initial stocking shrimp number}) \times 100. \quad (5)$$

2.7. Statistical Analysis

The statistical analysis was performed using SAS v9.4 for windows (Cary, NC, USA). The normality of distribution of data was checked using the Kolmogorov–Smirnov test ($p > 0.05$). The Levene test ($p > 0.05$) was applied for assessing the homogeneity of variance. One-way analysis of variance (ANOVA) was used to compare the average concentration of all the water quality parameters, concentrations of NR, TAN, NO_2^--N, NO_3^--N and TN on each sampling day, microbial alpha diversity indexes, shrimp growth indexes, the composition of microbial phylum and genus among treatments. Differences were considered significant at $p < 0.05$. When significant differences were found, Tukey's test was used to identify differences between treatments. The indices of Chao1, Abundance-based Coverage Estimator (ACE), Shannon and Simpson were calculated using Qiime 2.

The linear discriminant analysis (LDA) effect size (LEfSe) method (http://huttenhower.sph.harvard.edu/lefse/, 29 April 2021) was used for determining the features of bacterial communities between different treatments [30]. The LDA score was set as 3.5 to identify the significant features. The redundancy analysis (RDA) was carried on using the vegan package (v. 2.3) in R (v. 4.0). The functional prediction of bacterial community was performed by FAPROTAX [31] and BugBase [32].

3. Results

3.1. Water Quality

The descriptive statistics for the water quality data are given in Table 1. No significant differences in average temperature or DO were observed among the treatments (Table 1, $p > 0.05$). The pH in the CW system was significantly higher than that in the other treatments and pH in the P-BF system was significantly higher than that in the M-BF system (Table 1, $p < 0.05$). The average TA concentration in the CW system was significantly higher than that in the other systems ($p < 0.05$) and the biofloc volume in the M-BF systems was significantly higher than that in the CW and P-BF systems. The Chlorophyll a in the P-BF system was significantly lower than that in the other two systems.

Table 1. The mean and standard deviations of water quality parameters in the three systems during the experiment.

Parameters	CW	M-BF	P-BF
Temperature (°C)	25.5 ± 1.56 [a]	25.5 ± 1.56 [a]	25.5 ± 1.56 [a]
DO (mg L^{-1})	5.27 ± 0.74 [a]	4.83 ± 0.88 [a]	4.95 ± 0.84 [a]
pH	7.7 ± 0.08 [a]	7.1 ± 0.07 [c]	7.4 ± 0.05 [b]
TA (mg L^{-1} CaCO$_3$)	102.67 ± 0.44 [a]	81.69 ± 4.98 [b]	86.14 ± 1.81 [b]
TOC (mg L^{-1})	10.65 ± 3.29 [c]	27.01 ± 2.93 [a]	18.95 ± 3.15 [b]
Biofloc volume (mL L^{-1})	0.00 ± 0.00 [b]	3.35 ± 1.10 [a]	0.00 ± 0.00 [b]
Chl (mg L^{-1})	13.55 ± 1.19 [a]	10.48 ± 2.11 [a]	3.95 ± 0.57 [b]
TAN ((mg L^{-1})	0.13 ± 0.16 [a]	0.17 ± 0.04 [a]	0.23 ± 0.05 [a]
NO$_2^-$-N (mg L^{-1})	0.82 ± 0.26 [c]	6.44 ± 1.70 [a]	3.38 ± 0.49 [b]
NO$_3^-$-N (mg L^{-1})	0.94 ± 0.09 [c]	4.51 ± 0.90 [a]	2.79 ± 0.20 [b]
TN (mg L^{-1})	5.69 ± 0.30 [c]	13.65 ± 0.58 [a]	9.75 ± 0.39 [b]
Nitrification rate (mg L^{-1} h^{-1})	0.39 ± 0.06 [a]	0.10 ± 0.02 [b]	0.12 ± 0.01 [b]

CW, M-BF and P-BF represented the three systems of this study. TA: total alkalinity; TOC: total organic carbon; Chl: chlorophyll a; TAN: total ammonia nitrogen; TN: total nitrogen. In each row, different superscript letters indicate significant differences at the $p < 0.05$ level (one-way ANOVA and Tukey's test).

The average TAN concentration was not significantly different among the three systems (Table 1, $p > 0.05$). The concentration of TAN in the BFT systems (M-BF and P-BF) reached peak levels on day 7 of the experiment and then decreased sharply from day 7 to 14 (Figure 1a). Then, the concentration of TAN in the BFT systems remained relatively stable. The concentration of TAN in the CW system declined from days 1 to 14 and then increased slowly over that in the other systems (Figure 1a). No significant difference in TAN was observed among the treatments at the end of the study (Figure 1a, $p > 0.05$).

The average concentration of NO$_2^-$-N in the CW system was significantly lower than that in the BFT systems ($p < 0.05$) and it was relatively stable. The NO$_2^-$-N concentration in the M-BF system increased sharply from days 0 to 19, then, the concentration of NO$_2^-$-N decreased from days 19 to 24 and was stable after day 24 (Figure 1b).

The concentration of NO$_3^-$-N in all systems was stable from days 1 to 19 but increased sharply in the BFT systems from days 19 to 29 (Figure 1c). After day 29, the NO$_3^-$-N concentration in the BFT systems decreased slightly. The NO$_3^-$-N concentration in the CW system remained relatively constant during the study. The concentration of NO$_3^-$-N in the CW system was not significantly different from the other two systems during the first 19 days ($p > 0.05$). After day 19, the concentration of NO$_3^-$-N in the CW system was significantly lower than that in the carbon added systems ($p < 0.05$).

The average concentration of TN in the M-BF was 13.65 mg L^{-1}, which was significantly higher than that in the CW and P-BF systems (5.69 and 9.75 mg L^{-1}, respectively) (Table 1, $p < 0.05$). The TN concentration increased during the first 7 days in all systems but began to fluctuate after day 7 (Figure 1d). The concentration of TN in the M-BF system was significantly higher than that in the CW system after day 7.

The average nitrification rate in the CW system was significantly higher than that in the BFT systems (Table 1, $p < 0.05$). No significant differences were observed in nitrification rate among the three treatments on the first sampling date (d 14) (Figure 2). Then, the nitrification rate was consistently higher in the CW system than in the other systems (Figure 2). The nitrification rate in the BFT systems decreased dramatically after the first sampling and approached zero after day 19 (Figure 2).

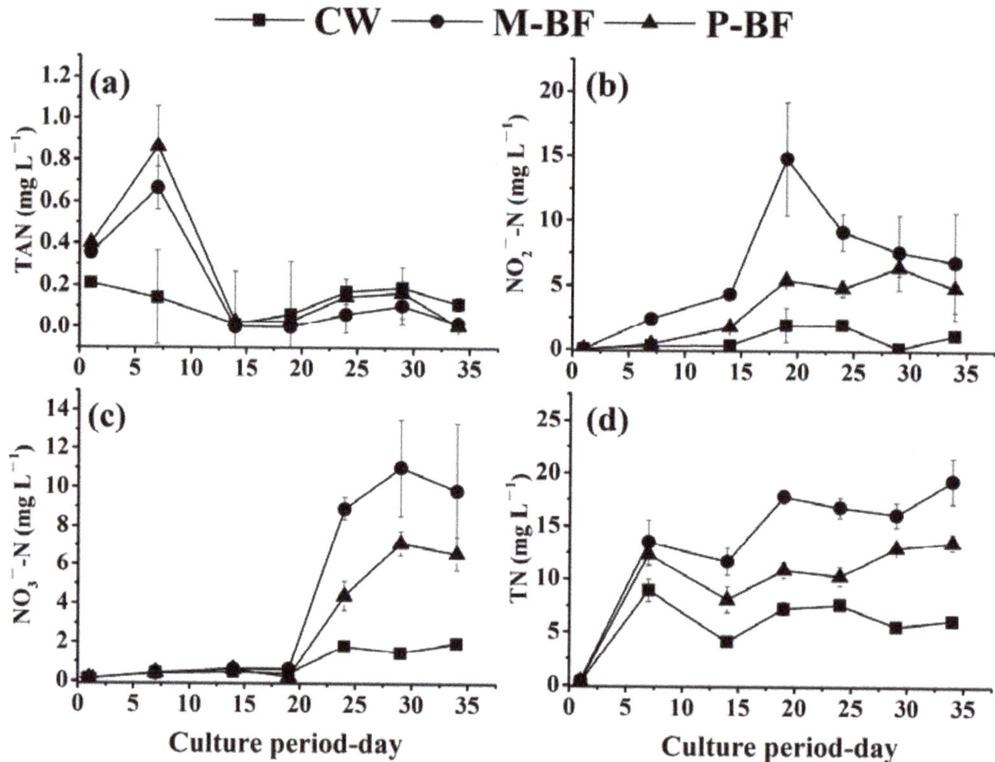

Figure 1. Fluctuations of (**a**) total ammonia nitrogen (TAN), (**b**) NO$_2$$^-$-N, (**c**) NO$_3$$^-$-N and (**d**) total nitrogen (TN) during the period of the experiment. CW, M-BF and P-BF represented the three systems of this study. Values are mean ± standard deviation of three replications on each sampling day.

3.2. Characterization of the Bacterial Community

A total of 446,015 high-quality reads were obtained and 3,219 operational taxonomic units (OTUs) were obtained in the water samples by clustering the OTUs at a 97% similarity level. The Chao1, ACE, Shannon, and Simpson indices, which were used to represent the richness and diversity of the bacterial community, were not significantly different among the treatments. The phyla Bacteroidetes, Proteobacteria and Actinobacteria were the dominant bacteria (48.76%, 38.20%, and 3.27%, respectively) in the three treatments (Figure 3a). The abundance of Cyanobacteria decreased in the M-BF system and increased in the P-BF system compared to the CW system, which was 0.44%, 7.32% and 2.83% in the M-BF, P-BF, and CW systems, respectively. Differences in the phyla among treatments were compared and the top 20 phyla with the smallest p-values were listed in Figure 3c. Cyanobacteria were included in the top 20 phyla. Planctomycetes was one of the dominant bacteria in the M-BF system and the abundance of this phylum was significantly higher than that in the other treatments ($p < 0.01$, Figure 3c). The abundance of Chloroflexi was higher in the CW systems (5.58%) than in the BFT systems (0.69% in M-BF and 0.46% in P-BF) and was included in the top 20 phyla with the smallest p-value (Figure 3c).

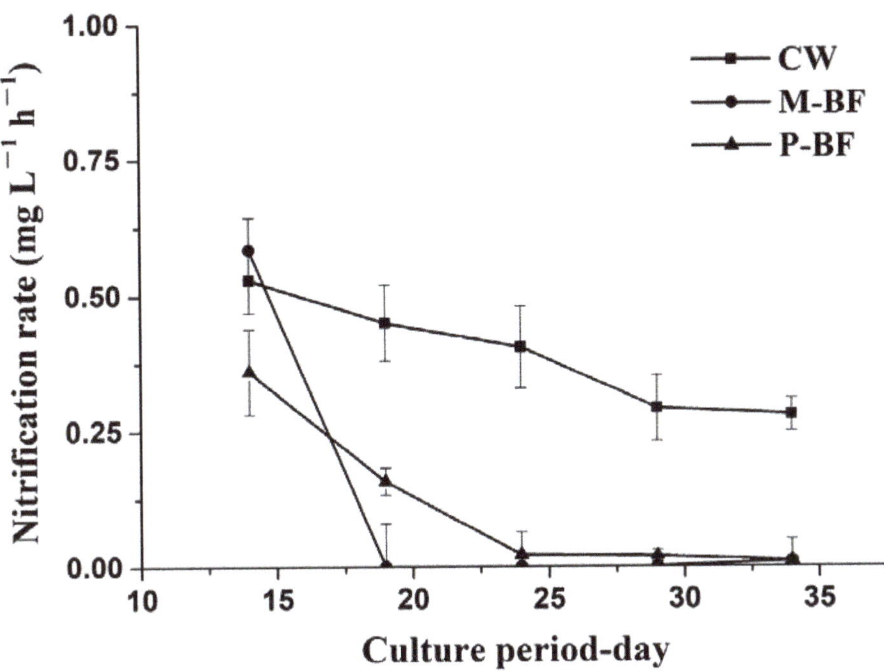

Figure 2. Nitrification rate during the period of the experiment. CW, M-BF and P-BF represented the three systems of this study. Values are mean ± standard deviations of three replications on each sampling day.

As for genera of the bacterial communities, an uncultured_bacterium_f_ Flavobacteriaceae, uncultured_bacterium_f_Rhodobacteraceae, and *Leucothrix* were the more abundant genera in the M-BF treatment compared to the other two treatments (Figure 3b). *Tamlana* and an uncultured _bacterium_o_Ardenticatenales were more abundant in the CW than the other two systems. Five genera were significantly different among the treatments in the top 20 genera with the smallest *p*-values (Figure 3d). *Vibrio* was significantly more abundant in the CW than the BFT systems (Figure 3d).

Linear discriminant analysis effect size (LEfSe) was used to identify the most differentially abundant microbial taxa in the three systems (Figure 4a). The taxa higher than the LDA significance threshold of 3.5 were scored in each system (Figure 4b). Various taxa were enriched in the BFT systems (Figure 4). For example, Oxyphotobacteria (Class), Cyanobacteria (Phylum), and Chloroplast (Order) were enriched in the *p*-BF treatment, whereas Gemmatimonadetes (Phylum) and planctomycetes (Class, Order, Family and Genus) were enriched in the M-BF treatment (Figure 4). Furthermore, Vibrionaceae (Family), Vibrionales (Order) and *Vibrio* (Genus) were enriched in the CW system, which was consistent with the analysis of variance in genus level (Figure 3d).

3.3. Correlation Analysis between Environmental Factors and Microbial Community

Redundancy analysis (RDA) was used to discern the possible correlations between the bacterial genera, the carbon varieties, and the environmental variables (Figure 5). Samples from the M-BF system were distributed on the left of the first canonical axis, whereas samples from the P-BF and CW systems were distributed on the right. Most environmental factors had a strong negative effect on the distribution of the samples on the first axis, whereas TAN and TA had a positive effect. Moreover, *Leucothrix* was strongly positively correlated with the M-BF treatment.

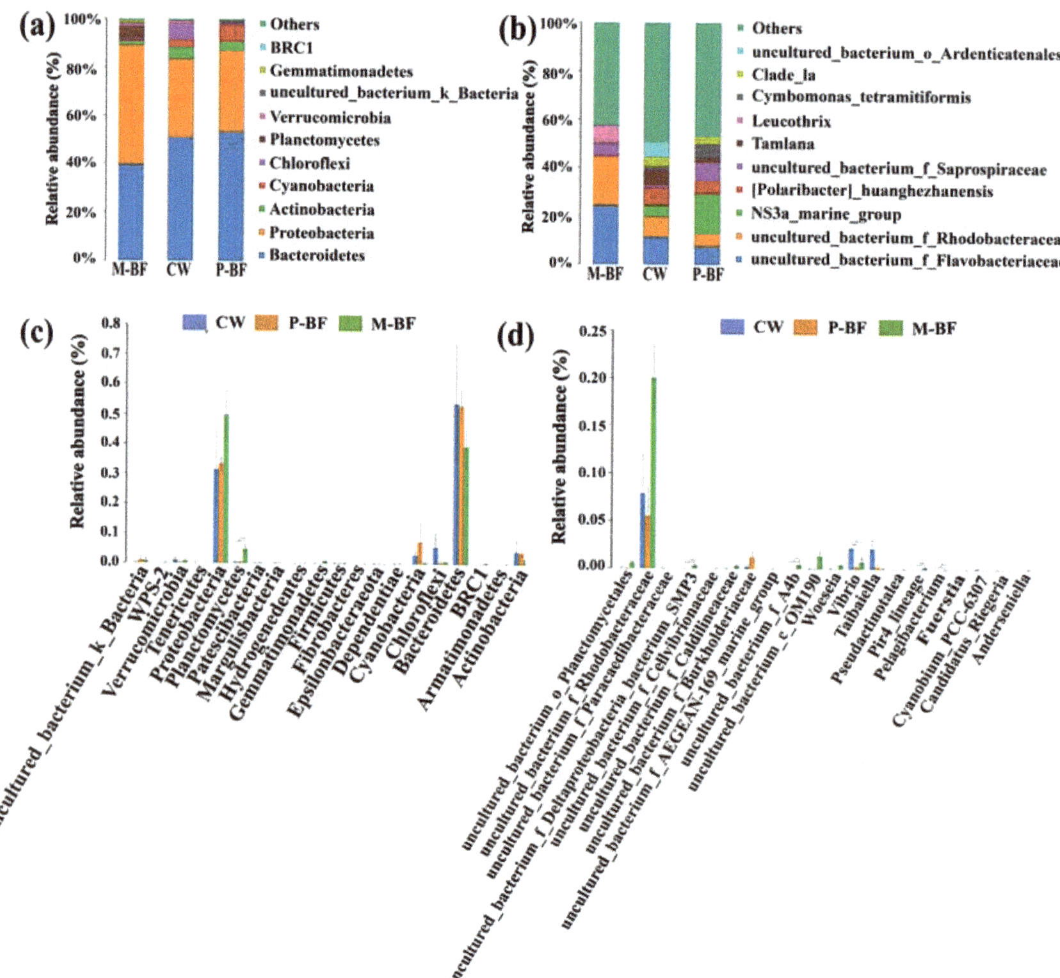

Figure 3. Bacterial community compositions and abundance at phyla (**a**) and genus (**b**) levels across all samples. The Analysis of Variance (ANOVA) shows the top 20 phyla (**c**) and genera (**d**) with the smallest *p*-value among treatments. *, $p < 0.05$; **, $p < 0.01$. CW, M-BF and P-BF represented the three systems of this study.

3.4. Functional Prediction in Carbon Adding Environments

FAPROTAX was adopted to annotate and screen the key ecological functions of the bacterial communities in the three treatments. The top 10 functions are shown in Figure 6a. The functions of chemoheterotrophy and aerobic chemoheterotrophy were abundant in all treatments. Increased relative abundances of chloroplasts and intracellular parasites were observed in the P-BF treatment, while fermentation increased in the M-BF treatments. Based on the 16s rRNA gene sequences, the pairwise functional comparison was investigated using BugBase (Figure 6b–d). The relative abundance of several functions, such as biofilm formation (i.e., forms biofilms), anaerobic functions (i.e., anaerobic), stress tolerance (i.e., stress-tolerant), and mobile elements (i.e., contains mobile elements), increased in the M-BF than the CW system (Figure 6b). The relative abundance of the function of Gram-negative increased in both carbon adding systems compared to the CW system

(Figure 6b,c). The relative abundance of the functions related to Gram-negative, aerobic, anaerobic, forms biofilms and stress-tolerant were all higher in M-BF than P-BF (Figure 6d).

3.5. Growth Performance

The final weight, specific growth rate (SGR), growth rate, and survival rate of *L. vannamei* were highest in the P-BF system and lowest in the CW system. The three parameters in the P-BF system were significantly higher than those in the CW system (Table 2). The feed conversion ratios (FCR) were significantly lower in M-BF and P-BF treatments than in the CW treatment ($p < 0.05$).

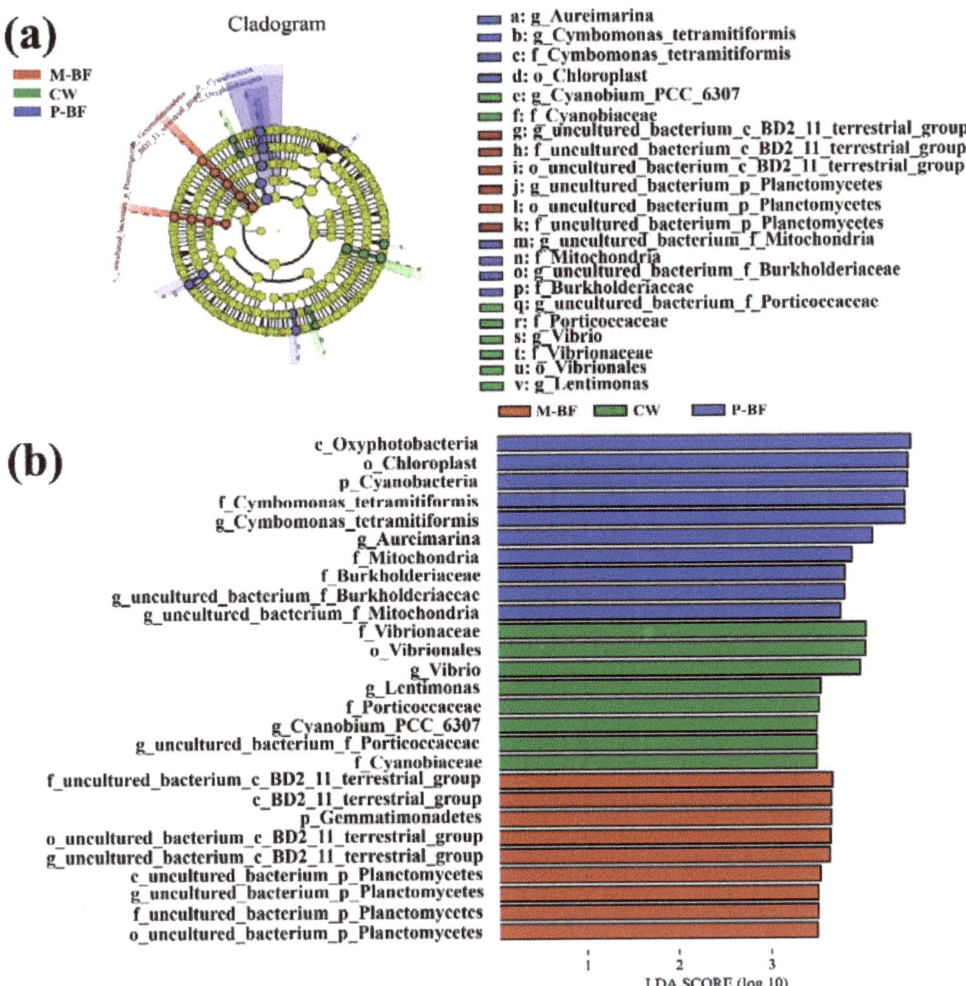

Figure 4. Linear discriminant analysis effect size (LEfSe) analysis showing the abundance of three treatments. (**a**) LEfSe analysis identified the most differentially abundant taxa among three systems. The six rings of the cladogram stand for domain (innermost) phylum, class, order, family, and genus. (**b**) Linear discriminant analysis score of the three systems with a threshold value of 3.5. CW, M-BF and P-BF represented the three systems of this study.

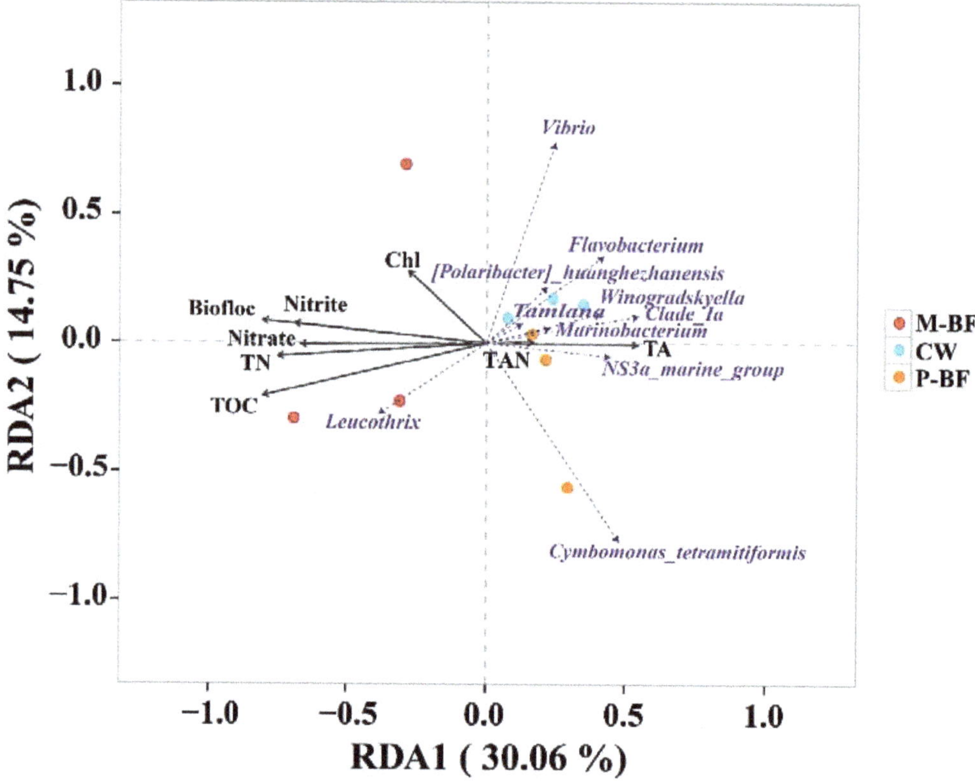

Figure 5. Redundancy analysis (RDA) identified the correlation of different treatments, bacterial communities, and environmental factors. Biofloc: biofloc volume; TN: total nitrogen; TAN: total ammonia nitrogen; TA: total alkalinity; Chl: chlorophyll a; TOC: total organic carbon. Details of all environmental factors were shown in Table 1. CW, M-BF and P-BF represented the three systems of this study.

Table 2. Growth parameters and survival of *Litopenaeus vannamei* of different experimental systems at the end of the 34-d trial.

Parameters	CW	M-BF	P-BF
final weight	5.17 ± 0.81 [b]	6.94 ± 1.17 [ab]	8.04 ± 0.94 [a]
Specific growth rate (%/day)	1.84 ± 0.00 [b]	2.67 ± 0.00 [ab]	3.11 ± 0.00 [a]
Growth rate (g/wk)	0.49 ± 0.16 [b]	0.85 ± 0.23 [ab]	1.07 ± 0.19 [a]
Feed conversion ratio (kg/kg)	1.95 ± 0.54 [a]	1.12 ± 0.30 [b]	0.86 ± 0.17 [b]
Survival rate (%)	28.33 ± 0.04 [b]	45.56 ± 0.11 [ab]	61.11 ± 0.29 [a]

Data are presented as mean ± standard deviation (n = 3). CW, M-BF and P-BF represented the three systems of this study. In each row, different superscript letters indicate significant differences at the $p < 0.05$ level (one-way ANOVA and Tukey's test).

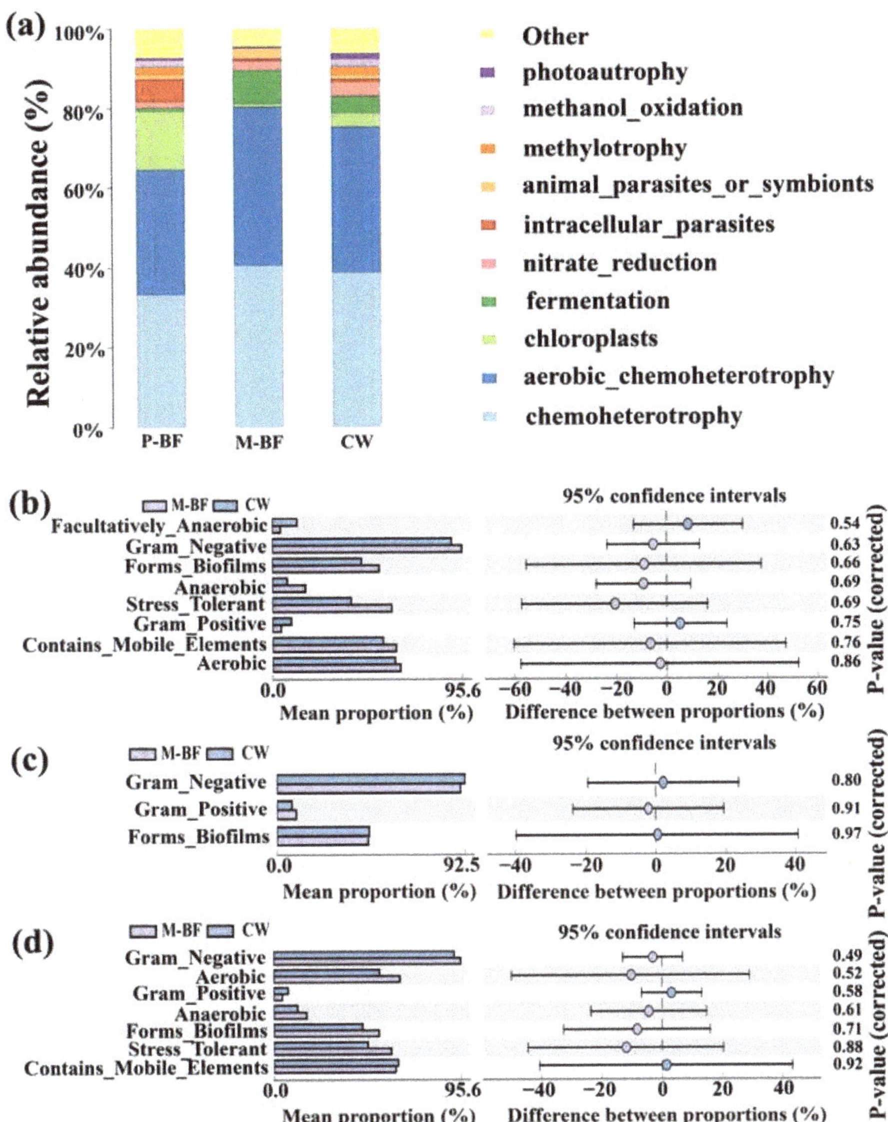

Figure 6. Functional prediction in different environments. (**a**) The predicted functions using FAPROTAX among different treatments. (**b**–**d**) Pairwise comparisons of functions between the three systems according to BugBase. The left part of each figure shows the mean proportion of two systems in different functions. The right part of each figure shows the difference between proportions. CW, M-BF and P-BF represented the three systems of this study.

4. Discussion

4.1. TAN Removal Pathways in Different Systems

4.1.1. Nitrification in Different Systems

In previous studies, apparent nitrification was observed in heterotrophic systems, where heterotrophic assimilation was promoted [7,33]. This agrees well with the findings of the present study, in which the nitrification rate was relatively high in the BFT treatments during early culture but was not significantly different from that in the CW treatment.

Several water quality variables demonstrated nitrification in the BFT systems. NO_2^--N is the product of the first step in the nitrification process. The accumulation of NO_2^--N and a decrease in TAN were observed during the early days of culture, suggesting that the first nitrification step ($NH_4^+ + 1.5\,O_2 \rightarrow NO_2^- + 2H^+ + H_2O$) progressed at a greater rate than the second step ($NO_2^- + 0.5\,O_2 \rightarrow NO_3^-$). Schveitzer, et al. [33] made a similar observation in BFT systems and arrived at the same conclusion. A dramatic increase in NO_3^--N was observed from days 19 to 29, indicating complete oxidation of nitrite to nitrate during the period. Nitrite increased to a high level in the middle of the experiment and Emilie, et al. [34] also observed this trend after 20 d. However, nitrification in the BFT systems became inhibited as the experiment progressed. The nitrification rates in the BFT systems decreased and almost reached zero at the end of the study. Previous studies have demonstrated that heterotrophic bacteria efficiently compete with nitrifying bacteria for nutrients and space resulting in inhibited nitrification in *L. vannamei* and *Oreochromis mossambicus* culture systems [7,8,12].

4.1.2. Heterotrophic Bacteria Assimilation

The heightened ammonia concentrations in BFT systems during week 1 of this study may indicate that heterotrophic bacterial communities require considerable time to develop. The water had not received added carbohydrates to stimulate the growth of heterotrophic bacteria before it was added to the culture tanks. The addition of carbohydrates began after the shrimp were placed into the tanks, and the bacterial community required time to become functional in the systems [6]. Gerardi [35] reported that the heterotrophic bacterial community takes 14 days to reach a stable stage. In this study, although the average TAN concentration was not significantly different among the treatments, the dominant TAN removal pathway may have been different. The low TAN concentration in the CW system was maintained by water exchange and nitrification. This was supported by the significantly higher nitrification rate in the CW system during the experiment period. TAN in the BFT systems was mainly removed through heterotrophic bacterial assimilation, and nitrification was greatly inhibited after 14 d.

4.2. Microbial Composition

Previous studies have shown that different carbon sources can affect the structure and function of bacterial communities in aquaculture systems [8,10,36,37]. In this study, Proteobacteria and Bacteroidetes were the dominant phyla of bacteria in all three systems. This finding agrees with Cardona, et al. [34] who reported that these two phyla represent more than 90% of the total bacteria present in BFT systems. Proteobacteria are widely dispersed in the aquatic environment and play an important role in nutrient cycling and the mineralization of organic compounds [34,38]. Bacteroidetes are an important part of the heterotrophic bacterial community of many water bodies [39]. Other phyla with different abundance among treatments, including Cyanobacteria, Planctomycetes, Chloroflexi, and Verrucomicrobia were found in *L. vannamei* culture systems [40]. Significant growth and proliferation of Cyanobacteria were observed in the P-BF system in this study, and this was reported by Miranda-Baeza, et al. [41] in BFT systems with added molasses. This phylum has developed a variety of ecological and physiological adaptation strategies to grow under poor conditions with high organic matter loads [41]. The phylum Planctomycetes, which was abundant in the M-BF system, performs anaerobic ammonium oxidation (i.e., oxidizing ammonium with nitrite as the electron acceptor to yield dinitrogen gas) to maintain good water quality [36,42].

Leucothrix was the most abundant genus in the M-BF system, and this was consistent with our previous studies [43]. One study reported that *Leucothrix* forms filaments and is usually found as an epiphyte on algae and invertebrates [43]. Species of this genus promote high mortality rates in shrimp [44,45], and this may be the reason for the decreased survival rate in the M-BF system. In addition, *Vibrio*, which are also bacterial pathogens to aquatic organisms, increased significantly in the CW system compared to the BFT systems [46,47]. Inhibition of a population of pathogenic *Vibrio* was reported in BFT systems [46].

4.3. The Correlation between Carbon Sources, Microbial Communities and Environmental Factors

According to the RDA, a close correlation was observed between bacterial composition and the water quality parameters. This result agrees with the findings of previous studies, which reported that bacterial community structure in the aquatic environment is affected by the abiotic environment [48]. Zhang, et al. [49] revealed that abiotic environmental factors, such as TAN and TN, have large impacts on bacterial populations in *L. vannamei* culture ponds, and Li, et al. [50] reported that TAN and Chl strongly affect the bacterial composition in the sediment of intertidal regions used for mariculture. Many of the environmental factors changed in the culture systems after adding the carbon sources. (Table 1). Different bacteria perform different ecological functions and the changing ecology and microbiota co-affected shrimp culture [51].

The changes in the functional profiles involved in the bacterial community provide valuable information from a functional perspective [31]. FAPROTAX is a gene functional annotation tool based on 16S rRNA sequencing that is appropriate for environmental samples. The results of the functional predictions indicated that biofilms formation and stress tolerance increased in the M-BF system. Biofilm contributes to the metabolism of the nitrogenous compounds generated within biofloc culture systems [52]. This finding indicates that adding molasses changed the environment to a more stressful condition but improved nitrogen metabolism in another aspect. The functional comparison using BugBase showed that adding PHBV produced a less stressful environment for bacteria compared to that of molasses. Our previous study also showed that shrimp are subject to be stressed in BFT systems where molasses serves as the carbon source other than PHBV [53].

4.4. Practical Value

Carbon sources are often applied in aquaculture systems to maintain a high C/N ratio and control the concentrations of nitrogenous metabolites in the system [54]. The carbon sources used are often by-products of the food processing and animal feed industries that are inexpensive and locally available [54]. Water exchange is greatly reduced in those systems to favor heterotrophic nitrogen assimilation pathways [33,55–57]. In the present study, the water exchange rate in the BFT systems was only one-fifth of that applied to the CW system, but the average TAN concentrations were not significantly different among the three systems. This agrees with the findings of Hargreaves [12] and Avnimelech [58], who demonstrated the usefulness of BFT in reducing water use while maintaining adequate water quality.

The growth rate of shrimp was better in the BFT systems than in the CW system, suggesting the benefits of BFT systems for improving the growth rate, decreasing the FCR of shrimp, and reducing feed costs. Avnimelech [59] also observed that the growth rate of overwintering tilapia fingerlings in BFT systems was significantly higher than in the control. Shrimp may utilize bioflocs as a source of protein and additional feed for growth [4,60]. The mechanism for the promoting effect of biofloc on the growth of aquatic animals is unclear. Some researchers have proposed that biofloc contains fungal proteins, and their amino acids facilitate absorption by aquatic animals. In addition, biofloc is a rich source of proteins and lipids continuously available in situ for consumption by shrimp [54].

Water-soluble carbon sources, such as molasses, must be added many times to a culture system, which increases the difficulty of system management. Such carbon sources are also limited by excessive DOC and color problems [61]. In the present study, water from the M-BF system was more turbid than that from the CW and P-BF systems. The biofloc in the P-BF system was almost in the PVC tube, which caused low biofloc volume in this system. Moreover, PHBV was added as a carbon source only once, so this was not a major management effort. Boley, et al. [61] demonstrated that solid carbon sources are better alternatives to water-soluble carbon sources because they reduce supervision and management efforts. Adding PHBV had a similar function for improving water quality and producing less stress compared with the molasses-added treatments. Taken together, our results suggest that insoluble biodegradable polymers could be good alternatives for improving water quality and promoting production in intensive aquaculture systems.

5. Conclusions

The carbon sources-added group maintained water quality with reduced water exchange compared to the control. The dominant ammonia removal pathways changed after adding a carbon source, where the TAN concentration was mainly removed through heterotrophic bacterial assimilation, and nitrification was greatly inhibited after day 14. Moreover, this study confirmed the effects of adding carbon sources on bacterial community composition in culture tanks, and *Leucothrix* was closely related to molasses treated systems. Results from the present study also indicated that the environment of the molasses-added systems tended to be more stressful to shrimp as the relative abundance of several functions, such as biofilms formation and stress tolerance increased in the treatment. The relatively better shrimp performance and less stressful culture environment in the PHBV-added systems suggest that PHBV may be a better alternative as a carbon source in BFT systems.

Author Contributions: Conceptualization, methodology, and writing—original draft preparation, L.L. and Y.X.; writing—review and editing, L.L.; visualization, Y.X.; data curation, project administration, L.L., S.D. and X.T.; supervision, L.L., S.D., Q.G. and X.T.; data curation, project administration, Y.X.; funding acquisition, L.L., X.T. and S.D.; investigation, Y.X.; validation and resources, L.L. and X.T.; formal analysis: Y.X. All authors have read and agreed to the published version of the manuscript.

Funding: This research was funded by the National Key R&D Program of China (No. 2020YFD0900200 and No. 2017YFE0122100) and the Natural Science Foundation of Shandong Province, China (ZR2020MC194).

Institutional Review Board Statement: Not applicable.

Informed Consent Statement: Not applicable.

Data Availability Statement: The data from this study are available from the corresponding author upon reasonable request.

Conflicts of Interest: The authors declare no conflict of interest.

References

1. Boyd, C.E.; McNevin, A.A. Land Use in Shrimp Aquaculture. *World Aquacult.* **2018**, *49*, 28–34.
2. Haslun, J. Characterization of Bioflocs in a No Water Exchange Super-intensive System for the Production of Food Size Pacific White Shrimp *Litopenaeus vannamei*. *Int. J. Aquac.* **2012**, *2*, 29–38. [CrossRef]
3. Li, Y.; Boyd, C.E. Laboratory tests of bacterial amendments for accelerating oxidation rates of ammonia, nitrite and organic matter in aquaculture pond water. *Aquaculture* **2016**, *460*, 45–58. [CrossRef]
4. Abu Bakar, N.S.; Mohd Nasir, N.; Lananan, F.; Abdul Hamid, S.H.; Lam, S.S.; Jusoh, A. Optimization of C/N ratios for nutrient removal in aquaculture system culturing African catfish, (*Clarias gariepinus*) utilizing Bioflocs Technology. *Int. Biodeterior. Biodegrad.* **2015**, *102*, 100–106. [CrossRef]
5. Azim, M.E.; Little, D.C. The biofloc technology (BFT) in indoor tanks: Water quality, biofloc composition, and growth and welfare of Nile tilapia (*Oreochromis niloticus*). *Aquaculture* **2008**, *283*, 29–35. [CrossRef]
6. Ray, A.J.; Lotz, J.M. Comparing a chemoautotrophic-based biofloc system and three heterotrophic-based systems receiving different carbohydrate sources. *Aquac. Eng.* **2014**, *63*, 54–61. [CrossRef]
7. Ebeling, J.M.; Timmons, M.B.; Bisogni, J.J. Engineering analysis of the stoichiometry of photoautotrophic, autotrophic, and heterotrophic removal of ammonia–nitrogen in aquaculture systems. *Aquaculture* **2006**, *257*, 346–358. [CrossRef]
8. Xu, W.-J.; Morris, T.C.; Samocha, T.M. Effects of C/N ratio on biofloc development, water quality, and performance of *Litopenaeus vannamei* juveniles in a biofloc-based, high-density, zero-exchange, outdoor tank system. *Aquaculture* **2016**, *453*, 169–175. [CrossRef]
9. Zhu, S.; Chen, S. Effects of organic carbon on nitrification rate in fixed film biofilters. *Aquac. Eng.* **2001**, *25*, 1–11. [CrossRef]
10. Avnimelech, Y. Carbon nitrogen ratio as a control element in aquaculture systems. *Aquaculture* **1999**, *176*, 227–235. [CrossRef]
11. Hargreaves, J.A. Photosynthetic suspended-growth systems in aquaculture. *Aquac. Eng.* **2006**, *34*, 344–363. [CrossRef]

12. Hargreaves, J.A. *Biofloc Production Systems for Aquaculture*; Southern Regional Aquaculture Center Stoneville: Stoneville, MS, USA, 2013; Volume 4503, pp. 1–11.

13. Ren, W.; Li, L.; Dong, S.; Tian, X.; Xue, Y. Effects of C/N ratio and light on ammonia nitrogen uptake in *Litopenaeus vannamei* culture tanks. *Aquaculture* **2019**, *498*, 123–131. [CrossRef]

14. Qiu, T.; Xu, Y.; Gao, M.; Han, M.; Wang, X. Bacterial community dynamics in a biodenitrification reactor packed with polylactic acid/poly (3-hydroxybutyrate-co-3-hydroxyvalerate) blend as the carbon source and biofilm carrier. *J. Biosci. Bioeng.* **2017**, *123*, 606–612. [CrossRef] [PubMed]

15. Wu, W.; Yang, L.; Wang, J. Denitrification performance and microbial diversity in a packed-bed bioreactor using PCL as carbon source and biofilm carrier. *Appl. Microbiol. Biotechnol.* **2013**, *97*, 2725–2733. [CrossRef] [PubMed]

16. Schneider, O.; Sereti, V.; Eding, E.H.; Verreth, J.A.J. Molasses as C source for heterotrophic bacteria production on solid fish waste. *Aquaculture* **2006**, *261*, 1239–1248. [CrossRef]

17. Fatimah, N.; Pande, G.S.J.; Natrah, F.M.I.; Meritha, W.W.; Widanarni; Sucipto, A.; Ekasari, J. The role of microbial quorum sensing on the characteristics and functionality of bioflocs in aquaculture systems. *Aquaculture* **2019**, *504*, 420–426. [CrossRef]

18. Coyte, K.Z.; Schluter, J.; Foster, K.R. The ecology of the microbiome: Networks, competition, and stability. *Science* **2015**, *350*, 663–666. [CrossRef]

19. Zhang, N.; Luo, G.; Tan, H.; Liu, W.; Hou, Z. Growth, digestive enzyme activity and welfare of tilapia (*Oreochromis niloticus*) reared in a biofloc-based system with poly-β-hydroxybutyric as a carbon source. *Aquaculture* **2016**, *464*, 710–717. [CrossRef]

20. Kim, M.S.; Min, E.; Kim, J.H.; Koo, J.K.; Kang, J.C. Growth performance and immunological and antioxidant status of Chinese shrimp, Fennerpenaeus chinensis reared in bio-floc culture system using probiotics. *Fish Shellfish Immunol.* **2015**, *47*, 141–146. [CrossRef]

21. Eaton, A. *Standard Methods for the Examination of Water and WasteWater*; AWWA: Washington DC, USA, 2005.

22. Gilcreas, F.W. Standard methods for the examination of water and waste water. *Am. J. Public Health Nations Health* **1966**, *56*, 387–388. [CrossRef]

23. Bratvold, D.; Browdy, C.L. Simple electrometric methods for estimating microbial activity in aquaculture ponds. *Aquac. Eng.* **1998**, *19*, 29–39. [CrossRef]

24. Murray, M.G.; Thompson, W.F. Rapid isolation of high molecular weight plant DNA. *Nucleic Acids Res.* **1980**, *8*, 4321–4325. [CrossRef] [PubMed]

25. Bolger, A.M.; Lohse, M.; Usadel, B. Trimmomatic: A flexible trimmer for Illumina sequence data. *Bioinformatics* **2014**, *30*, 2114–2120. [CrossRef] [PubMed]

26. Martin, M. Cutadapt removes adapter sequences from high-throughput sequencing reads. *EMBnet J.* **2011**, *17*. [CrossRef]

27. Edgar, R.C. UPARSE: Highly accurate OTU sequences from microbial amplicon reads. *Nat. Methods* **2013**, *10*, 996–998. [CrossRef] [PubMed]

28. Haas, B.J.; Gevers, D.; Earl, A.M.; Feldgarden, M.; Ward, D.V.; Giannoukos, G.; Ciulla, D.; Tabbaa, D.; Highlander, S.K.; Sodergren, E.; et al. Chimeric 16S rRNA sequence formation and detection in Sanger and 454-pyrosequenced PCR amplicons. *Genome Res.* **2011**, *21*, 494–504. [CrossRef]

29. Bolyen, E.; Rideout, J.R.; Dillon, M.R.; Bokulich, N.A.; Abnet, C.C.; Al-Ghalith, G.A.; Alexander, H.; Alm, E.J.; Arumugam, M.; Asnicar, F. Reproducible, interactive, scalable and extensible microbiome data science using QIIME 2. *Nat. Biotechnol.* **2019**, *37*, 852–857. [CrossRef]

30. Segata, N.; Izard, J.; Waldron, L.; Gevers, D.; Miropolsky, L.; Garrett, W.S.; Huttenhower, C. Metagenomic biomarker discovery and explanation. *Genome Biol.* **2011**, *12*, R60. [CrossRef]

31. Louca, S.; Parfrey, L.W.; Doebeli, M. Decoupling function and taxonomy in the global ocean microbiome. *Science* **2016**, *353*, 1272–1277. [CrossRef]

32. Ward, T.; Larson, J.; Meulemans, J.; Hillmann, B.; Lynch, J.; Sidiropoulos, D.; Spear, J.R.; Caporaso, G.; Blekhman, R.; Knight, R.; et al. BugBase Predicts Organism Level Microbiome Phenotypes. *BioRxiv* **2017**, 133462. [CrossRef]

33. Schveitzer, R.; Arantes, R.; Costódio, P.F.S.; do Espírito Santo, C.M.; Arana, L.V.; Seiffert, W.Q.; Andreatta, E.R. Effect of different biofloc levels on microbial activity, water quality and performance of *Litopenaeus vannamei* in a tank system operated with no water exchange. *Aquac. Eng.* **2013**, *56*, 59–70. [CrossRef]

34. Emilie, C.; Yannick, G.; Kevin, M.; Bénédicte, L.; David, P.; Fabien, P.; Florian, N.; Denis, S. Bacterial community characterization of water and intestine of the shrimp Litopenaeus stylirostrisin a biofloc system. *BMC Microbiol.* **2016**, *16*, 157. [CrossRef]

35. Gerardi, M.H. *Wastewater Bacteria*; Wiley-Interscience: Hoboken, NJ, USA, 2006.

36. Martínez-Córdova, L.R.; Vargas-Albores, F.; Garibay-Valdez, E.; Ortíz-Estrada, Á.M.; Porchas-Cornejo, M.A.; Lago-Lestón, A.; Martínez-Porchas, M. Amaranth and wheat grains tested as nucleation sites of microbial communities to produce bioflocs used for shrimp culture. *Aquaculture* **2018**, *497*, 503–509. [CrossRef]

37. Yu, E.; Xie, J.; Wang, J.; Ako, H.; Wang, G.; Chen, Z.; Liu, Y. Surface-attached and suspended bacterial community structure as affected by C/N ratios: Relationship between bacteria and fish production. *World J. Microbiol. Biotechnol.* **2016**, *32*, 116. [CrossRef]

38. Kirchman, D.L. The ecology of *Cytophaga-Flavobacteria* in aquatic environments. *FEMS Microbiol. Ecol.* **2002**, *39*, 91–100. [CrossRef]

39. Woebken, D.; Fuchs, B.M.; Kuypers, M.M.; Amann, R. Potential interactions of particle-associated anammox bacteria with bacterial and archaeal partners in the Namibian upwelling system. *Appl. Environ. Microbiol.* **2007**, *73*, 4648–4657. [CrossRef]

40. Yang, W.; Zhu, J.; Zheng, C.; Qiu, H.; Zheng, Z.; Lu, K. Succession of bacterioplankton community in intensive shrimp (*Litopenaeus vannamei*) aquaculture systems. *Aquaculture* **2018**, *497*, 200–213. [CrossRef]

41. Miranda-Baeza, A.; Mariscal-López, M.d.l.A.; López-Elías, J.A.; Rivas-Vega, M.E.; Emerenciano, M.; Sánchez-Romero, A.; Esquer-Méndez, J.L. Effect of inoculation of the cyanobacteria *Oscillatoria* sp. on tilapia biofloc culture. *Aquacult. Res.* **2017**, *48*, 4725–4734. [CrossRef]

42. Kim, Y.-S.; Kim, S.-E.; Kim, S.-J.; Jung, H.-K.; Park, J.; Jeon, Y.J.; Kim, D.-H.; Kang, J.-H.; Kim, K.-H. Effects of wheat flour and culture period on bacterial community composition in digestive tracts of *Litopenaeus vannamei* and rearing water in biofloc aquaculture system. *Aquaculture* **2021**, *531*, 735908. [CrossRef]

43. Goffredi, S.K. Indigenous ectosymbiotic bacteria associated with diverse hydrothermal vent invertebrates. *Environ. Microbiol. Rep.* **2010**, *2*, 479–488. [CrossRef]

44. Hansen, G.H.; Olafsen, J.A. Bacterial Interactions in Early Life Stages of Marine Cold Water Fish. *Microb. Ecol.* **1999**, *38*, 1–26. [CrossRef] [PubMed]

45. Gutiérrez-Salazar, G.J.; Molina-Garza, Z.J.; Hernández-Acosta, M.; García-Salas, J.A.; Mercado-Hernández, R.; Galaviz-Silva, L. Pathogens in Pacific white shrimp (*Litopenaeus vannamei* Boone, 1931) and their relationship with physicochemical parameters in three different culture systems in Tamaulipas, Mexico. *Aquaculture* **2011**, *321*, 34–40. [CrossRef]

46. Barcenal, A.; Traifalgar, R.; Jr, V. Anti-Vibrio harveyi Property of Micrococcus luteus Isolated from Rearing Water under Biofloc Technology Culture System. *Curr. Res. Bacteriol.* **2015**, *8*, 26–33. [CrossRef]

47. Dias Schleder, D.; Jatobá, A.; Silva, B.; Ferro, D.; Seiffert, W.; do Nascimento Vieira, F. Soybean protein concentrate in Pacific white shrimp reared in bioflocs: Effect on health and vibrio challenge. *Acta Sci. Anim. Sci.* **2018**, *40*, 1–6. [CrossRef]

48. Jiang, W.; Ren, W.; Li, L.; Dong, S.; Tian, X. Light and carbon sources addition alter microbial community in biofloc-based *Litopenaeus vannamei* culture systems. *Aquaculture* **2020**, *515*, 734572. [CrossRef]

49. Zhang, H.; Sun, Z.L.; Liu, B.; Xuan, Y.M.; Jiang, M.; Pan, Y.S.; Zhang, Y.M.; Gong, Y.P.; Lu, X.P.; Yu, D.S.; et al. Dynamic changes of microbial communities in *Litopenaeus vannamei* cultures and the effects of environmental factors. *Aquaculture* **2016**, *455*, 97–108. [CrossRef]

50. Li, J.; Li, F.; Yu, S.; Qin, S.; Wang, G. Impacts of mariculture on the diversity of bacterial communities within intertidal sediments in the Northeast of China. *Microb. Ecol.* **2013**, *66*, 861–870. [CrossRef]

51. Bunse, C.; Pinhassi, J. Marine Bacterioplankton Seasonal Succession Dynamics. *Trends Microbiol.* **2017**, *25*, 494–505. [CrossRef]

52. Lara, G.; Furtado, P.; Hostins, B.; Poersch, L.; Wasielesky Jr, W. Addition of sodium nitrite and biofilm in a *Litopenaeus vannamei* biofloc culture system. *Lat. Am. J. Aquat. Res.* **2016**, *44*, 760–768. [CrossRef]

53. Xue, Y.; Wei, F.; Jiang, Y.; Li, L.; Dong, S.; Tian, X. Transcriptome signatures of the Pacific white shrimp *Litopenaeus vannamei* hepatopancreas in response to stress in biofloc culture systems. *Fish Shellfish Immunol.* **2019**, *91*, 369–375. [CrossRef]

54. Emerenciano, M.G.C.; Martínez-Córdova, L.R.; Martínez-Porchas, M.; Miranda-Baeza, A. Biofloc technology (BFT): A tool for water quality management in aquaculture. *Water Qual.* **2017**, *5*, 92–109. [CrossRef]

55. Deng, M.; Chen, J.; Hou, J.; Li, D.; He, X. The effect of different carbon sources on water quality, microbial community and structure of biofloc systems. *Aquaculture* **2017**, *482*. [CrossRef]

56. Wang, G.J.; Yu, E.M.; Xie, J.; Yu, D.G.; Li, Z.F.; Luo, W.; Qiu, L.J.; Zheng, Z.L. Effect of C/N ratio on water quality in zero-water exchange tanks and the biofloc supplementation in feed on the growth performance of crucian carp, *Carassius auratus*. *Aquaculture* **2015**, *443*, 98–104. [CrossRef]

57. Wei, Y.; Liao, S.-A.; Wang, A.-l. The effect of different carbon sources on the nutritional composition, microbial community and structure of bioflocs. *Aquaculture* **2016**, *465*, 88–93. [CrossRef]

58. Avnimelech, Y. *Biofloc Technology: A Practical Guide Book*; The World Aquaculture Society: Baton Rouge, LA, USA, 2012; pp. 73–91.

59. Avnimelech, Y. Feeding with microbial flocs by tilapia in minimal discharge bio-flocs technology ponds. *Aquaculture* **2007**, *264*, 140–147. [CrossRef]

60. Ahmad, I.; Rani, A.M.B.; Verma, A.K.; Maqsood, M. Biofloc technology: An emerging avenue in aquatic animal healthcare and nutrition. *Aquacult. Int.* **2017**, *25*, 1215–1226. [CrossRef]

61. Boley, A.; Muller, W.R.; Haider, G. Biodegradable polymers as solid substrate and biofilm carrier for denitrification in recirculated aquaculture systems. *Aquac. Eng.* **2000**, *22*, 75–85. [CrossRef]

Article

Effects of Desiccation, Water Velocity, and Nitrogen Limitation on the Growth and Nutrient Removal of *Neoporphyra haitanensis* and *Neoporphyra dentata* (Bangiales, Rhodophyta)

Jingyu Li [1,2], Guohua Cui [1], Yan Liu [1,2], Qiaohan Wang [1,2], Qingli Gong [1,2] and Xu Gao [1,2,*]

1   Fisheries College, Ocean University of China, Qingdao 266003, China; qdlijingyu@ouc.edu.cn (J.L.); 15063973601@163.com (G.C.); qd_liuyan@ouc.edu.cn (Y.L.); wangqiaohan@ouc.edu.cn (Q.W.); qingli@vip.sina.com (Q.G.)
2   Key Laboratory of Mariculture, Ministry of Education, Ocean University of China, Qingdao 266003, China
*   Correspondence: gaoxu@ouc.edu.cn; Tel.: +86-532-8203-2377

Abstract: Seaweeds have been verified to effectively reduce the nutrients of aquaculture wastewater, and to increase the economic output when commercially valuable species are utilized. *Pyropia/Porphyra/Neopyropia/Neoporphyra* species are important seafood resources globally, and their growth and bioremediation capacities are affected by diverse biotic and abiotic stressors. In this study, we investigated the effects of desiccation (0, 1, 2, 4, and 6 h of air exposure), water velocity (0.1, 0.2, and 0.5 m s^{-1}), and the nitrogen limitation period (1, 2, and 3 d) on the relative growth rates (RGR) and nutrient removal rates of *Neoporphyra haitanensis* and *Neoporphyra dentata*. The RGRs and NO$_3$-N removal rates of the two species decreased significantly with increasing desiccation periods. A higher water velocity of 0.5 m s^{-1} had a greater negative impact on the RGRs and NO$_3$-N and PO$_4$-P removal rates than 0.1 and 0.2 m s^{-1}. *N. haitanensis* exhibited a greater tolerance to water motion than *N. dentata*. Additionally, the RGRs and NO$_3$-N and PO$_4$-P removal rates were significantly different among the nitrogen limitation periods. *N. haitanensis* and *N. dentata* exhibited different nitrogen usage strategies after nitrogen limitation and recovery. These results provide valuable information relating to the excessive nutrient removal from aquaculture wastewater by *Neoporphyra* species.

Keywords: aquaculture wastewater; desiccation; growth; *Neoporphyra*; nitrogen limitation; NO$_3$-N and PO$_4$-P removal; water velocity

Citation: Li, J.; Cui, G.; Liu, Y.; Wang, Q.; Gong, Q.; Gao, X. Effects of Desiccation, Water Velocity, and Nitrogen Limitation on the Growth and Nutrient Removal of *Neoporphyra haitanensis* and *Neoporphyra dentata* (Bangiales, Rhodophyta). *Water* **2021**, *13*, 2745. https://doi.org/10.3390/w13192745

Academic Editor: Jesus Gonzalez-Lopez

Received: 10 September 2021
Accepted: 27 September 2021
Published: 2 October 2021

Publisher's Note: MDPI stays neutral with regard to jurisdictional claims in published maps and institutional affiliations.

1. Introduction

Aquaculture was one of the fastest-growing commercial activities in the last few decades. The production of marine organisms has broken historical records, reaching 114.5 million tonnes in 2018 [1]. In terms of the increasing storage of natural resources, the recirculating aquaculture system (RAS) has become one of the most sustainable models of marine animal aquaculture [2–4]. Due to high density cultivation with limited volumes of seawater, the wastewater from RAS usually contains high concentrations of nutrients, posing a potential risk to the surrounding environment [3,5,6]. Much research has been performed to look for bioremediation technologies that could solve this problem and ensure its environmental sustainability [7–9]. Recently, due to their low cost and high uptake efficiency, seaweeds have become a feasible alternative in the bioremediation of eutrophic wastewater [10–12]. It is very critical to select appropriate seaweed species with great nutrient demands and high economic value for RAS.

Intertidal seaweeds are subjected to cyclical immersion and emersion because of their periodic exposure to tidal fluctuations. During low tide, the intertidal seaweeds are exposed to air and experience various environmental stresses, such as drastic temperature shifts, high osmotic pressure, and desiccation [13–15]. Desiccation with different periods and frequencies is unavoidable for seaweeds growing in different vertical zones, and their physiological and metabolic activities, including growth and nutrient assimilation, are significantly affected [16–18]. Besides desiccation, water velocity is another important abiotic factor affecting the metabolism of seaweeds [15,19]. Adverse hydrodynamic conditions in the sea can produce different degrees of influences on the productivity of natural and cultivated seaweeds, which depends on their different environmental tolerances and nutrient utilization performances [20–23]. Similarly, under laboratory conditions, the growth and nutrient uptake of diverse seaweed species have been identified to be significantly inhibited by immoderate water velocity [24,25].

Moreover, the nutrient uptake and growth of seaweeds are greatly affected by diverse biotic factors, including the species, competition, growth phase, and nitrogen level in the algae [15,26,27]. Nitrogen and phosphorus are two essential nutrient components to be incorporated into physiological compounds that are crucial for seaweed growth and development [28]. In recent years, ambient nitrogen deficiency has been reported to cause aggravated damage to the seaweed mariculture systems due to the reduced environmental tolerance of algae caused by an undesirable internal nutrient status [29–31]. Nevertheless, as a physiological strategy in response to nutrient deprivation, it was found that the nitrogen uptake ability of seaweeds was enhanced under the condition of nitrogen limitation [32–34], and was positively associated with the degree of nitrogen limitation [35]. Therefore, the assessment of the potential of nitrogen limitation in aquaculture water purification with seaweeds is considered to have significant ecological and commercial values.

Macroalgae of the genus *Neoporphyra* are considered edible, delicious and nutritious, and were recently separated from the genus *Porphyra* [36]. *Neoporphyra* species are important commercially available marine crops in China, and have been massively cultivated because of great demand [17,37,38]. Due to their extremely high surface area to volume ratio, they are capable of the rapid assimilation of nutrients, which promotes high rates of growth in these algae [39–41]. As a result, they have been proven to contribute to the purification of seawater in aquaculture ponds and nearshore farming areas in China, with total nitrogen and PO_4-P removal rates of 65.8–80.2% and 71.1–84.6% [42,43]. These facts suggest that this genus is one of the most promising candidates for bioremediation and integrated aquaculture [40,41,44].

Neoporphyra haitanensis and *Neoporphyra dentata* are two common *Pyropia* species inhabiting the intertidal zone of rocky shores along the coast of southern China [45]. Their optimal temperature ranges for growth are 19–23 °C and 17–23 °C, respectively [46,47], which are generally consistent with the cultivation temperature of economic marine animals such as turbot and grouper [48,49]. These provide advantageous conditions for aquaculture wastewater purification using these two species. As one of the major commercial species, *N. haitanensis* has received extensive attention for its physiological and metabolic responses to biotic and abiotic stressors for aquaculture production optimization [50–54]. *N. dentata* is a promising species for cultivation in South China, and has been farmed on an experimental scale over the past few years [46,47,55]. Nevertheless, the nutrient removal capacities and potential application in aquaculture wastewater purification have been rarely investigated in these species. Furthermore, it is very vital to understand the correlation between the nutrient removal capacities and diverse biotic and abiotic factors.

In the present study, three short-term laboratory experiments were conducted to investigate the respective effects of desiccation, water velocity and nitrogen limitation on the growth and NO_3-N and PO_4-P removal of *N. haitanensis* and *N. dentata*. The results of this study are expected to provide valuable information to improve aquaculture wastewater management and to assess the bioremediation potential of these two high-valued cultivars in China.

2. Materials and Methods

2.1. Sample Collection and Maintenance

Gametophytic thalli of *N. haitanensis* and *N. dentata* were collected from cultivated populations on Nanao Island, Guangdong, China (23°28′ N, 117°06′ E) in December 2018. These samples were rinsed several times with filtered seawater to remove epiphytic organisms and detritus. The surface seawater temperature at the sampling site was measured at the same time. The samples were promptly transported to the laboratory under low-temperature conditions. Healthy thalli were then selected and cultured in several plastic tanks containing sterilized seawater. For the subsequent experiments, these thalli were maintained at 23 °C (the surface seawater temperature at the sampling site), with an irradiance of 100 μmol photon m^{-2} s^{-1} and a 12:12-h light/dark cycle for 2 days.

2.2. Desiccation Experiment

A culture experiment was conducted over a period of 4 days after five periods of desiccation: 0, 1, 2, 4, and 6 h of air exposure. The water loss percentages of *N. haitanensis* and *N. dentata* were 34.4% and 41.4% after 1 h of desiccation, 55.2% and 59.2% after 2 h of desiccation, 70.5% and 71.9% after 4 h of desiccation, and 77.6% and 81.2% after 6 h of desiccation, respectively. There were a total of 10 experimental treatments for each species, and each treatment was performed in three replicates. Before the culturing, 5 g thalli were randomly selected for each replicate. After being blotted dry, those thalli at 1–6 h of desiccation treatments were transferred into autoclaved Petri dishes (10 cm in diameter) containing a layer of gauze soaked with a small amount of culture medium (NO_3-N: 50 mg L^{-1}; PO_4-P: 5 mg L^{-1}), which was made using a nutrient solution and sterilized seawater from the coast of Taipingjiao, Qingdao, with a salinity of approximately 31 psu. Next, these Petri dishes were placed into incubators at 23 °C for 1, 2, 4, and 6 h, respectively. After desiccation, the thalli of each replicate were moved into a side-arm flask with 500 mL culture medium and GeO_2, which were then gently aerated. During this experiment, a temperature of 23 °C, a 12:12-h light/dark cycle, and an irradiance of 100 μmol photon m^{-2} s^{-1} were maintained.

The fresh weights of all of the thalli before and after the experiment were measured after removing excess seawater on the surface. The relative growth rate (RGR; % day^{-1}) of each replicate was calculated using the following Equation (1):

$$RGR\ (\%\ day^{-1}) = 100 \times (\ln W_t - \ln W_o)/t \tag{1}$$

where W_o is the initial fresh weight, W_t is the final fresh weight, and t is the time of the culture in days.

For all of the treatments, the culture media before and after the experiment were separately collected, and the concentrations of NO_3-N and PO_4-P were analyzed using the cadmium column reduction method and the phosphomolybdenum blue spectrophotometric method, respectively [56,57]. The removal rates of NO_3-N and PO_4-P were estimated using the following Equation (2):

$$R_{N,\,P} = (C_0 - C_4)/C_0 \times 100\% \tag{2}$$

where $R_{N,\,P}$ are the removal rates of NO_3-N and PO_4-P (%); C_0 is the initial concentration of NO_3-N and PO_4-P (mg L^{-1}); and C_4 is the final concentration of NO_3-N and PO_4-P (mg L^{-1}) after 4 days.

2.3. Water Velocity Experiment

In order to examine the effect of the water velocity on the growth and nutrient removal of these two species, they were cultured for 4 days at three water velocities (0.1, 0.2, and 0.5 m s^{-1}) with three replicates. For this experiment, a total of 18 side-arm flasks were prepared, and each contained 500 mL culture medium with GeO_2 and 5 g thalli. During the experimental period, a temperature of 23 °C, a 12:12-h light/dark cycle, and an irradiance of 100 μmol photon m^{-2} s^{-1} were maintained. At the end of this experiment, the calculations of the RGR and the removal rates of NO_3-N and PO_4-P were the same as for the desiccation experiment.

2.4. Nitrogen Limitation Experiment

In order to investigate the effect of nitrogen limitation on the growth and nutrient removal of these two species, a total of 18 side-arm flasks including three replicates for each treatment were prepared. Each flask contained 500 mL culture medium with GeO_2 and 5 g thalli. These two species were incubated for 4 days after three different periods of nitrogen limitation (1, 2, and 3 day). We used nitrogen-deficient seawater (NO_3-N: 5 mg L^{-1}) to achieve different nitrogen levels in the algae. During the 4-day experiment, a temperature of 23 °C, a 12:12-h light/dark cycle, and an irradiance of 100 μmol photon m^{-2} s^{-1} were maintained. The RGR and removal rates of NO_3-N and PO_4-P were calculated in the same way as described above.

2.5. Statistical Analysis

Two-way analysis of variances (ANOVA) were used to analyze the effects of desiccation and the species, water velocity and the species, and nitrogen limitation and the species on the RGR and NO_3-N and PO_4-P removal rates. Prior to the ANOVA tests, all of the data were confirmed to show a normal distribution and homogeneity of variance. When a significant difference was identified by the ANOVA, Tukey's multiple comparisons test was used to determine which levels of each factor produced significant differences ($p < 0.05$). All of the analyses were performed using STATISTICA version 7.0 software.

3. Results

3.1. Effect of Desiccation on Growth and Nutrient Removal

The results of the two-way ANOVA showed that the RGRs of *N. haitanensis* and *N. dentata* were significantly affected by desiccation, but they did not significantly differ between the two species (Figure 1; Table 1). A significant interaction between desiccation and the species was not detected. The RGRs of the two species decreased significantly with increasing desiccation periods from 0 to 6 h.

Table 1. Analysis of variance (two-way ANOVA) examining the effects of desiccation on the *RGR* and NO_3-N and PO_4-P removal rates of *N. haitanensis* and *N. dentata*, and the comparison of these parameters between the two species.

Factors	df	F	P
RGR			
Desiccation (D)	4	5.973	<0.01
Species (S)	1	1.187	0.346
Interaction (D × S)	4	0.851	2.061
NO_3-N removal rate			
Desiccation (D)	4	16.177	<0.001
Species (S)	1	1.465	0.239
Interaction (D × S)	4	2.935	<0.05
PO_4-P removal rate			
Desiccation (D)	4	0.834	2.947
Species (S)	1	1.181	0.385
Interaction (D × S)	4	0.655	6.421

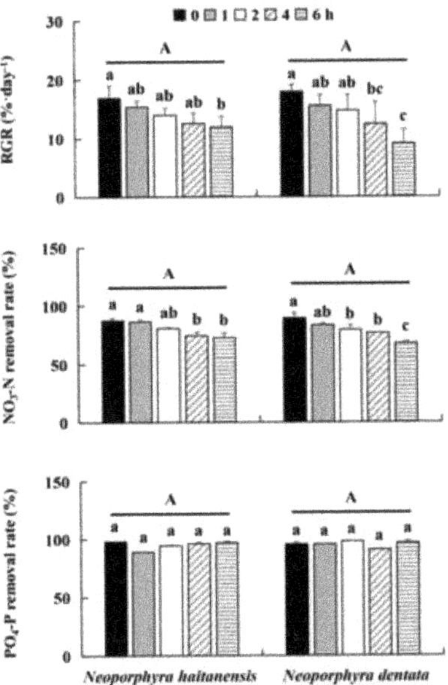

Figure 1. The RGRs and NO_3-N and PO_4-P removal rates of *N. haitanensis* and *N. dentata* cultured for 4 days under five desiccation periods. Different capital and small letters indicate statistical significance ($P < 0.05$) among species and desiccation periods, respectively. The data represent the mean $\pm$ SE (n = 3 replicates).

The NO_3-N removal rates of *N. haitanensis* and *N. dentata* were significantly affected by desiccation, but they did not significantly differ between the two species. A significant interaction between desiccation and the species was detected. The NO_3-N removal rates of *N. haitanensis* and *N. dentata* decreased significantly with increasing desiccation periods, with ranges of 87.2–72.4% and 89.2–67.1%, respectively.

The PO_4-P removal rates of *N. haitanensis* and *N. dentata* were not significantly affected by desiccation, and also did not significantly differ between the two species. A significant interaction between desiccation and the species was not detected. The PO_4-P removal rates of *N. haitanensis* and *N. dentata* ranged from 88.9 to 98.0%, and from 90.9 to 98.2%, respectively.

3.2. Effect of the Water Velocity on Growth and Nutrient Removal

The RGRs of *N. haitanensis* and *N. dentata* were significantly affected by the water velocity, and also differed significantly between the two species (Figure 2; Table 2). A significant interaction between the water velocity and species was detected. The RGRs of *N. haitanensis* and *N. dentata* at 0.1 and 0.2 m s^{-1} were significantly greater than those at 0.5 m s^{-1}. The RGR of *N. dentata* at 0.2 m s^{-1} was significantly greater than that at 0.1 m s^{-1}. The RGRs of *N. haitanensis* were significantly greater than those of *N. dentata* at all three water velocities.

The NO_3-N removal rates of *N. haitanensis* and *N. dentata* were significantly affected by the water velocity, but they did not significantly differ between the two species. A significant interaction between the water velocity and species was detected. The NO_3-N removal rates of *N. haitanensis* and *N. dentata* at 0.1 (*N. haitanensis*: 86.2%; *N. dentata*: 80.5%) and 0.2 m s^{-1} (*N. haitanensis*: 83.7%; *N. dentata*: 89.9%) were significantly greater than those at 0.5 m s^{-1} (*N. haitanensis*: 44.6%; *N. dentata*: 48.4%). There were no significant differences in the NO_3-N removal rates between 0.1 and 0.2 m s^{-1} for both species.

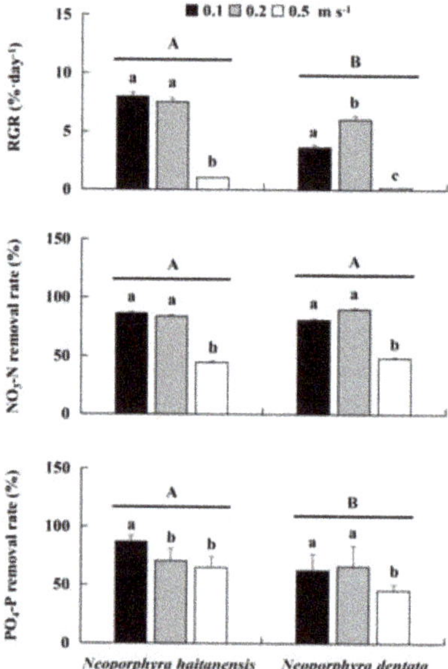

Figure 2. The RGRs and NO_3-N and PO_4-P removal rates of *N. haitanensis* and *N. dentata* cultured for 4 days under three water velocities. Different capital and small letters indicate statistical significance ($P < 0.05$) among species and water velocities, respectively. The data represent the mean $\pm$ SE (n = 3 replicates).

Table 2. Analysis of variance (two-way ANOVA) examining the effects of the water velocity on the *RGR* and NO_3-N and PO_4-P removal rates of *N. haitanensis* and *N. dentata*, and the comparison of these parameters between the two species.

Factors	df	F	P
RGR			
Water velocity (W)	2	29.933	<0.001
Species (S)	1	16.025	<0.001
Interaction (W × S)	2	27.251	<0.001
NO_3-N removal rate			
Water velocity (W)	2	20.070	<0.001
Species (S)	1	0.773	4.197
Interaction (W × S)	2	5.773	<0.01
PO_4-P removal rate			
Water velocity (W)	2	18.191	<0.001
Species (S)	1	9.966	<0.001
Interaction (W × S)	2	11.555	<0.001

The PO_4-P removal rates of *N. haitanensis* and *N. dentata* were significantly affected by the water velocity, and also differed significantly between the two species. A significant interaction between the water velocity and species was detected. The PO_4-P removal rate of *N. haitanensis* at 0.1 m s^{-1} (86.9%) was significantly greater than those at 0.2 (70.3%) and 0.5 m s^{-1} (64.8%). The PO_4-P removal rates of *N. dentata* at 0.1 (62.4%) and 0.2 m s^{-1} (65.9%) were significantly greater than that at 0.5 m s^{-1} (45.4%). The PO_4-P removal rate of *N. haitanensis* was significantly greater than those of *N. dentata* at 0.1 and 0.5 m s^{-1}.

3.3. Effect of Nitrogen Limitation on Growth and Nutrient Removal

The RGRs of *N. haitanensis* and *N. dentata* were significantly affected by nitrogen limitation, and also differed significantly between the two species (Figure 3; Table 3). A significant interaction between nitrogen limitation and the species was detected. After

4 day nitrogen recovery, the RGRs of *N. haitanensis* at 2 and 3 day nitrogen limitation were significantly greater than that at 1 day nitrogen limitation. There were no significant differences in the RGRs among the three nitrogen limitation levels for *N. dentata*. The RGR of P. dentata was significantly greater than that of *N. haitanensis* at each nitrogen limitation level.

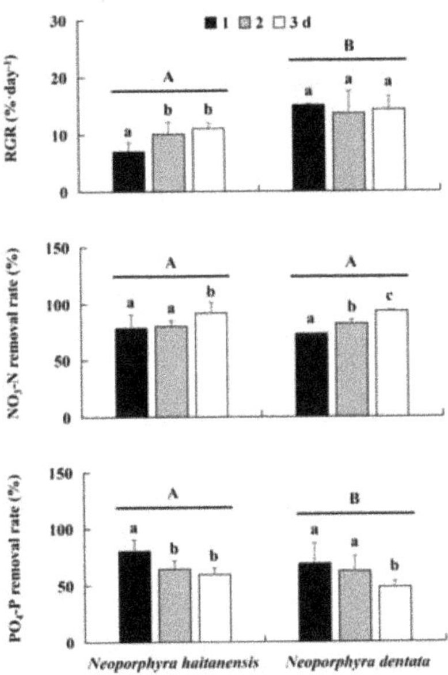

Figure 3. The RGRs and NO$_3$-N and PO$_4$-P removal rates of *N. haitanensis* and *N. dentata* cultured for 4 days after three different periods of nitrogen limitation. Different capital and small letters indicate statistical significance ($P < 0.05$) among species and nitrogen limitation periods, respectively. The data represents the mean $\pm$ SE (n = 3 replicates).

Table 3. Analysis of variance (two-way ANOVA) examining the effects of nitrogen limitation on the *RGR* and NO$_3$-N and PO$_4$-P removal rates of *N. haitanensis* and *N. dentata*, and the comparison of these parameters between the two species.

Factors	df	F	P
RGR			
Nitrogen limitation (N)	2	5.190	<0.01
Species (S)	1	25.021	<0.001
Interaction (N × S)	2	19.201	<0.001
NO$_3$-N removal rate			
Nitrogen limitation (N)	2	23.876	<0.001
Species (S)	1	0.482	9.110
Interaction (N × S)	2	15.766	<0.001
PO$_4$-P removal rate			
Nitrogen limitation (N)	2	28.171	<0.001
Species (S)	1	8.960	<0.001
Interaction (N × S)	2	12.095	<0.001

The NO$_3$-N removal rates of *N. haitanensis* and *N. dentata* were significantly affected by nitrogen limitation, but they did not significantly differ between the two species. A significant interaction between nitrogen limitation and species was detected. After 4 day nitrogen recovery, the NO_3-N removal rate of *N. haitanensis* at 3 day of nitrogen limitation (91.8%) was significantly greater than those at 1 day (78.4%) and 2 day of nitrogen limitation (80.0%). The NO_3-N removal rate of *N. dentata* increased significantly from 1 to 3 day nitrogen limitation (1 day: 72.9%; 2 day: 82.0%; 3 day: 93.3%).

The PO_4-P removal rates of *N. haitanensis* and *N. dentata* were significantly affected by nitrogen limitation, and differed significantly between the two species. A significant interaction between nitrogen limitation and the species was detected. After 4 day of nitrogen recovery, the PO_4-P removal rate of *N. haitanensis* at 1 day of nitrogen limitation (80.3%) was significantly greater than those at 2 (64.4%) and 3 day of nitrogen limitation (59.5%). The PO_4-P removal rates of *N. dentata* at 1 (68.9%) and 2 day of nitrogen limitation (62.4%) were significantly greater than that at 3 day of nitrogen limitation (48.3%). The PO_4-P removal rates of *N. haitanensis* were significantly greater than those of *N. dentata* at 1 and 3 day of nitrogen limitation.

4. Discussion

Intertidal seaweeds usually experienced the stress of desiccation with different periods originating from tidal alternation because they are essentially marine organisms [15,16]. Several studies have demonstrated that desiccation significantly affects the growth and nutrient uptake of *Neoporphyra* species under laboratory conditions [17,18]. Similarly, in the present study, we found that different degrees of desiccation exhibited a significant inhibitory effect on the growth and nitrogen removal of *N. haitanensis* and *N. dentata*. This inhibitory effect intensified significantly with the increase of the desiccation period. This finding is supported by Cao et al. [58], showing that, due to the artificial reduction of the air exposure period, the thalli of *Neoporphyra yezoensis* (formerly *Pyropia yezoensis*) from cultivated populations were significantly larger than those from wild populations in the surrounding area. Li et al. [18] also documented that periodical dehydration could reduce the relative growth rate of *N. yezoensis* by 7–10% in different salinity conditions. However, the effects of desiccation on the growth and nutrient uptake of seaweeds varied from species to species, which is coordinated with their vertical distribution patterns [16,59,60]. Thomas et al. [61] found that the upper-shore species *Pelvetiopsis limitata* and *Fucus distichus* (Phaeophyceae) achieved their maximum nitrate uptake following severe desiccation that inhibited the nitrate uptake in the low-shore species *Gracilaria pacififica* (Rhodophyta). Kim et al. [16] suggested that species in the intertidal zone that have longer exposure times may have a higher time-use efficiency than subtidal species in terms of their nitrate uptake and growth. On the other hand, the impact of desiccation on seaweeds appears to be associated with their water-retaining abilities caused by structure features. For example, desiccation appeared to have no significant influences on the growth and photosynthesis of *Ulva linza* (Chlorophyta) and *Gloiopeltis furcata* (Rhodophyta) because of their internal hollow cavities being conductive to water retention [60,62]. The two *Neoporphyra* species used in this study are monolayer or bilayer membranous thalli with a weak water-retaining capacity, which may partially contribute to the inhibition of nutrient removal under desiccation stress.

Although the water loss percentages of the two experimental species in this study were as high as approximately 80% after 6 h of desiccation, they were still alive. After a 96-h culture, both species effectively recovered the capacity of nutrient absorption. Particularly, the removal rates of PO_4-P were not significantly different among the desiccation periods from 0 to 6 h. These results suggest that *N. haitanensis* and *N. dentata* possess extremely strong resilience to high desiccation stress, and thus can adapt well to changeable oceanic conditions. Similarly, Gao and Wang [63] examined the effect of single dehydration on the physiological features of *N. yezoensis*, and found that the photosynthetic activity of *N. yezoensis* that lost 86% of its cellular water could be fully restored after 30 min of rehydration. Therefore, the periodical emersion of seeded nets has been widely applied in *Pyropia/Neopyropia/Neoporphyra* field aquaculture to kill most fouling organisms, including diatoms and macroalgal spores [64]. Additionally, *N. dentata* appears to be slightly more sensitive to desiccation stress than *N. haitanensis*, even though their vertical distribution ranges are similar [45]. This may be closely associated with the fact that *N. haitanensis* has a thicker thallus (65–110 µm) with local double-layer cell tissue, whereas *N. dentata* only has a very thin single-layer thallus (30–55 µm) [45,65].

Water motion is a critical physical process in the marine environment that can transport the inorganic carbon and nutrients required for the growth and survival of marine macrophytes [66]. It has been demonstrated that an increased water velocity can reduce the diffusion boundary layer along the algae surface, and can thereby enhance its nutrient uptake and growth until a saturating velocity is reached [67,68]. Nevertheless, in this study, the growth of *N. dentata* exhibited a significant increase from 0.1 to 0.2 m s^{-1}, but was significantly inhibited at 0.5 m s^{-1}. Similarly, *N. haitanensis* had a significant lower RGR at 0.5 m s^{-1} compared with lower water velocities. Their removal capacities of NO$_3$-N and PO$_4$-P were also inhibited at a high water velocity of 0.5 m s^{-1}. Analogous growth and physiological inhibition caused by an over-high water velocity has also been reported for the brown algae *Laminaria digitata* and *Sargassum siliquastrum* (Phaeophyceae) [68,69]. Furthermore, Yang [70] found a clear inverse correlation between the cell growth and chlorophyll synthesis of *Chlorella* (Chlorophyta) species and water velocity. All of these results suggested that over-high water velocity can greatly restrict the photosynthetic and nutrient accumulation activities of algae, and can even cause direct damage to algae membranes. Furthermore, *N. haitanensis* appeared to exhibit a greater tolerance and less sensitivity to variable hydrodynamic conditions than *N. dentata*. This may still be correlated with their different structure traits. It is conceivable that thicker and more solid algae can better resist the negative impact of external forces.

In seaweed species, nitrogen can be accumulated for growth demands and stored in its inorganic form in the cellular reserve pool [15,19]. As an ecological adaption to nitrogen deficiency stress, stored nitrogen can be remobilized to support seaweed survival without an ongoing inorganic nitrogen supply [33,71]. Their physiological and metabolic activities, including growth, photosynthesis, and protein and nucleic acid synthesis, will be suppressed by nitrogen deficiency below the critical concentration [72,73]. However, nitrogen replenishment quickly reverted the nitrogen accumulation and metabolism of algae, with the reestablishment of the nitrogen reserve pool and growth promotion [33,50,74]. In this study, the nitrogen removal capacities of both species after nitrogen recovery were positively correlated with the nitrogen limitation period, exhibiting a compensatory response in nitrogen accumulation. Similar phenomena have been documented for floating *Sargassum horneri* (Phaeophyceae) and *Agarophyton tenuistipitatum* (formerly *Gracilaria tenuistipitata*) (Rhodophyta), showing that nitrogen limitation and then recovery significantly increased their ammonium uptake rates in comparison with a control without nitrogen limitation [34,75]. These results suggest that nitrogen limitation can lead to a deficit status of the nitrogen reserve pool. In an effort to maintain life and restore health, the algae in worse conditions must possess a stronger ability to replenish nitrogen when subjected to nitrogen resupply. Curiously, the phosphorus removal in both species did not show the compensatory responses consistent with nitrogen removal. As we known, in seaweed species, there are interactions among different nutrients, which are required in a certain ratio [19]. Perini and Bracken [76] confirmed that nitrogen availability may mediate the ability of marine primary producers to access phosphorus. Therefore, we suspect that particularly rapid nitrogen accumulation may have a negative effect on phosphorus accumulation, which is supported by a finding regarding the phosphorus absorption inhibition by an ambient high nitrogen concentration in *Skeletonema costatum* (Bacillariophyta) [77]. Further studies should be carried out to test our hypothesis.

Under the same nitrogen limitation and recovery conditions, *N. haitanensis* and *N. dentata* showed similar NO$_3$-N removal rates, whereas the RGRs of *N. haitanensis* were significantly lower than those of *N. dentata*. Liu et al. [74] demonstrated that nitrogen deprivation and then recovery notably promoted the growth of *Asparagus schoberioides* (formerly *Gracilaria lemaneiformis*) (Rhodophyta), but significantly inhibited the accumulation of phycoerythrin and chlorophyll a. In contrast, Friedlander et al. [78] reported that after nitrogen limitation and recovery, although the external nitrogen concentration was sufficient, *Gracilaria conferta* (Rhodophyta) did not grow significantly, probably due to more internal organic synthesis. All of these findings suggest that seaweeds under nitrogen starvation show a species-specific nitrogen utilization strategy when encountering nitrogen resupply. As mentioned above, in comparison with *N. dentata*, *N. haitanensis* with a thicker thallus and more complex structure appears to require more nitrogen resources for the synthesis of the necessary organic compounds rather than growth, resulting in the significantly lower RGR and deeper color observed in our experiment.

5. Conclusions

Dealing with eutrophic wastewater is a major challenge for aquaculture production and water environment protection. According to the data of this study, *N. haitanensis* and *N. dentata* appear to play an important role in the removal of inorganic nitrogen and phosphorus from aquaculture wastewater, and therefore are likely to be used as efficient and environmentally friendly remediation tools. *N. dentata* is a more ideal candidate because of its greater environmental resistance. Our results show that desiccation, water velocity and nitrogen limitation are three significant variables affecting the growth and nutrient removal of both species. *N. haitanensis* and *N. dentata* showed different nitrogen usage strategies after nitrogen limitation and recovery. These findings provide valuable information for developing and improving wastewater treatment technologies for RAS. Moreover, apart from being a biological filter, these two *Neoporphyra* species can be harvested as aquaculture byproducts to increase economic income owing to their excellent growth capacities. Due to the limited physiological data in this study, further experiments measuring multiple parameters are warranted to comprehensively assess the application feasibility of wastewater purification using *Neoporphyra* species.

Author Contributions: Conceptualization, methodology, and writing—original draft preparation, J.L. and X.G.; writing—review and checking, X.G.; data curation, project administration, and funding acquisition, J.L.; investigation, G.C. and Y.L.; validation and resources, Q.G.; formal analysis, J.L. and Q.W. All authors have read and agreed to the published version of the manuscript.

Funding: This work was financially supported by the National Key R&D Program of China (No. 2020YFD0900201).

Institutional Review Board Statement: Not applicable.

Informed Consent Statement: Not applicable.

Data Availability Statement: The data of this study are available from the corresponding author upon reasonable request.

Acknowledgments: We are grateful to Weizhou Chen of Shantou University for providing *N. haitanensis* and *N. dentata* as the experimental materials.

Conflicts of Interest: The authors declare no conflict of interest.

References

1. FAO. *The State of World Fisheries and Aquaculture 2020*; FAO: Rome, Italy, 2020.
2. Martins, C.; Eding, E.; Verdegem, M.; Heinsbroek, L.; Schneider, O.; Blancheton, J.; Roque d'Orbcastel, E.; Verreth, J. New developments in recirculating aquaculture systems in Europe: A perspective on environmental sustainability. *Aquac. Eng.* **2010**, *43*, 83–93. [CrossRef]
3. Baloo, L.; Azman, S.; Said, M.; Ahmad, F.; Mohamad, M. Biofiltration potential of macroalgae for ammonium removal in outdoor tank shrimp wastewater recirculation system. *Biomass Bioenerg.* **2014**, *66*, 103–109. [CrossRef]
4. Badiola, M.; Basurko, O.; Piedrahita, R.; Hundley, P.; Mendiola, D. Energy use in recirculating aquaculture systems (RAS): A review. *Aquac. Eng.* **2018**, S0144860917302327. [CrossRef]
5. Marinho-Soriano, E.; Azevedo, C.; Trigueiro, T.; Pereira, D.; Carneiro, M.; Camara, M. Bioremediation of aquaculture wastewater using macroalgae and Artemia. *Int. Biodeterior Biodegrad.* **2011**, *65*, 253–257. [CrossRef]
6. Liu, K.; Fang, T.; Feng, Z.; Zong, S. Research progress on algae treatment to aquaculture wastewater. *J. Huaihai Inst. Technol.* **2016**, *25*, 74–79, (In Chinese with English abstract).

7.   Lin, Y.F.; Jing, S.; Lee, D. The potential use of constructed wetlands in a recirculating aquaculture system for shrimp culture. *Environ. Pollut.* **2003**, *123*, 107–113. [CrossRef]
8.   Burgin, A.; Yang, W.; Hamilton, S.; Silver, W. Beyond carbon and nitrogen: How the microbial energy economy couples elemental cycles in diverse ecosystems. *Front. Ecol. Environ.* **2011**, *9*, 44–52. [CrossRef]
9.   Castine, S.; McKinnon, A.; Paul, N.; Trott, L.; de Nys, R. Wastewater treatment for land-based aquaculture: Improvements and value-adding alternatives in model systems from Australia. *Aquac. Environ. Interact.* **2013**, *4*, 285–300. [CrossRef]
10. Brito, L.; Cardoso Junior, L.; Lavander, H. Bioremediation of shrimp biofloc wastewater using clam, seaweed and fish. *Chem. Ecol.* **2018**, *34*, 901–913. [CrossRef]
11. Marinho-Soriano, E. Historical context of commercial exploitation of seaweeds in Brazil. *J. Appl. Phycol.* **2017**, *29*, 665–671. [CrossRef]
12. Shin, S.; Kim, S.; Kim, J.; Han, T.; Yarish, C.; Kim, J. Effects of stocking density on the productivity and nutrient removal of Agarophyton vermiculophyllum in Paralichthys olivaceus biofloc effluent. *J. Appl. Phycol.* **2020**, *32*, 2605–2614. [CrossRef]
13. Burritt, D.; Larkindale, J.; Hurd, K. Antioxidant metabolism in the intertidal red seaweed Stictosiphonia arbuscula following desiccation. *Planta* **2002**, *215*, 829–838. [CrossRef]
14. Kumar, M.; Gupta, V.; Trivedi, N.; Kumari, P.; Bijo, A.; Reddy, C.; Jha, B. Desiccation induced oxidative stress and its biochemical responses in intertidal red alga Gracilaria corticata (Gracilariales, Rhodophyta) Environ. *Exp. Bot.* **2011**, *72*, 194–201. [CrossRef]
15. Hurd, C.L.; Harrison, P.; Bischof, K.; Lobban, C. *Seaweed Ecology and Physiology*; Cambridge University Press: Cambridge, UK, 2014.
16. Kim, J.; Kraemer, G.; Yarish, C. Comparison of growth and nitrate uptake by New England Porphyra species from different tidal elevations in relation to desiccation. *Phycol. Res.* **2009**, *57*, 152–157. [CrossRef]
17. Wang, W.; Sun, X.; Liu, F.; Liang, R.; Zhang, J.; Wang, F. Effect of abiotic stress on the gameophyte of Pyropia katadae var. hemiphylla (Bangiales, Rhodophyta). *J. Appl. Phycol.* **2016**, *28*, 469–479. [CrossRef]
18. Li, X.; Sun, X.; Gao, L.; Xu, J.; Gao, G. Effects of periodical dehydration on biomass yield and biochemical composition of the edible red alga Pyropia yezoensis grown at different salinities. *Algal Res.* **2021**, *56*, 102315. [CrossRef]
19. Roleda, M.Y.; Hurd, C. Seaweed nutrient physiology: Application of concepts to aquaculture and bioremediation. *Phycologia* **2019**, *58*, 552–562. [CrossRef]
20. Peteiro, C.; Freire, Ó. Biomass yield and morphological features of the seaweed Saccharina latissima cultivated at two different sites in a coastal bay in the Atlantic coast of Spain. *J. Appl. Phycol.* **2013**, *25*, 205–213. [CrossRef]
21. Kregting, L.; Blight, A.; Elsäßer, B.; Savidge, G. The influence of water motion on the growth rate of the kelp Laminaria digitata. *J. Exp. Mar. Biol. Ecol.* **2016**, *478*, 86–95. [CrossRef]
22. Mols-Mortensen, A.; Jacobsen, C.; Holdt, S. Variation in growth, yield and protein concentration in Saccharina latissima (Laminariales, Phaeophyceae) cultivated with different wave and current exposures in the Faroe Islands. *J. Appl. Phycol.* **2017**, *29*, 2277–2286. [CrossRef]
23. Visch, W.; Nylund, G.; Pavia, H. Growth and biofouling in kelp aquaculture (Saccharina latissima): The effect of location and wave exposure. *J. Appl. Phycol.* **2020**, *32*, 3199–3209. [CrossRef]
24. Msuya, F.; Neori, A. Effect of water aeration and nutrient load level on biomass yield, N uptake and protein content of the seaweed Ulva lactuca cultured in seawater tanks. *J. Appl. Phycol.* **2008**, *20*, 1021–1031. [CrossRef]
25. Rula, N.; Ganzon-Fortes, E.T.; Pante, M.; Trono, J. Influence of light, water motion, and stocking density on the growth and pigment content of Halymenia durvillei (Rhodophyceae) under laboratory conditions. *J. Appl. Phycol.* **2021**, *33*, 2367–2377. [CrossRef]
26. Liu, H.; Wang, F.; Wang, Q.; Dong, S.; Tian, X. A comparative study of the nutrient uptake and growth capacities of seaweeds Caulerpa lentillifera and Gracilaria lichenoides. *J. Appl. Phycol.* **2016**, *28*, 3083–3089. [CrossRef]
27. Gao, X.; Endo, H.; Nagaki, M.; Agatsuma, Y. Interactive effects of nutrient availability and temperature on growth and survival of different size classes of Saccharina japonica (Laminariales, Phaeophyceae). *Phycologia* **2017**, *56*, 253–260. [CrossRef]
28. Mouritsen, O.; Mouritsen, J.; Johansen, M. *Seaweeds: Edible, Available, and Sustainable*; University of Chicago Press: Chicago, IL, USA, 2013.
29. Du, R.; Liu, L.; Wang, A.; Wang, Y. Effects of temperature, algae biomass and ambient nutrient on the absorption of dissolved nitrogen and phosphate by Rhodophyte Gracilaria asiatica. *Chin. J. Oceanol. Limnol.* **2013**, *31*, 353–365. [CrossRef]
30. Gao, X.; Endo, H.; Taniguchi, K.; Agatsuma, Y. Combined effects of seawater temperature and nutrient condition on growth and survival of juvenile sporophytes of the kelp Undaria pinnatifida (Laminariales; Phaeophyta) cultivated in northern Honshu, Japan. *J. Appl. Phycol.* **2013**, *25*, 269–275. [CrossRef]
31. Martín, L.; Rodríguez, M.; Matulewicz, M.; Fissore, E.; Gerschenson, L.; Leonardi, P. Seasonal variation in agar composition and properties from Gracilaria gracilis (Gracilariales, Rhodophyta) of the Patagonian coast of Argentina. *Phycol. Res.* **2013**, *61*, 163–171. [CrossRef]
32. Xu, Z. The Physiological Responses of Two Economic Marine Macroalga Species to Nutrients Supplies in Seawater. Master's Thesis, Shantou University, Shantou, China, 2007. (In Chinese with English abstract).
33. Cha, Y. Study on the Nutrients Uptake and Photosynthesis Based on Bioremediation in Two Species of Economic Marine Macroalgae. Master's Thesis, South China University of Technology, Guangzhou, China, 2013. (In Chinese with English abstract).

34. Li, D.P.; Ma, Z.; Li, H.; Ding, G.; Xin, M.; Wu, H.; Guo, W. NH4-N over-compensatory uptake of Sargassum horneri under the stress of nutrients deficiency. *Oceanol. ET Limnol. Sin.* **2018**, *49*, 904–909, (In Chinese with English abstract).

35. McGlathery, K.; Pedersen, M.; Borum, J. Changes in intracellular nitrogen pools and feedback controls on nitrogen uptake in Chaetomorpha linum (Chlorophyta). *J. Phycol.* **1996**, *32*, 393–401. [CrossRef]

36. Sutherland, J.; Lindstrom, C.; Nelson, W.; Brodie, J.; Lynch, M.; Hwang, M.; Choi, H. A new look at an ancient order: Generic revision of the Bangiales (Rhodophyta). *J. Phycol.* **2011**, *47*, 1131–1151. [CrossRef]

37. Zhang, T.; Li, J.; Ma, F.; Lu, Q.; Shen, Z.; Zhu, J. Study of photosynthetic characteristics of the Pyropia yezoensis thallus during the cultivation process. *J. Appl. Phycol.* **2014**, *26*, 859–865. [CrossRef]

38. He, P.; Zhang, Z.; Zhang, X.; Ma, J. *Seaweed Cultivation*; Science Press: Beijing, China, 2018. (In Chinese)

39. Neori, A.; Chopin, T.; Troell, M.; Buschmann, A.; Kraemer, G.; Halling, C.; Shpigel, M.; Yarish, C. Integrated aquaculture: Rational, evolution and state of the art emphasizing seaweed biofiltration in modern mariculture. *Aquaculture* **2004**, *231*, 361–391. [CrossRef]

40. Carmona, R.; Kraemer, G.; Yarish, C. Exploring Northeast American and Asian species of Porphyra for use in an integrated finfish–alga aquaculture system. *Aquaculture* **2006**, *252*, 54–65. [CrossRef]

41. Kang, Y.; Kim, S.; Lee, J.; Chung, I.; Park, S. Nitrogen biofiltration capacities and photosynthetic activity of Pyropia yezoensis Ueda (Bangiales, Rhodophyta): Groundwork to validate its potential in Integrated Multi-Trophic Aquaculture (IMTA) *J. Appl. Phycol.* **2014**, *26*, 947–955. [CrossRef]

42. Tian, J.B.; Wang, Y.; Sun, X.; Zhang, H.; Huang, B. Design and engineering of mariculture effluents purification system. *Fish. Mod.* **2008**, *35*, 1–5, (In Chinese with English abstract).

43. Sun, Q. Nutrients Absorption of Aquaculture Wastewater and Bioremediation of Sea Area by Macroalgae. Master's Thesis, Fujian Normal University, Fuzhou, China, 2013. (In Chinese with English abstract).

44. He, P.; Xu, S.; Zhang, H.; Wen, S.; Dai, Y.; Lin, S.; Yarish, C. Bioremediation efficiency in the removal of dissolved inorganic nutrients by the red seaweed, Porphyra yezoensis, cultivated in the open sea. *Water Res.* **2008**, *42*, 1281–1289. [CrossRef] [PubMed]

45. Zheng, B.; Li, J. *Marine Algae of China*; Science Press: Beijing, China, 2017. (In Chinese)

46. Xu, J.; Chen, W.; Song, Z.; Jiang, H.; Zhu, J.; Lu, Q. Effects of different culture conditions on growth and physiological response of Porphyra dentata thallus. *J. Fish. China* **2013**, *37*, 1319–1327, (In Chinese with English abstract). [CrossRef]

47. Chen, P.; Huang, Z.; Zhu, J.; Lu, Q.; Chen, W. Effect of environmental factor on conchospore releasing, attachment and germination in Pyropia dentata. *South. China Fish. Sci.* **2015**, *11*, 55–61, (In Chinese with English abstract).

48. Cheng, Q.; Zheng, B. *Fish. System Retrieval of China*; Science Press: Beijing, China, 1987. (In Chinese)

49. Lei, L.; Liu, X. A primary study on culture of turbot Scophthalmus maeoticus. *Mod. Fish. Inform.* **1995**, *10*, 1–3, (In Chinese with English abstract).

50. Xu, K.; Chen, H.; Wang, W.; Xu, Y.; Ji, D.; Chen, C.; Xie, C. Responses of photosynthesis and CO_2 concentrating mechanisms of marine crop Pyropia haitanensis thalli to large pH variations at different time scales. *Algal Res.* **2017**, *28*, 200–210. [CrossRef]

51. Jiang, H.; Zou, D.; Lou, W.; Deng, Y.; Zeng, X. Effects of seawater acidification and alkalization on the farmed seaweed, Pyropia haitanensis (Bangiales, Rhodophyta), grown under different irradiance conditions. *Algal Res.* **2018**, *31*, 413–420. [CrossRef]

52. Xiong, Y.; Yang, R.; Sun, X.; Yang, T.; Chen, H. Effect of the epiphytic bacterium Bacillus sp. WPySW2 on the metabolism of Pyropia haitanensis. *J. Appl. Phycol.* **2018**, *30*, 1225–1237. [CrossRef]

53. Wang, W.; Xing, L.; Xu, K.; Ji, D.; Xu, Y.; Chen, C.; Xie, C. Salt stress-induced H_2O_2 and Ca^{2+} mediate K^+/Na^+ homeostasis in Pyropia haitanensis. *J. Appl. Phycol.* **2020**, *32*, 4199–4210. [CrossRef]

54. Huang, L.; Peng, L.; Yan, X. Multi-omics responses of red algae Pyropia haitanensis to intertidal desiccation during low tides. *Algal Res.* **2021**, *58*, 102376. [CrossRef]

55. Zhong, Z.; Wang, W.; Sun, X.; Liu, F.; Liang, Z.; Wang, F.; Chen, W. Developmental and physiological properties of Pyropia dentata (Bangiales, Rhodophyta) conchocelis in culture. *J. Appl. Phycol.* **2016**, *28*, 3435–3445. [CrossRef]

56. Li, S.; Liu, D. Improvement of spectrophotometric determination of phosphorus in water by phosphomolybdenum blue. *Environ. Prot. Chem. Ind.* **2006**, *26*, 78–80, (In Chinese with English abstract).

57. Sun, X.; Hong, L.; Ye, H. Experiment determining nitrate nitrogen in water samples by on-line cadmium column reduction-flow injection method. *Water Resour. Prot.* **2010**, *26*, 75–77, (In Chinese with English abstract).

58. Cao, Y.; Wang, W.; Liang, Z.R.; Liu, F.; Sun, X.; Yao, H.; Li, X.; Wang, F. Genetic and nutrient analysis of new Pyropia yezoensis strain "Huangyou No. 1". *Guangxi. Sci.* **2016**, *23*, 131–137, (In Chinese with English abstract).

59. Davison, R.; Pearson, G. Stress tolerance in intertidal seaweeds. *J. Phycol.* **1996**, *32*, 197–211. [CrossRef]

60. Guo, G.L.; Dong, S. Effects of desiccation on the growth and photosynthetic rate of four intertidal macroalgae from different vertical locations. *Trans. Oceanol. Limnol.* **2011**, *4*, 78–84, (In Chinese with English abstract).

61. Thomas, T.; Turpin, D.H.; Harrison, P. Desiccation enhanced nitrogen uptake rates in intertidal seaweeds. *Mar. Biol.* **1987**, *94*, 293–298. [CrossRef]

62. Ji, Y.; Tanaka, J. Effects of desiccation on the photosynthesis of seaweeds from the intertidal zone in Honshu, Japan. *Phycol. Res.* **2002**, *50*, 145–153. [CrossRef]

63. Gao, S.; Wang, G. The enhancement of cyclic electron flow around photosystem I improves the recovery of severely desiccated Porphyra yezoensis (Bangiales, Rhodophyta). *J. Exp. Bot.* **2012**, *63*, 4349–4358. [CrossRef]

64. Ma, J.; Cai, S. *The Cultivation and Processing of Pyropia yezoensis*; Science Press: Beijing, China, 1996. (In Chinese)

65. Yoshida, T. *Marine Algae of Japan*; Uchida Roukakuho Publishing: Tokyo, Japan, 1998. (In Japanese)
66. Raven, J.; Hurd, C. Ecophysiology of photosynthesis in macroalgae. *Photosynth. Res.* **2012**, *113*, 105–125. [CrossRef]
67. Hurd, C. Shaken and stirred: The fundamental role of water motion in resource acquisition and seaweed productivity. *Perspect. Phycol.* **2017**, *4*, 73–81. [CrossRef]
68. Kregting, L.; Hepburn, C.D.; Savidge, G. Seasonal differences in the effects of oscillatory and uni-directional flow on the growth and nitrate-uptake rates of juvenile Laminaria digitata (Phaeophyceae) *J. Phycol.* **2015**, *51*, 1116–1126. [CrossRef]
69. Inoue, Y.; Terada, R.; Belleza, D.; Nishihara, G. Effect of water velocity on the physiology of a collapsing Sargassum siliquastrum canopy under a controlled environment. *Phycol. Res.* **2020**, *68*, 313–322. [CrossRef]
70. Yang, W. Hydrodynamic, illumination, nitrogen and phosphorus on the response mechanism of Chlorella. sp. Master's Thesis, Xi'an University of Architecture and Technology, Xi'an, China, 2018. (In Chinese with English abstract).
71. Naldi, M.; Wheeler, P. Changes in nitrogen pools in Ulva fenestrata (Chlorophyta) and Gracilaria pacifica (Rhodophyta) under nitrate and ammonium enrichment. *J. Phycol.* **1999**, *35*, 70–77. [CrossRef]
72. Yarish, C.; Redmond, S.; Kim, J. *Gracilaria Culture Handbook for New England*; Wrack Lines; University of Connecticut Sea Grant: Groton, CT, USA, 2012.
73. Wang, Y.; Feng, Y.; Liu, X.; Zhong, M.; Chen, W.; Wang, F.; Du, H. Response of Gracilaria lemaneiformis to nitrogen deprivation. *Algal Res.* **2018**, *34*, 82–96. [CrossRef]
74. Liu, X.; Wen, J.; Zheng, C.; Jia, H.; Chen, W.; Du, H. The impact of nitrogen deficiency and subsequent recovery on the photosynthetic performance of the red macroalga Gracilariopsis lemaneiformis. *J. Appl. Phycol.* **2019**, *31*, 2699–2707. [CrossRef]
75. Liu, J.; Dong, S.; Ma, S. Effects of temperature and salinity on growth and NH4-N uptake of Gracilaria tenuistipitata var. liui, Ulva pertusa, Gracilaria filicina and NH4-N uptake of Gracilaria tenuistipitata var. liui. *Acta. Oceanol. Sin.* **2001**, *23*, 109–116.
76. Perini, V.; Bracken, M. Nitrogen availability limits phosphorus uptake in an intertidal macroalga. *Oecologia* **2014**, *175*, 667–676. [CrossRef] [PubMed]
77. Liu, D.; Sun, J.; Chen, Z.; Wei, T. Effect of N/P ratio on the growth of a red tide diatom Skeletonema costatum. *Trans. Oceanol. Limnol.* **2002**, *2*, 39–44, (In Chinese with English abstract).
78. Friedlander, M.; Krom, M.; Ben-Amotz, A. The effect of light and ammonium on growth, epiphytes and chemical constituents of Gracilaria conferta in outdoor cultures. *Bot. Mar.* **1991**, *34*, 161–166. [CrossRef]

water

Review

Advances in Ecology Research on Integrated Rice Field Aquaculture in China

Jing Yuan [1], Chuansong Liao [1,*], Tanglin Zhang [1,2], Chuanbo Guo [1,2] and Jiashou Liu [1,2]

[1] State Key Laboratory of Freshwater Ecology and Biotechnology, Institute of Hydrobiology, Chinese Academy of Sciences, Wuhan 430072, China; yuanj@ihb.ac.cn (J.Y.); tlzhang@ihb.ac.cn (T.Z.); guocb@ihb.ac.cn (C.G.); jsliu@ihb.ac.cn (J.L.)

[2] University of Chinese Academy of Sciences, Beijing 100049, China

* Correspondence: liaocs@ihb.ac.cn

Abstract: Integrated rice field aquaculture, a practice normally used by rural small-scale farmers, is not only supporting farms and livelihoods but is also reducing poverty and is playing a more and more important role in China. It is also becoming one of the main freshwater aquaculture systems, in addition to ponds, lakes, reservoirs, streams, and other aquaculture systems. During the past 40 years, both the production and areas of integrated rice field aquaculture in China have significantly increased from 0.13 million t and 0.74 million ha in 1990 to 3.25 million t and 2.56 million ha in 2020, respectively. Advances in ecology research on integrated rice–fish aquaculture were one of the main contributors to this achievement. In this paper, we systematically reviewed the advances in ecology research on three major integrated rice field aquaculture systems in China, namely rice–fish, rice–crab, and rice–crayfish coculture systems, the contribution of the research, and future prospects. We found that progress in ecology research on theories, biological studies, models, and eco-engineering techniques, coupled with policy support promoted the development of the rice field aquaculture industries. This review could assist individual small-scale farmers to make better use of rice field space to produce safer aquatic and rice products at a lower cost and help aquaculture scientists to further study the ecology of integrated rice field aquaculture systems.

Keywords: rice field; integrated aquaculture; ecology; review; prospects; coupling degree; eco-certification

Citation: Yuan, J.; Liao, C.; Zhang, T.; Guo, C.; Liu, J. Advances in Ecology Research on Integrated Rice Field Aquaculture in China. *Water* **2022**, *14*, 2333. https://doi.org/10.3390/w14152333

Academic Editors: Xiangli Tian and Li Li

Received: 26 May 2022
Accepted: 24 July 2022
Published: 28 July 2022

Publisher's Note: MDPI stays neutral with regard to jurisdictional claims in published maps and institutional affiliations.

1. Introduction

Integrated rice field aquaculture evolved from rice–fish coculture, and is playing a more and more important role in China, as the largest aquaculture producer in the world. It is also becoming one of the main freshwater aquaculture systems, in addition to ponds, lakes, reservoirs, streams, and other aquaculture systems, amounting to 10.52% of China's freshwater aquaculture production and 33.71% of the aquaculture area in 2020 [1] (Figure 1). China, the most populous country, is short of land resources and is also the biggest consumer of rice; therefore, food security is a priority of the Chinese government. The development of aquaculture by digging more ponds is strictly controlled by law. Therefore, integrated rice fields are getting more and more attention as a resource for increasing aquaculture production because of their properties of water conservation and using less land. This practice is normally adopted by individual small-scale farmers as an ideal use of land and an easy source of cheap, fresh, and convenient animal protein. It not only promotes sustainable agricultural and aquaculture development and farms and livelihoods, but also reduces poverty [2–4]. Consequently, during the past 40 years, both the production and area of integrated rice field aquaculture in China have significantly increased from 0.02 million t and 0.35 million ha in 1982 to 3.25 million t and 2.56 million ha in 2020, respectively [1,5] (Figure 2). Advances in ecology research, policy support, and aquaculture techniques for integrated rice–fish aquaculture are probably the main contributors to this achievement.

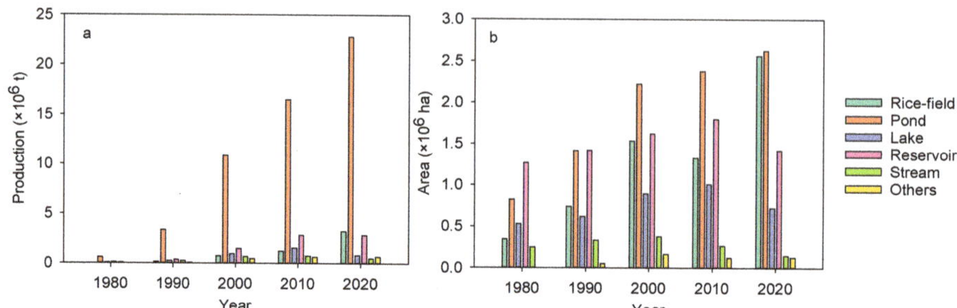

Figure 1. Comparison of freshwater aquaculture production (**a**) and area (**b**) between rice fields and other main water bodies in China [1].

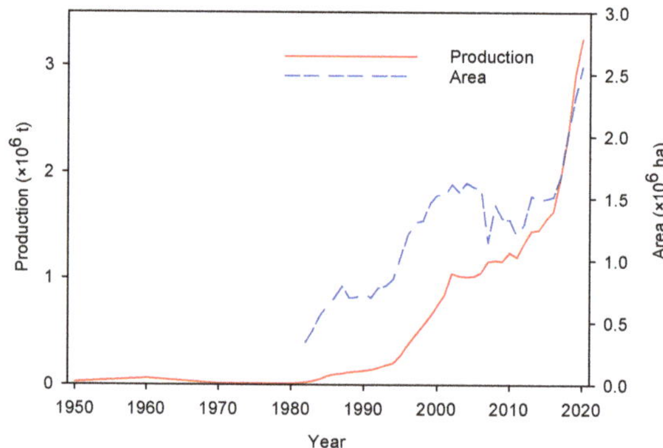

Figure 2. Changes in the aquaculture production and area in rice fields in China from 1950 to 2020 [1,5].

Integrated rice field aquaculture in China includes 25 different models [4]. In this paper, we systematically reviewed the advances in ecology research for the three most important systems in China, namely the rice–fish, rice–crab, and rice–crayfish coculture systems [6]. We also introduced the prospects for ecology research on integrated rice field aquaculture. This information could assist individual small-scale farmers to make better use of rice field space to produce safer aquatic and rice products at a lower cost and help aquaculture scientists to further study the ecology of this industry.

For this review, we searched for Chinese articles in the China National Knowledge Infrastructure database and English articles in the Web of Science database in 2022. We collected as many relevant pieces of literature as possible, and we only selected and referred to important published articles and books. Most of the information that was extracted included productive data (such as density, area, yield, survival rate, feed, and pesticide usage), aquaculture models, and advances in ecological research.

2. Advances in Ecology Research on the Rice–Fish Coculture System

The rice–fish coculture system is the most ancient integrated rice field aquaculture system in the world [2], and many fish species have been chosen for coculture, such as grass carp (*Ctenopharyngodon idella*), common carp (*Cyprinus carpio*) and its diverse strains, goldfish (*Carassius auratus*) and its diverse strains, swamp loach (*Misgurnus anguillicaudatus*), silver carp (*Hypophthalmichthys molitrix*), bighead carp (*Aristichthys nobilis*), and tilapias (*Oreochromis* sp.) [7,8]. Furthermore, ecology research in this system not only accelerates the development of the rice–fish aquaculture industry, but also lays the foundation for the formulation and development of other rice field integrated aquaculture systems.

2.1. Establishment of the Rice–Fish Symbiosis Theory

Ni [9] studied the population interactions between rice and fish in the coculture system based on biological and ecological traits in the late 1970s and early 1980s and put forward the Rice–Fish Symbiosis Theory in 1981. Ni and Wang suggested that rice is the main body in the rice field ecosystem and is also the dominant population, which absorbs solar energy, carbon dioxide, water, and other nutrients, and produces organic materials by photosynthesis and rice and straw for stakeholders. Moreover, plenty of weeds, phytoplankton, and photosynthetic bacteria in the field also conduct similar energy conversion, transportation, and storage as rice, but they also compete with rice for fertilizers, water, space, and solar energy without providing useful products for people [10]. Thus, the clearance of weeds in rice fields, which results in the loss of fertility and solar energy for rice, is imperative. Furthermore, plankton, bacteria, and other microorganisms are usually flushed away by irrigation, directly or indirectly causing the loss of fertility and solar energy [10]. Fish culture in rice fields can partially compensate for this loss. Fish in rice fields can directly or indirectly utilize weeds, zoobenthos, plankton, and detritus, decreasing competition with weeds for fertilizers, and utilizing the food and energy that cannot be utilized by rice [10]. In addition, nutrient loadings from fish can also provide nutrients for rice and plankton, and the CO_2 released by fish can also be utilized by rice, weeds, and algae. Additionally, the fish may loosen the surface soil and improve the oxygen condition of the soil to promote the mineralization of organic matter and the release of nutrients [9]. In this ecosystem, rice and fish complement each other and both play an active role, promoting the cycle of materials inside the field, directing beneficial energy flow to both the rice and fish, and efficiently recycling the materials and energy in this coculture ecosystem [10].

The Rice–Fish Symbiosis Theory, although a qualitative description in theory, provides a theoretical base for the Chinese government to adopt specific extension policies for the development of this traditional practice with a long history. In 1983, the Health Commission of China listed fish culture in rice fields as an important measure for killing mosquitoes, and the Ministry of Agriculture, Husbandry, and Fisheries held the first nationwide "On Site Experience Exchange Conference on Fish Culture in Rice Fields" in Wenjiang County, Sichuan Province. Additionally, in 1984, the National Economic Commission of China listed fish culture in rice fields as a national technical development project and extended this technique to 18 provinces; and in 1987, the technical extension of fish culture in rice fields was accepted into the National Harvest Project and State Key Agricultural Extension Project. Then, in 1990, the Ministry of Agriculture held the second nationwide "On Site Experience Exchange Conference on Fish Culture in Rice Fields" in Chongqing [2]. Thus, the breakthrough of ecological theories for integrated rice field aquaculture won national policy support and accelerated the development of fish culture in rice fields in the 1980s (Figure 2).

2.2. Quantitative Determination of Material Cycling and Energy Flow in the Rice–Fish System

2.2.1. Rice–Fish Coculture Remains Rice Production

With the rapid development of integrated rice field aquaculture in the 1980s and 1990s, one of the biggest concerns for agricultural administration officers was whether this system would affect rice production since China has the largest population in the world and food security is a priority for the government. A systematic large-scale in situ experiment, including 155 rice–fish coculture fields and 93 rice monoculture fields in 31 villages in Qingtian County, Zhejiang Province (where its rice–fish culture system was designated by the FAO-GEF as one of the five first Globally Important Agricultural Heritage System (GIAHS) pilot sites in the world), was conducted to compare the difference between rice–fish coculture and rice monoculture from 2006 to 2010. Studies showed that there was no significant difference in rice production between the rice–fish coculture system and the rice monoculture system ($p > 0.05$). The average rice production over five years in the rice–fish coculture system was 6190 ± 150 kg/ha/crop, while that for the monoculture was

6520 ± 390 kg/ha/crop [11]. This result eliminated the concerns about a decrease in rice production due to fish culture in rice fields.

2.2.2. Rice–Fish Coculture Decreases the Application of Chemical Fertilizers and Agricultural Pesticides

Fertilizers and agricultural pesticides are the main inputs of rice fields, affecting economic income and food safety. Experiments from 31 villages in Qingtian County indicated that the application amount of agricultural pesticides in the rice monoculture system was significantly higher than that in the rice–fish coculture system ($p < 0.05$) [11]. The application amount of chemical fertilizers was 158.39 ± 16.97 kg/ha/growing season in the rice–fish coculture system, which was much lower than 266.28 ± 18.15 kg/ha/growing season in the rice monoculture system. In addition, the application amount of agricultural pesticides was 1.81 ± 0.15 kg/ha/growing season in the rice–fish coculture system, which was also much lower than the 4.22 ± 0.16 kg/ha/growing season in the rice monoculture system [12]. During the five-year experiments, no herbicides were used in the rice–fish coculture system [12]. These results support that fish can control weeds and thus reduce the need for chemical fertilizers and decrease the application of agricultural pesticides. No application or much lower application of agricultural pesticides provides security for rice quality safety; thus, many rice products from integrated rice field aquaculture can be certificated and labeled as green and organic food [13]. Consequently, they have become famous brands that can be sold at a much higher price than rice from rice monoculture systems [5].

Further studies on the reason for the maintenance of rice production with low or zero pesticide application showed that rice planthoppers (including *Nilaparvata lugen*, *Sogatella furcifera*, and *Laodelphax striatellus*) were more abundant during their outbreak period (from late August to early September in each year, $p < 0.05$), the incidence of rice sheath blight caused by *Thanatephorus cucumeris* was higher ($p < 0.01$), and the weed biomass was significantly higher ($p < 0.01$) in the rice monoculture system than those in the rice–fish coculture system [14]. In addition, a lower cost and better price are the drivers for farmers to extend rice fish coculture, and better food safety increases consumers' confidence.

2.2.3. Rice–Fish Coculture Increases Nitrogen Use Efficiency

In the rice monoculture system, nitrogen in chemical fertilizers can only be used by rice, while nitrogen in feeds can be used only by fish in the fish monoculture system. However, nitrogen in chemical fertilizers and feeds can be mutually used in the coculture system. Xie et al. [14] confirmed that the rice–fish coculture system increases nitrogen use efficiency when compared to the rice and fish monoculture systems, and Hu et al. [15] and Zhang et al. [16] confirmed that it decreases the nutrient loadings to the water environment. Experiments showed that 41.02% of the food source of fish in the rice–fish coculture system was from natural food produced in the rice fields [11]. Nitrogen intake from feed was 37.6 kg/ha and that from natural food sources in the rice fields was 42.2 kg/ha [15]. Among them, 17.99% (14.36 kg/ha) of nitrogen was assimilated and 82.01% (65.44 kg/ha) was released by the fish, while 84.38% (55.22 kg/ha) of the released nitrogen was dissolved into the water in the form of ammonia excreted by the fish, which can be directly absorbed by the rice. The remaining nitrogen, 15.62% (10.22 kg/ha), was released in the form of feces, which returns to the soil [15]. Thus, fish can provide nutrients through feeding, excretion, and fecal production, compensating for the nitrogen needed for rice growth, and feces added to the soil can also provide nitrogen for the rice.

Quantitative studies on material cycling and energy flow in the rice–fish coculture system, together with other ecological research achievements in the rice–crab, rice–crayfish, and rice–turtle coculture systems led to the leapfrog development of the integrated rice field aquaculture industry both in terms of the amount produced and the area used (Figure 2).

3. Advances in Ecology Research on the Rice–Crab Coculture System

The Chinese mitten crab (*Eriocheir sinensis*) is used for traditional seafood in China and its aquaculture production reached 775,887 t in 2020 [1]. Fish–crab coculture systems started in the late 1980s [17] but developed very fast with a coculture area of 13,869 ha and a production of 61,800 t in 2019 [18]. As a young aquaculture sector, ecology research on rice–crab coculture received attention from the start of the industry [19–21]. Promoted by the Rice–Fish Symbiosis Theory, symbiosis was also studied in the rice–crab coculture and it was found that crabs can utilize the weeds in rice fields, loosen the soil, prey on pests, and provide fertilization through excretion and fecal production, while rice can purify water and protect crabs from their natural enemies [22]. Modern studies on rice–crab coculture ecology have focused on two aspects. One is the crab stocking density in relation to growth, yield, and environmental improvements including the larvae and juvenile crabs. The other is the progress in ecological engineering research.

For megalopa, a stocking density of 15 ind./m^2 can produce the highest crab yield and profit [23], while a stocking density of 120 ind./m^2 with an average weight of 0.005 g/ind. can promote nitrogen use efficiency [24]. For juveniles, a stocking density of 0.75 ind./m^2 with an average weight of 7.04 g/ind. has the highest profit without affecting the soil chemical indexes [25], and weeds can be effectively controlled at a stocking density of 0.50 ind./m^2 with an average weight of 12 g/ind. [26]. This quantitative determination of different stocking densities for different sizes under different ecological conditions lays the foundation for rice–crab coculture development [27–29].

Progress in eco-engineering techniques was the other driver for the development of the rice–crab coculture. Enlightened by the edge effect of rice growth, the trench structure in rice fields has changed from line-shaped trenches to cross-shaped trenches, and then to circular-shaped trenches composed of peripheral and cross trenches (Figure 3). The compensation rates for rice yield loss because of trench building were 95.89%, 85.58%, and 58.02% for line-shaped, cross-shaped, and circular-shaped trenches, respectively [30]. Thus, the new trench shape helps to maintain rice yield without increasing the trench area. Normally, the total trench area accounts for approximately 10% of rice fields. Another reform is the building of new style ridges [13]. The most successful method of ecological ridge engineering is the Panshan model, which is applied in Liaoning Province. In this model, rice planting is based on large single ridges between double rows, i.e., one wide row and a narrow row, with narrow row spacing within the ridge and wider row spacing between the ridges [31]. Edge row densification reduces the plant spacing in the rice row on the field's edge, where there is a marginal improvement in light, water, and fertilizer supply, which can improve rice yield [22]. Because of the improvements from this ecological engineering project, many provinces and autonomous regions in northern China have referenced the Panshan model when developing rice–crab coculture systems.

Fence building is another important eco-engineering project for rice–crab coculture. Vertical fences were designed in recent years for surrounding rice fields based on understanding the ecological traits of the mitten crab (Figure 4). Preventing the escape of the mitten crab ensures a better crab production. The fences are established using rigid plastics, metal leaves, bricks, or other materials, which are usually fixed into the soil and reached a 30 cm height above the soil.

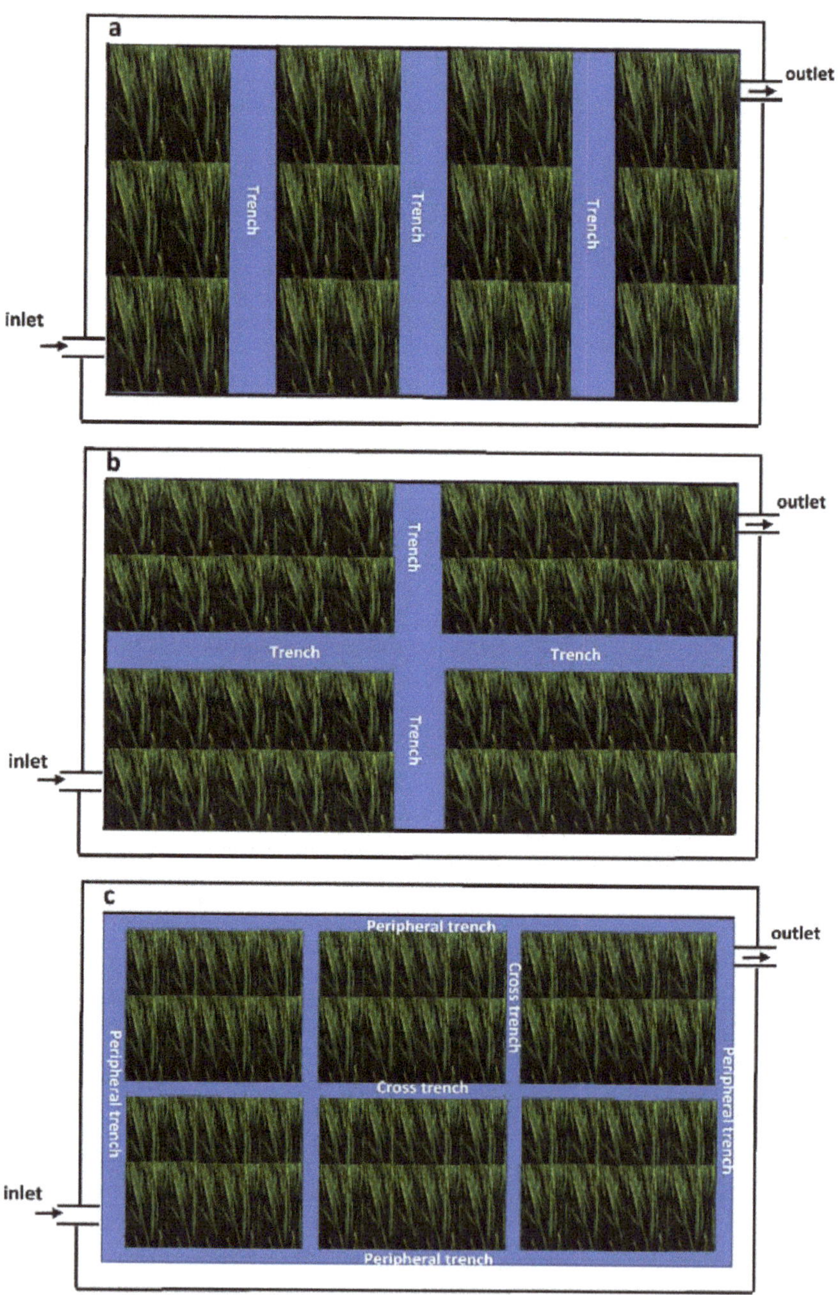

Figure 3. The evolution of trench designs in the rice–crab coculture system: (**a**) line-shaped trenches; (**b**) cross-shaped trenches; (**c**) and circular-shaped trenches.

Figure 4. A plastic fence for the prevention of escaping crabs in the rice–crab coculture system.

4. Advances in Ecology Research on the Rice–Crayfish Coculture System

The red swamp crayfish (*Procambarus clarkii*) is an alien crustacean that was introduced to China from Japan as early as 1929, although it is natively distributed in North America [32]. Since its introduction, the fighting against its invasion and effects on biodiversity and the environment, especially for the destruction of water conservation projects has never stopped until today [33–36]. However, it has also become the most successful exotic aquaculture species with a production increase from 0.05 million t in 2003 to 2.39 million t in 2020, and an area increase from 0.6 million ha in 2016 to 1.46 million ha in 2020 (Figure 5) [1]. Therefore, it is playing a more and more important role in China's aquaculture industry [37]. It has almost overtaken the production of 1.66 million t for tilapias and its production is 7.8 times the production of channel catfish (*Ictalurus punctatus*), which was 0.31 million t in 2020 [1]. Among all the crayfish aquaculture systems, the rice–crayfish coculture has become the most important model and it has developed very fast. The percentage of the rice–crayfish coculture area to the total crayfish aquaculture area increased from 70.83% in 2017 to 86.61% in 2020, while that of rice–crayfish coculture production to total crayfish aquaculture production increased from 72.40% in 2018 to 86.16% in 2020 [1] (Figure 6).

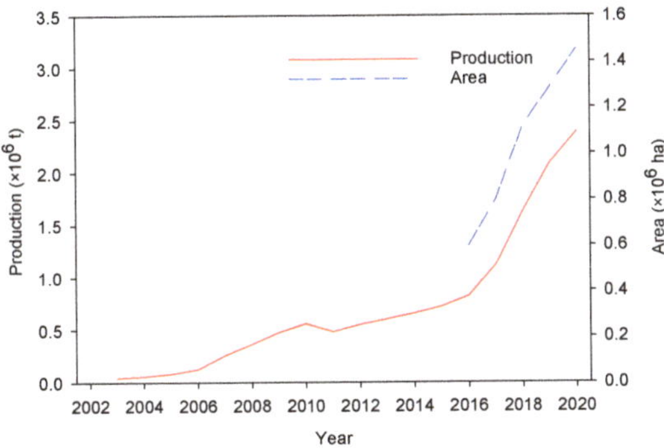

Figure 5. The red swamp crayfish aquaculture area and production from 2003 to 2020 in China [1].

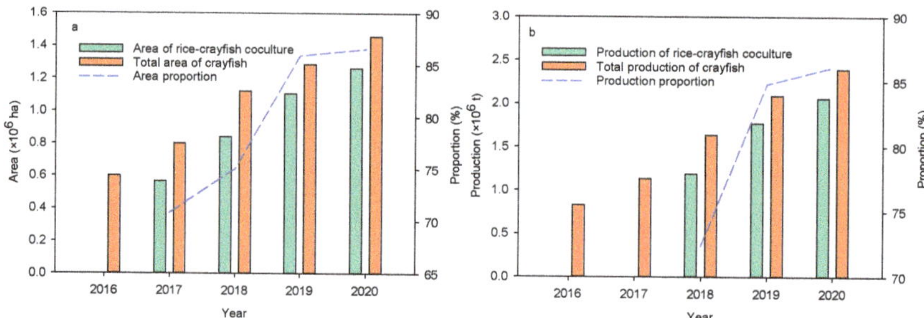

Figure 6. The red swamp crayfish area (**a**) and production (**b**) in the rice–crayfish coculture system and total crayfish aquaculture system in China [1].

Red swamp crayfish aquaculture in rice fields originated from fish farmers' practices and its aquaculture techniques were later summarized and extended by scientists and the government. The biggest barrier to developing this aquaculture is the concern for crayfish burrowing. However, this phenomenon is rarely observed in rice fields. The main reason is likely that the rice in the fields provides shelter for the crayfish. For instance, Cheng et al. confirmed that shadows can act as a solid shelter for red swamp crayfish, and more shadow partitioning, with a 60% shadow area, resulted in a decreased agonistic behavior, a lower mortality, and a higher body weight gain [38]. Yu et al. [39] also reported that an 80% artificial macrophyte coverage significantly increased the total biomass, the molting frequency, the total weight gain, the specific growth rate, and the daylight shelter occupancy when compared to those of a 20% coverage. Rice fields have more shadow areas and even partitioning conditions, which could be the ideal environment for red swamp crayfish growth. Moreover, an increased structural complexity could decrease the predation of native fishes by red swamp crayfish [38]. Additionally, dense rice and circular trenches in rice fields are beneficial for biodiversity protection from red swamp crayfish.

Red swamp crayfish have a small fecundity of about 500 eggs per female and normally reproduce in autumn when the water temperature is low and the hatching time is long [40], which affects the large-scale supply of juveniles. Research on the reproductive biology of red swamp crayfish is focused on solving this problem. For example, Jin et al. [41] systematically studied the spawning traits of red swamp crayfish and found that its spawning activities mainly took place from September to November with a mean fecundity of 429 ± 9 eggs per female, and there were two recruitments yearly, a major one from October to November and a minor one from March to May. They inferred that some eggs, prevented from hatching by a lower water temperature in winter, were more likely to hatch in the next spring, suggesting that reducing the fishing intensity on immature crayfish and avoiding sex selection during the reproductive period could improve the overall sustainability. Jin et al. [42] further indicated that the optimal temperature is 21–25 °C for adult reproduction and 25 °C for the embryonic development of red swamp crayfish. Consequently, this study provides evidence that manipulating the water temperature is an effective alternative for the mass production of juvenile red swamp crayfish.

The determination of the optimal stocking density for red swamp crayfish is very important since they have agonistic behaviors and fight for shelter. Many experimental and in situ studies have been conducted on this topic [39,43–45]. In summary, the optimal stocking densities were 90,000–120,000 juvenile crayfish at 3–4 cm in length per hectare and 75,000–90,000 juvenile crayfish at 4–5 cm in length per hectare [30].

Overall, progress in the ecology research on rice–fish coculture systems contributes to increased economic benefits, decreased chemical fertilizer and agricultural pesticides usage [4], and improved environmental conditions [44,46,47]. Using Hubei Province as an example, with the largest amount of rice–fish coculture in China, the average production was 1.80 t/ha of crayfish with a benefit of about 7000 USD and the use of chemical fertilizers and agricultural pesticides decreased by more than 30% when compared to the rice monoculture system [4].

5. Future Prospects

During the past two decades, integrated rice field aquaculture has become the fastest developing sector of all the aquaculture systems in China, partially because of the progress in ecology research, policy support, and aquaculture techniques in this field. The basic principle of integrated rice field aquaculture is the coupling of aquatic animals, including fishes, crustaceans, and turtles, with rice production similar to that of the aquaponics to some degree from the viewpoint of coupling [48]. The present study systematically reviewed the advances in ecology research for three major integrated rice field aquaculture systems, the development process of each system, and the contribution of the ecology research. Moreover, we discussed new trends that have appeared in the integrated rice field aquaculture systems and industry, and we have introduced some prospects below.

Some qualitative and quantitative ecological theories of symbiosis, material cycling, and energy flow have been established for the aquaculture systems that were discussed above. However, there are as many as 25 types of integrated rice field aquaculture systems and more than 30 aquatic animals have been included in the industry [4]. Therefore, it is necessary to establish a standard coupling degree comprising a set of parameters to quantitatively determine the coupling efficiency of the nutrient flow in the different coculture systems. This will maximize nutrient use and minimize the input of feeds, fertilizers, and agricultural pesticides, contributing to the extension of specific coculture techniques and the sustainability of integrated rice field aquaculture. Currently, the trench area is forbidden to exceed 10% of the total area of the rice fields [30]; therefore, establishing a standard coupling degree that is suitable for different integrated rice field aquaculture systems will increase the efficiency of land use and reduce environmental impacts.

Although a green aquatic food and organic aquatic food certification system, and an aquatic products quality and safety traceability system have been established for food safety in China [13,49], it is necessary to build an eco-certification system for integrated rice field aquaculture in China since this industry is related with both agriculture and aquaculture. Eco-certification in integrated rice field aquaculture should be based on producing positive sustainability outcomes through the process of certification and a market for certified sustainable products from this system. This would rely on producer compliance with eco-certification criteria that effectively addresses sustainability issues and the acceptance of eco-certification amongst the stakeholders including the consumers, producers, and harvesters [50].

The development of integrated rice field aquaculture is also closely related to brand building, which relies on eco-culture development. Because of the success of eco-culture building in Qianjiang (a medium-sized city in Hubei Province), in this city, red swamp crayfish produced through the rice–crayfish coculture system has promoted the development of the tourism and catering industry, and it has become a model city for red swamp crayfish aquaculture and trade [51]. Similar cases have also been reported in Qingtian, a GIAHS pilot site [11], and many other places [30]. Additionally, aquaculture products from eco-culture-famous sites normally demand much higher prices than those from normal producing areas. Thus, eco-culture can ensure a much better income for stakeholders.

Author Contributions: Conceptualization, J.Y., C.L. and J.L.; validation, J.Y., C.L., T.Z., C.G. and J.L.; data curation, C.L., J.Y. and J.L.; writing—original draft preparation, C.L. and J.Y.; writing—review and editing C.G., and J.L.; visualization, C.L. and J.L.; funding acquisition, C.L. and J.L. All authors have read and agreed to the published version of the manuscript.

Funding: This research was funded by the National Key Research and Development Program of China: 2020YFD0900304, 2019YFD0900304; the Key Research and Development Program of Hubei Province: 2020BBB077.

Institutional Review Board Statement: Not applicable.

Informed Consent Statement: Not applicable.

Data Availability Statement: Not applicable.

Conflicts of Interest: The authors declare no conflict of interest.

References

1. FDMARA (Fisheries Department of Ministry of Agriculture Rural Affairs). *China Fisheries Year Book, 1983–2021*; China Agriculture Press: Beijing, China, 2021.
2. Liu, J.; Wang, Q.; Yuan, J.; Zhang, T.; Ye, S.; Li, W.; Li, Z.; Gui, J. Integrated rice-field aquaculture in China: A long-standing practice with recent leapfrog developments. In *Aquaculture in China: Success Stories and Modern Trends*; Gui, J., Tang, Q., Li, Z., Liu, J., De Silva, S.S., Eds.; John Wiley & Sons Ltd.: Canberra, Australia, 2018; pp. 174–183.
3. Prein, M. Integration of aquaculture into crop-animal systems in Asia. *Agric. Syst.* **2002**, *71*, 127–146. [CrossRef]
4. Tang, J.; Li, W.; Lu, X.; Wang, Y.; Ding, X.; Jiang, J.; Tang, Y.; Li, J.; Zhang, J.; Du, J.; et al. Development status and rethinking of the integrated rice-fish system in China. *China Rice* **2020**, *26*, 1–10. (In Chinese with English Abstract) [CrossRef]
5. Hu, L.; Tang, J.; Zhang, J.; Ren, W.; Guo, L.; Halwart, M.; Li, K.; Zhu, Z.; Qian, Y.; Wu, M.; et al. Development of rice-fish system: Today and tomorrow. *Chin. J. Eco-Agric.* **2015**, *23*, 268–275. (In Chinese with English Abstract) [CrossRef]
6. Luo, S. *Agroecological Rice Production in China: Restoring Biological Interactions*; Food and Agriculture Organization of the United Nations: Rome, Italy, 2018; pp. 1–115.
7. Cai, R. *Fish Culture in Rice Fields*; Science Press: Beijing, China, 1982; pp. 22–23. (In Chinese)
8. Ni, D.; Wang, J. Recent development of fish culture in the rice field in China. *Acta Hydrobiol. Sin.* **1988**, *12*, 364–375. (In Chinese)
9. Ni, D. Rice-fish mutualism theory. *Sci. Hum. Being* **1984**, *1*, 10–11. (In Chinese)
10. Ni, D.; Wang, J. *Theories and Practices of Fish Culture in Rice Fields*; China Agriculture Press: Beijing, China, 1990; pp. 1–291. (In Chinese)
11. Chen, X.; Tang, J.; Hu, L.; Wu, M.; Ren, W. *Rice-Fish System in Qingtian: Ecology, Conservation and Utilization*; Science Press: Beijing, China, 2021; pp. 1–232.
12. Chen, X.; Wu, X.; Li, N.; Ren, W.; Hu, L.; Xie, J.; Wang, H.; Tang, J. Globally important agricultural heritage system (GIHAS) rice-fish system in China: An ecological and economic analysis. In *Advances in Ecological Research*; Li, P.P., Ed.; Zhejiang University Press: Hangzhou, China, 2011; pp. 126–137.
13. Miao, W. Recent developments in rice-fish culture in China: A holistic approach for livelihood improvement in rural areas. In *Success Stories in Asian Aquaculture*; De Silva, S.S., Davy, F.B., Eds.; Springer: Ottawa, ON, Canada, 2010; pp. 15–40.
14. Xie, J.; Hu, L.; Tang, J.; Wu, X.; Li, N.; Yuan, Y.; Yang, H.; Zhang, J.; Luo, S.; Chen, X. Ecological mechanisms underlying the sustainability of the agricultural heritage rice–fish system. *Proc. Natl. Acad. Sci. USA* **2011**, *108*, 1381–1387. [CrossRef]
15. Hu, L.; Ren, W.; Tang, J.; Li, N.; Zhang, J.; Chen, X. The productivity of traditional rice-fish co-culture can be increased without increasing nitrogen loss to the environment. *Agric. Ecosyst. Environ.* **2013**, *177*, 28–34. [CrossRef]
16. Zhang, J.; Hu, L.; Ren, W.; Guo, L.; Wu, M.; Tang, J.; Chen, X. Effects of fish on field resource utilization and rice growth in rice-fish coculture. *Chin. J. Appl. Ecol.* **2017**, *28*, 299–307. (In Chinese with English Abstract) [CrossRef]
17. Xue, Y.; Yang, H.; Ge, X. Rice crab symbiosis technology. *Shanghai Agric. Sci. Technol.* **1991**, *6*, 32–33. (In Chinese)
18. Ministry of Agriculture and Rural Affairs of China. Development report of China's rice and fishery comprehensive planting and breeding industry. *China Fish.* **2020**, *10*, 12–19. (In Chinese)
19. Cheng, Y.; Wu, X.; Li, J. Chinese mitten crab culture: Current status and recent progress towards sustainable development. In *Aquaculture in China: Success Stories and Modern Trends*; Gui, J.-F., Tang, Q., Li, Z., Liu, J., De Silva, S.S., Eds.; John Wiley & Sons Ltd.: Canberra, Australia, 2018; pp. 197–217.
20. Chen, F.; Zhang, Z. Ecological economic analysis of a rice-crab model. *Chin. J. Appl. Ecol.* **2002**, *13*, 323–326. (In Chinese with English Abstract) [CrossRef]
21. Xu, Q.; Wang, X.; Xiao, B.; Hu, K. Rice-crab coculture to sustain cleaner food production in Liaohe River basin. *J. Clean. Prod.* **2019**, *208*, 188–198. [CrossRef]
22. Bao, J.; Jiang, H.; Li, X. Thirty years of rice-crab coculture in China—Research progress and prospects. *Rev. Aquac.* **2022**, *14*, 1597–1612. [CrossRef]
23. Wang, W.; Wang, C.; Ma, X. *Mitten Crab Eco-Culture*, 2nd ed.; China Agriculture Press: Beijing, China, 2013; pp. 1–357. (In Chinese)
24. Wang, A.; Ma, X.; Xu, J.; Lu, W. Methane and nitrous oxide emissions in rice-crab culture systems of northern China. *Aquac. Fish.* **2019**, *4*, 134–141. [CrossRef]
25. Lv, D.; Wang, W.; Ma, X.; Chen, Z.; Yu, Y.; Li, C.; Yan, B. The effect of stocking density of Chinese mitten crab on yields of rice and crab in rice-crab culture system. *Hubei Agric. Sci.* **2010**, *49*, 1677–1680. (In Chinese with English Abstract)
26. Lv, D.; Wang, W.; Ma, X.; Wang, Q.; Wang, A.; Chen, Z.; Tang, S. Ecological prevention and control of weeds in rice-crab polyculture field. *Hubei Agric. Sci.* **2011**, *50*, 1574–1578. (In Chinese with English Abstract)
27. Xu, M.; Wang, W.; Ma, X. Regularity of different cultivation patterns on the soil physical and chemical properties, nutrition changes in rice-carb culture system. *Guangdong Agric. Sci.* **2013**, *9*, 53–57. (In Chinese with English Abstract) [CrossRef]
28. Sun, W.; Zhang, Q.; Ma, X.; Wang, W.; Wang, A. A study on effects of different crab stocking density on water environment and rice yield. *J. Shanghai Ocean Univ.* **2014**, *23*, 366–373.

29. Shou, L.; Zhang, Z.; Chen, S.; Wu, W. Analysis of economic benefit of organic rice and rice-crab cropping patterns in rice planting area of northeast China. *Shaanxi J. Agric. Sci.* **2021**, *67*, 1–8. (In Chinese with English Abstract) [CrossRef]

30. National Fisheries Technology Extension Center. *Technical Modes for Integrated Rice-Field Aquaculture*; China Agriculture Press: Beijing, China, 2021; pp. 86–127. (In Chinese)

31. Dong, Y.; Jiang, H.; Yu, Y.; Sun, B. Water-temperature characteristics of rice-field-crab "Panshan mode". *J. Anhui Agri. Sci.* **2010**, *38*, 12483–12485. (In Chinese with English Abstract) [CrossRef]

32. Wu, T.; Gao, P. Status and development prospects of the freshwater red swamp crayfish. *Inland Fish.* **2008**, *2*, 15–17. (In Chinese)

33. Li, Z.; Xie, Y. *Invasive Alien Species in China*; China Forestry Publishing House: Beijing, China, 2002. (In Chinese)

34. Gong, S.; Li, L.; Lu, J.; Zhang, X.; He, X.; Xiong, C. A study on burrowing behavior of *Procambarus clarkii*. *Freshwater Fish.* **2007**, *37*, 3–7. (In Chinese with English Abstract) [CrossRef]

35. Wu, Z.; Cai, J.; Jia, Y.; Lu, J.; Jiang, Y.; Huang, C. Predation impact of *Procambarus clarkii* on *Rana limnocharis* tadpoles in Guilin area. *Biodivers. Sci.* **2008**, *16*, 150–155. (In Chinese with English Abstract) [CrossRef]

36. Cai, F.; Wu, Z.; He, N.; Ning, L.; Huang, C. Research progress in invasion ecology of *Procambarus clarkii*. *Chin. J. Ecol.* **2020**, *29*, 124–132. (In Chinese with English Abstract) [CrossRef]

37. Liu, J.; Li, Z. The role of exotics in Chinese inland aquaculture. In *Success Stories in Asian Aquaculture*; De Silva, S.S., Davy, F.B., Eds.; Springer: Ottawa, ON, Canada, 2010; pp. 173–186.

38. Huang, C.; Xiong, Q.; Tang, J.; Wu, M. Shadow area and partitioning influencing mortality, healthiness and growth of juvenile red swamp crayfish *Procambarus clarkii* (Decapoda). *Aquac. Res.* **2012**, *43*, 1677–1686. [CrossRef]

39. Yu, J.; Xiong, M.; Ye, S.; Li, W.; Xiong, F.; Liu, J.; Zhang, T. Effects of stocking density and artificial macrophyte shelter on survival, growth and molting of juvenile red swamp crayfish (*Procambarus clarkii*) under experimental conditions. *Aquaculture* **2020**, *521*, 735001. [CrossRef]

40. Xia, A. *Aquaculture of Red Swamp Aquaculture*; China Agriculture University Press: Beijing, China, 2007. (In Chinese)

41. Jin, S.; Jacquin, L.; Xiong, M.; Li, R.; Li, W.; Lek, S.; Tang, Z. Reproductive pattern and population dynamics of commercial red swamp crayfish (*Procambarus clarkii*) from China: Implications for sustainable aquaculture management. *PeerJ* **2019**, *7*, e6214. [CrossRef]

42. Jin, S.; Jacquin, L.; Huang, F.; Xiong, M.; Li, R.; Lek, S.; Li, W.; Liu, J.; Zhang, T. Optimizing reproductive performance and embryonic development of red swamp crayfish *Procambarus clarkii* by manipulating water temperature. *Aquaculture* **2019**, *510*, 32–42. [CrossRef]

43. Ren, Z. Technology for culture red swamp crayfish in rice fields. *Chin. Aquac.* **2012**, *3*, 22–25. (In Chinese) [CrossRef]

44. Yu, J.; Ren, Y.; Xu, T.; Li, W.; Xiong, M.; Zhang, T.; Li, Z.; Liu, J. Physicochemical water quality parameters in typical rice-crayfish integrated systems (RCIS) in China. *Int. J. Agric. Biol.* **2018**, *11*, 54–60. [CrossRef]

45. Huang, F.; Feng, Y.; Li, H.; Li, W.; Zhang, T. Effects of stocking of different sizes on growth performance and production of red swamp crayfish in rice fields. *Biol. Resour.* **2020**, *42*, 421–427. (In Chinese with English Abstract) [CrossRef]

46. Huang, J.; Zheng, X.; Wu, Z.; Liu, H.; Deng, F. Can increased structural complexity decrease the predation of an alien crayfish on a native fish? *Hydrobiologia* **2016**, *781*, 191–197. [CrossRef]

47. Ma, D.; Qian, J.; Liu, J.; Gui, J. Development strategy for integrated rice field aquaculture. *Strateg. Study CAE* **2016**, *18*, 96–100. (In Chinese with English Abstract) [CrossRef]

48. Baganz, G.F.M.; Junge, R.; Portella, M.C.; Goddek, S.; Keesman, K.J.; Baganz, D.; Staaks, G.; Shaw, C.; Lohrberg, F.; Kloas, W. The aquaponic principle—It is all about coupling. *Rev. Aquac.* **2021**, *14*, 252–264. [CrossRef]

49. Huang, L.; Song, Z.; Feng, Z.; Meng, T. The application of traceability system of aquatic products quality and safety in building the market access system. *Chin. Fish. Qual. Stand.* **2011**, *2*, 26–33. (In Chinese with English Abstract)

50. Amundsen, V.S.; Osmundsen, T.C. Virtually the reality: Negotiating the distance between standards and local realities when certifying sustainable aquaculture. *Sustainability* **2019**, *11*, 2603. [CrossRef]

51. Qin, Z. Propel the innovation of system and mechanism with "crayfish-rice cooperation" model—Observation and reflection of comprehensive reform in Qiangjiang as a nationwide small-medium. *Chin. Devel.* **2016**, *16*, 51–56.